Nicola Armaroli and
Vincenzo Balzani

Energy for a
Sustainable World

Related Titles

Olah, G. A., Goeppert, A., Prakash, G. K. S.

Beyond Oil and Gas: The Methanol Economy

350 pages
2010
Softcover
ISBN: 978-3-527-32422-4

Cocks, F. H.

Energy Demand and Climate Change

Issues and Resolutions

267 pages with 30 figures
2009
Softcover
ISBN: 978-3-527-32446-0

Wengenmayr, R., Bührke, T. (eds.)

Renewable Energy

Sustainable Energy Concepts for the Future

120 pages
2008
Hardcover
ISBN: 978-3-527-40804-7

Bührke, T., Wengenmayr, R. (eds.)

Erneuerbare Energie

Alternative Energiekonzepte für die Zukunft

108 pages
2007
Hardcover
ISBN: 978-3-527-40727-9

Paul, B.

Future Energy

How the New Oil Industry Will Change People, Politics and Portfolios

approx. 240 pages
2007
Hardcover
ISBN: 978-0-470-09642-0

Kruger, P.

Alternative Energy Resources

The Quest for Sustainable Energy

272 pages
2006
Hardcover
ISBN: 978-0-471-77208-8

Romm, J. J.

Der Wasserstoff-Boom

Wunsch und Wirklichkeit beim Wettlauf um den Klimaschutz

219 pages with 3 figures
2006
Softcover
ISBN: 978-3-527-31570-3

Nicola Armaroli and Vincenzo Balzani

Energy for a Sustainable World

From the Oil Age to a Sun-Powered Future

WILEY-VCH

WILEY-VCH Verlag GmbH & Co. KGaA

The Authors

Dr. Nicola Armaroli
Ist. ISOF/CNR
Molecular Photoscience Group
Via Gobetti 101
40129 Bologna
Italy

Prof. Vincenzo Balzani
Dept. of Chemistry G. Ciamician
University of Bologna
Via Selmi 2
40126 Bologna
Italy

Cover idea:

Fausto Puntoriero

Library of Congress Card No.: applied for

British Library Cataloguing-in-Publication Data
A catalogue record for this book is available from the British Library.

Bibliographic information published by the Deutsche Nationalbibliothek
The Deutsche Nationalbibliothek lists this publication in the Deutsche Nationalbibliografie; detailed bibliographic data are available on the Internet at <http://dnb.d-nb.de>.

© 2011 Wiley-VCH Verlag & Co. KGaA, Boschstr. 12, 69469 Weinheim, Germany

Cover   Formgeber, Eppelheim
Typesetting   Toppan Best-set Premedia Limited, Hong Kong
Printing and Binding   betz-druck GmbH, Darmstadt

Printed in the Federal Republic of Germany
Printed on acid-free paper

ISBN: 978-3-527-32540-5

To Claudia and Carla

Energy for a Sustainable World: From the Oil Age to a Sun-Powered Future. Nicola Armaroli and
Vincenzo Balzani
Copyright © 2011 WILEY-VCH Verlag GmbH & Co. KGaA, Weinheim
ISBN: 978-3-527-32540-5

Contents

Energy for a Sustainable World: From the Oil Age to a Sun-Powered Future. Nicola Armaroli and Vincenzo Balzani
© 2011 WILEY-VCH Verlag GmbH & Co. KGaA, Weinheim
ISBN: 978-3-527-32540-5

Preface

"With no foresight into the future
one is bound to find troubles at hand."

Ancient saying

In recent decades, by observing the Earth from space, we have fully realized that we live in a spaceship that cannot land and cannot dock anywhere to be refueled or repaired. We travel alone in the Universe and we can only rely on the resources available on the surface or in the hold of our planet, and on the energy coming from the Sun. We have also realized that the Earth is a system of intricately connected parts and that human activities can affect biogeochemical cycles. In fact, our 4.5 billion year old planet has entered a new epoch, Anthropocene, characterized by a dramatic increase of the size of human ecological footprint.

Energy is embedded in any type of goods and is needed to produce any kind of service. What makes the modern life of affluent people apparently so easy compared to that of our ancestors, or even to that of billions of individuals still living in poverty, is a steady flux of cheap and plentiful energy in the form of fossil fuels. We know, however, that these resources will not last forever and we have also learnt that their use has caused, and is still causing, severe damage to the Earth's atmosphere. Furthermore, fossil fuels have indirectly contributed to establish disparities and iniquities in human society: almost half of the total primary energy supply is consumed by about 10% of the population living in rich countries, while the poorest 25% of mankind consumes less than 3% of global energy.

Nowadays, everybody wants to have more and more energy, an attitude that poses a variety of entangled problems. When a blackout takes place in a country for whatever reason, the solution proposed by politicians who are seeking to be (re)elected is that of making new power plants. Is it the right solution? Many economists seem to believe that well-being correlates with energy consumption, that energy prices reflect all significant costs and that any societal problems can be solved by enhanced economic growth. Is it true? Several scientists are convinced that technology will solve the energy problem as well as the problems that technology itself is creating. Can we trust them?

The aim of this book is to show that we live in a fragile world and that the world's fragility can be strongly reduced or increased depending on how the energy

Energy for a Sustainable World: From the Oil Age to a Sun-Powered Future. Nicola Armaroli and Vincenzo Balzani
© 2011 WILEY-VCH Verlag GmbH & Co. KGaA, Weinheim
ISBN: 978-3-527-32540-5

problem is tackled. According to Stephen J. Gould, the fragility of the world is related to an intrinsic law of Nature that he called "the great asymmetry principle" (Gould, S.J., *Science*, 1998, **279**, 812): "The essential human tragedy, and the true source of science's potential misuse for destruction, lies in a great asymmetry in our universe of natural laws. We can only reach our pinnacles by laborious steps, but destruction can occur in a minute fraction of the building time, and can often be truly catastrophic. A day of fire destroyed a millennium of knowledge in the library of Alexandria and centuries of building in the city of London." Within this general principle, the destruction force depends on place and time. Leaving aside the menace coming from nuclear weapons, presently the biggest danger for spaceship Earth comes from too much consumption of natural resources, too much waste generation and too many disparities among the passengers. Energy plays a key role in controlling Earth's fragility because most of mankind's problems, including food, water, health, wealth, climate, heating, lighting, cooling, transportation, communication, and, of course, wars are strictly related to the energy issue. The way out of the difficulties and disparities generated during the fossil fuel era is a global problem: the supply of secure, clean, sustainable energy to *all* of the passengers of spaceship Earth is the most important scientific and technological challenge of the twenty-first century.

Fortunately, the energy crisis is not only a tough challenge, but also an unprecedented opportunity. It offers a unique chance to become more concerned about the world in which we live and the society we have built up. Whereas it used to be axiomatic that civilization would always progress over time, because science and technology would have solved any problem, now we are no longer sure about that. Human progress is neither automatic nor inevitable. We have to take urgent and responsible decisions right now: tomorrow might be too late. The quest for ecological and social sustainability requires every single citizen to become aware that consuming resources above a threshold of his/her real needs does not help to create a better world. Earth is in our hands: are we wise enough to develop, with the help of science and technology, an ecological sustainable civilization capable of reducing disparity and creating a more peaceful world?

An old Italian proverb says that the only difference between an optimist and a pessimist is that the latter is better informed. A short-sighted optimism based on unawareness will not allow mankind to move toward a real progress. Pessimism, which arises from the consciousness of the gravity of the situation, is the right starting point: to propose solutions, we must acknowledge that there are problems and we must know them in any possible detail. There is a great need for spreading information about the unsafe conditions of our planet.

Finding a solution to the energy problem is a challenge of utmost difficulty, but also an extraordinary opportunity. Perhaps we are still in time to change and create an Anthropocene epoch based on resource conservation, waste reduction, human relationships, and global solidarity. To achieve this epochal result, we need to educate public opinion and to find visionary leaders capable of looking far, over the planet and into the future.Our generation will ultimately be defined by how we live up to the energy challenge.

Acknowledgments

No book can be written in isolation and this one has, indeed, benefited from the work of the thousands of authors of books and articles that allowed us to gain a deeper understanding of the problems we have tried to illustrate and discuss comprehensively. We strived to acknowledge their work and we apologize beforehand if we have missed someone.

We are glad to thank the members of our research groups, including PhD students, for support, discussions, suggestions, and, even more, for their friendship. Special thanks are due to Gianluca Accorsi, Giacomo Bergamini, Francesco Barigelletti, Paola Ceroni, Sandra Monti John Mohanraj, and Margherita Venturi for their critical reading of several chapters of the manuscript. Public debates, many lectures in high schools and universities and intelligent questions by many students and colleagues have helped us to focus several topics better.

We also wish to thank Fausto Puntoriero for drawing the cover page of the book, Filippo Monti for preparing with great care and ability all the graphics and illustrations, and Andrea Listorti and Abdelhalim Belbakra for searching and gathering literature. We would also like to thank the staff of Wiley-VCH for their highly professional and valuable assistance.

Last but not least we wish to thank our families, and in particular our wives Claudia and Carla, who have provided inspiration, sustained encouragement, and, definitely, a great deal of patience during the writing of this book.

Bologna, August 2010

Nicola Armaroli and
Vincenzo Balzani

Notation

Prefixes

exa (E)	10^{18}
peta (P)	10^{15}
tera (T)	10^{12}
giga (G)	10^{9}
mega (M)	10^{6}
kilo (k)	10^{3}
milli (m)	10^{-3}
micro (μ)	10^{-6}
nano (n)	10^{-9}
pico (p)	10^{-12}
femto (f)	10^{-15}
atto (a)	10^{-18}

Abbreviations

bbl	barrel of oil
Dwt	deadweight ton
ppm	part per million
toe	ton of oil equivalent
W_{th}	thermal watt
W_p	watt peak
W_{el}	electric watt

Acronyms

AC	Alternating Current
AFC	Alkaline Fuel Cell
ASPO	Association for the Study of Peak Oil and Gas
ASTM	American Society for Testing and Materials

Energy for a Sustainable World: From the Oil Age to a Sun-Powered Future. Nicola Armaroli and Vincenzo Balzani
© 2011 WILEY-VCH Verlag GmbH & Co. KGaA, Weinheim
ISBN: 978-3-527-32540-5

BHJ	Bulk Heterojunction
BP	British Petroleum
bpd	barrel per day
bpy	barrel per year
BTU	British Thermal Units
CAES	Compressed Air Energy Storage
CBM	Coalbed Methane
CCS	Carbon Capture and Sequestration
CFC	Chlorofluorocarbons
CHP	Combined Heat and Power
CNG	Compressed Natural Gas
CPV	Concentrated Photovoltaics
CR	Concentration Ratio (in CSP)
CSP	Concentrating Solar Power
DC	Direct Current
DME	Dimethyl Ether
DOD	US Department of Defense
DOE	US Department of Energy
DSSC	Dye-Sensitized Solar Cell
DU	Depleted Uranium
EEA	European Environment Agency
EES	Earth Energy Systems
EI	Energy Intensity
EIA	US Energy Information Administration
ENI	Ente Nazionale Idrocarburi (Italy)
EPA	US Environmental Protection Agency
EROI (EROEI)	Energy Return on Investment
EU	European Union
EUROSTAT	Statistical Office of the European Communities
EV	Electric Vehicle
FAME	Fatty Acid Methyl Ester
FIT	Feed-in Tariffs
GDP	Gross Domestic Product
GHG	Greenhouse Gas
GWP	Global Warming Potential
HFC	Hydrofluorocarbon
IAEA	International Atomic Energy Agency
ICE	Internal Combustion Engine
ICF	Inertial Confinement Approach
ICT	Information and Communication Technology
IEA	International Energy Agency
IGCC	Integrated Gasification Combined Cycle
IPCC	International Panel on Climate Change
IR	Infrared (radiation)
KERS	Kinetic Energy Recovery Systems

LCA	Life-Cycle Analysis
LNG	Liquefied Natural Gas
LPG	Liquid Petroleum Gas
NASA	US National Aeronautics and Space Administration
NEA	Nuclear Energy Agency
NGO	Non-Governmental Organization
NIR	Near-Infrared (radiation)
NPT	Non-Proliferation Treaty
NREL	US National Renewable Energy Laboratory
OECD	Organization for Economic Cooperation and Development
OSC	Organic Solar Cell
OTEC	Ocean Thermal Energy Conversion
OWC	Oscillating Water Column
PCET	Proton-Coupled Electron Transfer
PEM	Proton Exchange Membrane
PM	Particulate Matter
PSII	Photosystem II
PV	Photovoltaic
QUAD	quadrillion BTU (10^{15} BTU)
RC	Reaction Center
RMFC	Reformed Methanol Fuel Cell
SEGS	Solar Energy Generating System
SHP	Small Hydropower
SI	International System of Units
SMES	Superconducting Magnetic Energy Storage
SUV	Sport Utility Vehicle
TPES	Total Primary Energy Supply
UCG	Underground Coal Gasification
UCTE	Union for the Coordination of the Transmission of Electricity
UNEP	United Nations Environment Programme
URFC	Unitized Regenerative Fuel Cell
USGS	US Geological Survey
UV	Ultraviolet (radiation)
Vis	Visible (radiation)
VOC	Volatile Organic Compound
WEC	World Energy Council
WHO	World Heath Organization
WNA	World Nuclear Association
WWII	World War II

Part One
Living on Spaceship Earth

Energy for a Sustainable World: From the Oil Age to a Sun-Powered Future. Nicola Armaroli and
Vincenzo Balzani
© 2011 WILEY-VCH Verlag GmbH & Co. KGaA, Weinheim
ISBN: 978-3-527-32540-5

1
The Energy Challenge

"Pay attention to the whispers,
so you won't have to listen to the screams."

Cherokee Proverb

1.1
Our Spaceship Earth

On Christmas Eve 1968, the astronauts of the Apollo 8 spacecraft, while in orbit around the Moon, had the astonishment to contemplate the Earthrise. William Anders, the crewmember who took what is considered one of the most influential photographs ever taken, commented: "We came all this way to explore the Moon, and the most important thing is that we discovered the Earth" [1] (Figure 1.1).

The image taken by the Cassini Orbiter spacecraft on September 15, 2006, at a distance of 1.5 billion kilometers (930 million miles) shows the Earth as a pale blue dot in the cosmic dark (Figure 1.2). There is no evidence of being in a privileged position in the Universe, no sign of our imagined self-importance. There is no hint that we can receive help from somewhere, no suggestion about places to which our species could migrate. Like it or not, Earth is a spaceship. It's the only home where we can live.

Spaceship Earth moves at the speed of $29\,\mathrm{km\,s^{-1}}$, apparently without any destination. It does not consume its own energy to travel, but it requires a huge amount of energy to make up for the needs of its 6.8 billion passengers who increase at a rate of 227 000 per day (the population of a medium-sized town), almost 83 million per year (the population of a large nation) [2]. Spaceship Earth cannot land and cannot dock anywhere to be refueled or repaired. Any damage has to be fixed and any problem has to be solved by us passengers, without disembarking. We travel alone in the Universe, and we can only rely on the energy coming from the Sun and on the resources available on the surface or stored in the hold of our spaceship.

Earth's civilization has always depended on the incessant flow of solar energy that sustains the biosphere and powers the photosynthetic production of food. Until a few centuries ago societies obtained their energy from sources that were almost immediate transformations of solar radiation (flowing water and wind) or that took relatively short periods of time to become available (wood) [3]. The feature

Energy for a Sustainable World: From the Oil Age to a Sun-Powered Future. Nicola Armaroli and Vincenzo Balzani
© 2011 WILEY-VCH Verlag GmbH & Co. KGaA, Weinheim
ISBN: 978-3-527-32540-5

Figure 1.1 Earthrise: a photograph of the Earth taken by astronaut William Anders on December 24, 1968, during the Apollo 8 mission while in orbit around the Moon. This picture is one of the *Life*'s 100 Photographs that Changed the World. Credit: NASA.

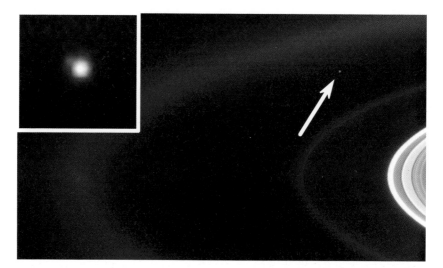

Figure 1.2 Photograph taken by the Cassini Orbiter spacecraft on September 15, 2006, at a distance of 1.5 billion kilometers from Earth. The dot to the upper left of Saturn's rings, indicated by the arrow, is the Earth. Saturn was used to block the direct light from the Sun, otherwise the Earth could not have been imaged. Inset: expanded image of the Earth which shows a dim extension (the Moon). Credit: NASA.

that distinguishes modern industrial society from all previous epochs is the exploitation of fossil fuel energy. Currently over 80% of the energy used by mankind comes from fossil fuels [4]. Harnessing coal, oil, and gas, the energy resources contained in the store of our spaceship, has prompted a dramatic expansion in energy use. Powering our spaceship Earth with fossil fuels has been very convenient, but now we know that this entails severe consequences [5, 6].

Firstly, fossil fuels are a nonrenewable resource that is going to exhaust. We have consumed 1 trillion barrels of oil in the last 140 years, and currently the world's growing thirst for energy amounts to almost 1000 barrels of oil, 93 000 cubic meters of natural gas, and 221 tons of coal per second [7]. How long can we keep running this road? Secondly, the use of fossil fuels causes severe damage to human health and the environment. It has been pointed out [8] that the energy challenge we face relates to "the tragedy of the commons" [9]: we treat fossil fuels as a resource that anyone anywhere can extract and use in any fashion, and Earth's atmosphere and oceans as a dump for their waste products, including more than 30 Gt per year of CO_2. A third critical aspect concerning fossil fuels is that their uneven allocation, coupled with the unfair distribution of wealth, leads to strong disparities in the quality of life among the Earth's passengers.

1.2
An Unsustainable Growth in an Unequal World

1.2.1
Population Growth and Carrying Capacity

In the last 100 years there has been a rapid population growth due to medical advances and massive increases in agricultural productivity. In 1950, the world population was 2.6 billion, with an increase of 1.5% per year [10]. In 2010, it was more than 6.8 billion, but with a lower rate of annual increase (1.1%), that is expected to decline further until 2050, when the Earth will be populated by about 9.2 billion people. At that time, the median age of the world population will be 37.3 years, up from 26.6 in 2000 [11].

The population size of a biological species that a given environment can sustain indefinitely is termed carrying capacity. Overpopulation may result from growth in population or reduction in capacity. The resources to be considered when assessing the carrying capacity of a given ecological system include clean water, clean air, food, shelter, warmth and other resources necessary to sustain life. In the case of humans, several additional resources must be considered, including medical care, education, sewage treatment, waste disposal, and, of course, energy.

Clearly, spaceship Earth has a limited carrying capacity, but it is quite difficult to assess the maximum number of humans who can live on it in satisfactory welfare conditions, also because "satisfactory welfare" is a somewhat subjective concept. An alarm bell, however, comes from the estimation of the ecological footprint, defined as the amount of biologically productive land and sea area needed to regenerate the resources a human population consumes and to absorb and render harmless the corresponding waste [12]. In global hectares per person, in 2006 the Earth's biocapacity was 1.8, while the average footprint was 2.5. In 2009, the Earth Overshoot Day, that is, the day when humanity begins living beyond its ecological means, was September 25 [13]. In other words, mankind uses biological services faster than the Earth can renew them.

1.2.2
Economic Growth and Ecologic Degradation

The expansion of the human enterprise in the twentieth century was phenomenal, particularly because of the availability of low-cost energy. Unfortunately, however, it has caused bad consequences that we have now to face. Ecologists emphasize that dominant patterns of production and consumption are causing environmental devastation and a massive extinction of species [14]. Climatologists warn about anthropogenic climate change [15]. Geologists point out that we will soon reach, or maybe we have already surpassed, the peak of oil production [16]. Seismologists wonder whether natural disasters, like the devastating earthquake which in May 2008 killed 80 000 people in China, are triggered by exaggerated human constructions [17]. International agencies inform us that about 6 million hectares of primary forest are lost each year [18]. People are worried about nuclear waste [19], and in affluent countries even disposal of electronic waste causes domestic and international problems [20, 21]. Last but not least, food security is a growing concern worldwide [22, 23].

Some scientists have pointed out that global effects of human activities, directly or indirectly related to the use of fossil fuels, are producing distinctive global signals. Accordingly it has been proposed that, since the beginning of the Industrial Revolution, we have entered a new epoch that can be called Anthropocene [24], in which the Earth has endured changes sufficient to leave a global stratigraphic signature distinct from that of the Holocene or of previous Pleistocene interglacial phases [25].

In spite of these alarm bells, growth remains the magic word of narrow-minded economists and politicians. They believe that the economic growth must continue indefinitely, and therefore they incessantly press for increasing production and consumption. In affluent countries, we live in societies where the concepts of "enough" and "too much" have been removed [26]. We do not take into account that the larger the rates of resource consumption and waste disposal, the more difficult it will be to reach sustainability and guarantee the survival of human civilization.

1.2.3
Inequalities

The goal of ecological sustainability is even more imperative if we consider the problem of disparity [27]: the passengers of spaceship Earth travel, indeed, in very different "classes." As an average, an American consumes about 7.11 toe of energy per year, a quantity approximately equal to that consumed by two Europeans, 4 Chinese, 17 Indians and 240 Ethiopians [28]. The uneven consumption of fossil fuels and the related generation of waste products are accompanied by uneven consumption and consequent uneven waste generation of any kind of nonrenewable store, for example, metals [29, 30].

Disparity is indeed the most worrying feature of our society. The poorest 40% of the world's population account for 5% of global income, and the richest 20% account for three-quarters of global income. According to the World Bank [31], the Gross Domestic Product (GDP) at purchasing power parity per capita is higher than $30000 in at least 25 countries ($46400 in the US), but it is below $3000 in more than 50 countries and less than $1000 in 15 African nations [32]. The three richest persons in the world have assets that exceed the combined GDP of the poorest 47 countries.

Income inequality is vast and is reflected in all aspects of life: health, education, food, energy, and so on. Life expectancy at birth is higher than 79 years in most of the affluent countries with a peak of 83 years in Japan, but in several African nations it is below 50 years, for example, 47 years in Nigeria [2]. The adult literacy rate is close to 100% in many countries, but it is below 50% for at least 15 nations, mostly African. In more than 45 nations at least 20% inhabitants do not have access to a reliable water source. Large differences are also found in the ecological footprint, that is 9.0 ha per capita in the US, 11 times higher than that in India [33]. It has been calculated that if all the world's 6.8 billion inhabitants were to live at current American ecological standards, we should look around for another four Earths to accommodate them [33].

There are also strong inequalities among citizens within each nation. The gap between rich and poor is larger in developing nations, but is increasing in almost all the affluent countries. The ratio between the household incomes of the richest 10% to the poorest 10% is 168 in Bolivia, 51 in Brazil, 16 in the US, and 11 in Italy [34]. In spite of the presence numerous billionaires (356 in 2009), poverty in the US is endemic, with roughly 13–17% of the people living below the federal poverty line [35]. In 2008, 11% of American households were food insecure, with 30% of African American minors living below the poverty threshold [36]. Health disparity is a big problem in several countries including the US [37]. Domestic disparity is a difficult problem to solve in a society where the way of life is based on consumerism, and international disparity is a problem set aside by politicians of affluent countries to please their supporters. In the long run, however, both problems have to be tackled because disparities destabilize human society. If things do not improve, sooner or later the poor will rise up against the rich. The boost of "illegal" immigration in affluent countries that lie at the boundary between the North and the South of the world (e.g., United States, Italy, Spain) is indeed a forewarning of what will happen in the international scene. Any action to restore equity should likely pass through lowering resource consumption (in particular, energy) by the rich while attempting to raise that of the poor.

Our time is characterized by an unsustainable growth in an unequal world. We should try to decrease disparity, while being aware that growth based on consumption of nonrenewable resources is poised to be an ephemeral illusion.

1.3
Energy and Climate Crisis

In the last 100 years the strong increase in population and the availability of large amounts of fossil fuels have led to an average primary energy consumption rate of almost 15 trillion watts (i.e., 15 TW) of power worldwide [7]. Current global trends in energy supply and consumption are environmentally, economically, and socially unsustainable. We need more energy to fill the gap between the industrialized and the developing countries, but at the same time we will not be able or allowed to consume more fossil fuels for several reasons: their limited amount, their increasing cost, and, above all, the need to reduce CO_2 emissions.

According to the fourth assessment report of the United Nations Intergovernmental Panel on Climate Change (IPCC) [15], an increase in carbon dioxide concentration leads to an increase in the greenhouse effect that, in turn, causes climate change. This will impact food security [38], water availability [39], fish production [40], and global forests [18]. Other not less dangerous effects will be ocean acidification [41] and permafrost melting [42]. Indeed, climate change caused by an increase in the CO_2 concentration in the atmosphere might result in much more than a simple stratigraphic signature of the Anthropocene epoch: it could lead to devastating effects on humanity.

Recently, two important steps in the right direction have been made: Europe signed its own climate agreement committing the region to a 20% cut in emissions by 2020 as well as a doubling of use of renewable energy and boosting energy efficiency by 20% over the same period, and US President Obama decided that combating climate change is a priority of his administration. These positive signals are counterbalanced by the great difficulties encountered when trying to set global and concerted policies to curb atmospheric carbon pollution [43].

1.4
Dealing with Change

In the last few decades we have become aware that we live in a fragile spaceship, with limited resources. We realize that we are in danger and that risks derive from two main features: too much consumption and too much disparity. In a restricted and perhaps overpopulated system like spaceship Earth, opportunities discovered and exploited by a generation can cause challenges to the subsequent ones. Fossil fuels have offered outstanding opportunities during the twentieth century in the rich countries of the western world, but now mankind has to face the challenges arising from fossil-fuel overexploitation. We need to reduce progressively the production of CO_2, while providing a suitable energy supply to allow a decent standard of life to all of the people around the world. This means that we have to learn to save energy, to find more intelligent ways of exploiting traditional energy sources and to develop new ones. From the supply side, there are several options, including (i) the use of fossil fuels (in particular, coal) with carbon capture and sequestration, (ii) expansion of nuclear energy with third- and fourth-generation power plants, and (iii) development of a variety of unexploited or underexploited renewable sources like solar energy, wind energy, geothermal energy, ocean energy, and biofuels. On the demand side, the opportunities include (i) saving energy in every action of our life, (ii) limiting energy wastage in heating and cooling buildings, (iii) increasing the efficiency of internal combustion engines and electric motors, (iv) moving from fossil fuel-powered automobiles to electric vehicles, and perhaps (v) developing a hydrogen economy. But energy supply and demand have implications that go far beyond technical features. In developing countries billions of people work hard to improve their standard of living and need much energy to succeed. They look forward to reaching a welfare level comparable to that enjoyed by the citizens of the countries that developed in the past century. To reach this goal, they constantly increase energy consumption and CO_2 production, but these parameters continue to increase also in affluent countries, where people falsely believe that the quality of life increases linearly with energy consumption.

In the last few decades the world has undergone big changes, and we should deal with them. We need new thinking and new ways of perceiving the world's problems.

1.5
Unavoidable Questions

For several reasons, we are deeply interested in finding solutions to the climate and energy crisis. As passengers of spaceship Earth, we need energy for satisfying our fundamental needs while living in a pleasant environment. As parents, we wish to leave our planet in a good shape for the benefit of future generations. As passengers traveling in the first class, we feel obliged to help passengers living in much worse compartments to find a better accommodation. As members of mankind, we have the moral duty of contributing to solving the energy problem as a decisive step towards creating a more peaceful world. And if we are scientists, we have a great responsibility that comes from our knowledge and educational duty [44].

We are at a crossroads, and decision on the path to be taken lies with policy makers who, unfortunately, usually look only at the interest of their own nation and at the next election. What we would need are politicians capable of looking far-off in space and time, statesmen thinking over the whole planet and the future generations. To play as statesmen, politicians need to be counseled; this is particularly true in the case of important and complex problems like energy that need to be tackled globally, with the wisdom deriving from an interdisciplinary approach. Advice should come from scientists of the various scientific and humanistic disciplines, who are less conditioned and better informed than politicians on the present state of spaceship Earth and on what will likely happen in the next few decades. It is indeed their duty to find answers to several entangled, fundamental questions like the following:

- Can we afford to stop burning fossil fuels or at least reduce their consumption?
- How long can we continue to treat the atmosphere and the oceans as a carbon dioxide sewer?
- Can scientists find any new energy source capable of replacing fossil fuels?
- Is it wise to develop nuclear energy?
- Can renewable energies supply us with all the energy we need?
- How can people living in poor countries improve their quality of life?
- Will it be possible for all Earth's inhabitants to reach the standard of living of developed countries without devastating the planet?
- Should the citizens of the affluent countries change their lifestyle and look for innovative social and economic paradigms?
- Is it possible to reach the goal of ecological sustainability?
- Will decreasing resources lead to a destructive collapse of economy or can we manage to descend without too much damage?
- To what extent is well-being, or even happiness, related to energy consumption?
- Is it possible to do more with less?
- Can we afford to wait for the end of the crises, that follow one after the other, before addressing the energy problem and the related climate change?
- Will science and technology alone take us to where we need to be in the next few decades?

These are, indeed, hard questions. History teaches that the pressures of the great, hard questions can bend and even break well-established principles, thereby transforming difficult challenges into unexpected, astonishing opportunities [45]. But we should not forget that the challenge of saving spaceship Earth and its passengers needs the engagement of all of us.

And we have to start right now.

2
Concepts and Misconcepts

"The laws of thermodynamics control, in the last resort,
the rise and fall of political systems, the freedom or bondage of nations,
the movements of commerce and industry,
the origins of wealth and poverty."

Frederick Soddy

2.1
The Elusive Definition of Energy

Energy (E) is not a primary concept. Its definition is based on another concept, namely work (W). Work is defined as the use of a force (F) to move something, where the force is the intensity with which we try to displace (push, pull, lift, kick, throw) an object. The amount of work to do depends on how much force is applied and what distance (d) is being covered. This is expressed mathematically with Equation 2.1:

$$W = Fd \qquad (2.1)$$

The most classic example of work is to lift up some weight (e.g., a book) against the force of gravity. The entity of our work depends on the mass m to move (this book or a big dictionary?), on the height we want to place it (which shelf on the bookcase?), and on the force of gravity (we are probably working on Earth, so this is constant everywhere).

Energy is the capacity of a given system (e.g., a liter of gasoline, an organism, a machine) of doing work. In broader practical terms, energy is the capacity to produce a change (of temperature, speed, location, chemical composition, etc.) in a given system. The SI unit of measure to quantify energy is the *joule*, which is the amount of energy exerted when a force of 1 newton (amounting to the force required to accelerate a mass of 1 kg at a rate of $1\,\mathrm{m\,s^{-1}}$) is applied over a displacement of 1 m. Energy must not be confused with *power*, which is the rate at which energy is used and is thus expressed as joules per second ($\mathrm{J\,s^{-1}}$), defined as the watt (W).

Although the above arguments allow to give a scientific definition of energy, we have to admit that energy continues to be an elusive concept, capable of defying

Energy for a Sustainable World: From the Oil Age to a Sun-Powered Future. Nicola Armaroli and Vincenzo Balzani
© 2011 WILEY-VCH Verlag GmbH & Co. KGaA, Weinheim
ISBN: 978-3-527-32540-5

attempts of rationalization and classification. This is true not only for ordinary people but also for illustrious scientists. In fact the physicist and Nobel Prize winner Richard Feynman wrote that "It is important to realize that in physics today, we have no knowledge of what energy is" [46]. To clarify this issue, Einstein's discovery that energy and mass are basically the same stuff packaged in forms that make them appear different, likewise water and ice, probably does not help too much (Section 8.1.2).

Indeed energy, like time, is something that everyone feels to be intimately familiar with but, when asked to define it, one gets stuck. Just be satisfied with this: energy changes and transformations make things happen.

2.2
A Taste of Basic Principles

Despite the fact that, generally, there is a superficial notion of the concept of energy, science has been capable of acquiring a deep understanding of the rules that govern its production and transformation. This was made possible thanks to the pioneering work of a few scientists and engineers who operated in the period ranging approximately between 1820 and 1940. Among many others, we just mention James Watt, Sadi Carnot, Justus von Liebig, James Joule, Rudolf Clausius, William Thompson (Lord Kelvin), Ludwig Boltzmann, Walther Nernst, Albert Einstein, and Edwin Schrödinger [3]. Curiously, the ambition of some of these scholars was just to set up or improve relatively rudimentary machines that transformed heat into mechanical work (a then apparently minor branch of physics termed "thermodynamics"). While doing this, however, they enabled the disclosure of some of the principles that drive the entire Universe [47]. Unfortunately, at that time, they often could not comprehend the importance of their discoveries, which stand among the greatest achievements of human ingenuity.

The first law of thermodynamics states that the energy of an isolated system is constant. Therefore the energy content of the Universe is invariable: energy can neither be created nor destroyed, it can be only transformed from one guise to another. This is a somewhat counterintuitive concept, in fact people are normally convinced that energy is destroyed upon use. After all the gas tank of the car is empty after traveling for a while and, apparently, there seems to be no traces anywhere of the valuable fuel we had carefully put in.

The second law of thermodynamics, which can be phrased in many ways, essentially states that the conservation of energy is inexorably accompanied by a loss of "quality": energy is transformed from useful and noble forms to increasingly useless forms. This qualitative argument can be quantified by using the quantity called *entropy*. The second principle is of central importance in the whole of science and is the key tool for the rational comprehension of the Universe because it provides the instruments to understand why any change occurs [47].

It is far beyond the scope of this book to discuss in detail the four (and not just two) principles of thermodynamics, which entail a deep understanding of the only apparently simple concepts of heat and temperature [47]. Our goal is not to ponder the ethereal concept of energy and the fascinating laws that govern it, but just to describe how we harness, transform, distribute, fight for, waste, and use this ubiquitous and essential stuff that pervades our existence.

And why we have to change all of this as soon as possible.

2.3
Converting Primary Energy into Useful Energy

Our very source of energy is Nature itself. The energy embodied in natural resources, prior to undergoing any human-made conversions or transformations, is defined as *primary energy*. Primary sources fall into two categories: flowing and stored. Examples of energy flows include sunlight, wind, and waves (e.g., water or sound). Stored energy includes fossil fuels, biomass, fissile atomic nuclei, and the heat stored in Earth's upper crust. The availability of each primary energy source offered by Nature is normally huge; however, there are a number of technological, economic and environmental constraints on converting them into useful energy.

Primary energy sources can be converted into *useful energy* which, in our everyday life, fall into three main categories: electricity, heat, and fuels. The last are the most versatile because they can be stored for an indefinite period of time, conveniently transported and used when needed. On the contrary electricity and heat can be stored only in limited quantities and for short periods of time; practically, they are used as soon as generated. On a broader perspective, there are essentially seven forms of energy, most of which are commonly encountered in our everyday life such as heat, electromagnetic (e.g., light), kinetic (or mechanical), and chemical (e.g., food, gasoline). Their conversions are summarized in Figure 2.1.

Conversion table of different types of energy

FROM: \ TO:	Thermal	Chemical	Electrical	Light	Kinetic	Nuclear	Gravitational
Thermal		Endothermic reactions	Thermionic processes	Incandescent light bulbs	Internal combustion engine		
Chemical	Combustion		Batteries	Fireflies	Muscles		
Electrical	Electrical resistances	Electrolysis		Electro-luminescence	Electric engine		Pumped reservoir storage
Light	Solar collectors	Natural photosynthesis	Photovoltaic panels		Solar sail		
Kinetic	Friction	Radiolytic reactions	Electric alternator	Accelerated charged			Rising objects
Nuclear	Nuclear fission and fusion	Ionization	Nuclear batteries	Nuclear weapons	Radioactive decay		
Gravitational					Water turbine		

Figure 2.1 Types of energy and some processes or devices that convert them into each other (adapted from [3]).

2.4
It Takes Energy to Make Energy: the EROI

It is crucial to emphasize that primary energies, as such, are hardly usable directly to provide useful energy services, perhaps with the only exception of biomass. In fact this is virtually the only energy source utilized by primitive civilization or, still today to a large extent, by the poorest people in the world.

Oil is a precious natural resource, but only upon refining does it become the most useful and versatile stuff in the world. To make useful energy for modern societies, a largely invisible infrastructure made of extraction/harvesting, transformation, and distribution facilities is needed. This has to be built, maintained, and ultimately decommissioned, and this requires an incommensurate amount of money. Even more important, perhaps, is the concept that all of this effort requires a great deal of energy to be accomplished.

The parameter that quantifies the above concept is the EROI, energy return on investment, sometimes indicated as EROEI, energy return on energy investment. EROI is the ratio of the amount of usable energy acquired from a particular energy resource to the amount of energy expended to make it usable. When the EROI of a resource is equal to or less than one, that energy source becomes an "energy sink" and its use as a primary source becomes pointless or at least questionable.

The estimation of EROI is extremely complex because it implies aggregation of rather different energy investments for completely diverse energy sources, which has prevented the implementation of a universally accepted methodology of calculation [3, 48]. Indeed reliable comparisons of EROI can be accomplished for the same energy source (e.g., oil coming from different geographic locations), but the same is not guaranteed if comparisons are made between completely different energy sources (e.g., oil versus wind). In the US, when oil was extracted at very shallow depth (1930s), the EROI of the oil extracted was around 100. Through a complicated pattern of ups and downs, it then progressively decreased to 15 in the late 1990s. Since the early 1970s, both the extraction rates and the extraction EROI have shown a decline [49].

EROI is a very useful parameter to underpin the physical limits of energy resources available on spaceship Earth. It ultimately explains why the oil age will finish well before we will run out of oil.

2.5
Embodied Energy

It also instructive to assess the energy that is used to make a product and thus turns out to be "embodied" in it. This exercise can be accomplished at different levels of analysis and would be (ideally) perfect only if all the materials and machinery utilized directly and indirectly to manufacture a given good were accounted for, from the primary energy resources initially extracted to dismantling and/or recycling [50].

Table 2.1 Embodied energy of some common materials (MJ kg^{-1}) [51].

Material	Energy cost (MJ/kg)	Made or extracted form
Aluminum	230–340	Bauxite
Bricks	2–5	Clay
Cement	5–9	Clay and limestone
Copper	60–125	Sulphide ore
Glass	18–35	Sand, etc.
Iron	20–25	Iron ore
Limestone	0.07–0.10	Sedimentary rock
Nickel	70–230	Concentrated ore
Paper	25–50	Standing timber
Polyethylene	87–115	Crude oil
Polystyrene	62–108	Crude oil
Polyvinylchloride	85–107	Crude oil
Sand	0.08–0.10	Riverbed
Silicon	200–250	Silica
Steel	20–50	Iron
Sulphuric acid	2–3	Sulphur
Titanium	900–950	Concentrated ore
Water	0.001–0.010	Streams, reservoirs
Wood	3–7	Standing timber

Clearly, it is more difficult to assess embodied energy for a complex object such as an aircraft, which is made of many different materials, than for a bulk material like steel or for an agricultural product. Normally, for a given item, an analysis that accounts for the energy used in its assembly and consumed to make its structural materials provides at least 80–90% of embodied energy. In Table 2.1 are reported the energy costs of some common materials.

In 1971 the American ecologist Howard Odum stated that "This is a sad hoax, for industrial man no longer eats potatoes made from solar energy, now he eats potatoes partly made of oil." Indeed, by analyzing the energy cost of harvested products in modern agriculture, this observation does not seem to be exaggerated. In fact harvested wheat, corn, and fruits require 0.25 tons of oil equivalent (toe) for every ton of product, but greenhouse tomatoes or peppers may require 1 toe per ton [50]. This means that their dietary energy content can be up to 50 times smaller than the energy needed to cultivate them. A sizeable fraction of the hidden energy subsidy in foodstuffs is represented by fertilizers, whose production is highly energy intensive [52].

Even so, in affluent countries the largest share of the energy cost of food is spent *after* harvesting, and is ascribable to processing, packaging, transport (sometimes overseas), refrigeration, retail, shopping trips, cooking, and finally washing of dishes. Taking all of this into account, it has been estimated that, on average, the amount of energy that goes into producing a week's supply of food in the UK is nearly five times greater than what the eater gets out of the final product [53].

A similar "embodied approach" is also being pursued to assess the hidden consumption of water, another indispensable resource for the production of food and materials [54].

2.6
Energy Units and Conversions

The *joule* (J), which amounts to $1\,kg\,m^2\,s^{-2}$, is a too small a unit to describe ordinary energy quantities, therefore it has to be typically utilized with the conventional prefixes. The daily adult food intake in affluent countries is about 10 million joules (10 MJ), the gasoline storable in the tank of an average car contains about two billion joules (2 GJ), and the current world primary energy consumption is about 500 billion billion joules (500 EJ).

Very often, larger energy units of measure are utilized. In the electrical sector, the multiples of the *watt-hour* (Wh) are very popular (kWh, GWh, TWh). 1 Wh is the work done by 1 W acting for 1 h and amounts to 3600 J.

In the fossil fuel industry the most popular unit is probably the *ton of oil equivalent* (toe), which is the amount of energy released by burning one metric ton of crude oil. Since different crude oils have different calorific values, the exact value of the toe can only be defined by convention. Unfortunately, there are several slightly different conventions: for the IEA and EIA 1 toe = 41.868 GJ, for the WEC 1 toe = 42.0 GJ. These small differences may become important when oil balances are made on the grand scale, by aggregating data from difference sources. This confusion in the use of conversion factors renders the accounting of oil trade, which has already a reputation of scarce transparency, even more obscure [55]. With a similar meaning as toe, a popular energy unit is also the *barrel of oil equivalent* (boe), which, as a unit of volume not belonging to the SI and not univocally defined, makes the conversion to energy units possibly even more arbitrary than for toe [55].

Other popular energy units of measure, particularly in the thermal energy sector, are the *calorie* (1 cal = 4.1868 J) and the *British thermal unit* (btu) (1 btu = 1055.055 85 J), the latter often expressed as *quad* (short for quadrillion btu, i.e., 10^{15} btu) when accounting for huge fluxes of energy. Finally, the *electronvolt* (eV) is a tiny unit employed to describe process at the atomic or molecular level, $1\,eV = 1.602\,176 \times 10^{-19}\,J$. In Table 2.2 are reported the principal energy units of measure and their conversion factors.

2.7
The Immense Energy and Power Scales

The quantities of energy at play in natural or man-made processes can vary immensely over powers of ten [51]. The box "A Journey Through Energy and Power" gives us an idea of the immense power of Mother Nature on tiny spaceship Earth.

Table 2.2 Conversion factors between some major units of energy, according to IEA/OECD/EUROSTAT [56].

From	To				
	TJ	Gcal	Mtoe	MBtu	GWh
Terajoule	1	238.8	2.388×10^{-5}	947.8	0.2778
Gigacalorie	4.187×10^{-3}	1	10^{7}	3.968	1.163×10^{-3}
Megatoe	4.187×10^{4}	10^{7}	1	3.968×10^{7}	11 630
Million Btu	1.055×10^{-3}	0.252	2.52×10^{-8}	1	2.931×10^{-4}
Gigawatt-hour	3.6	860	8.6×10^{-5}	3412	1

A Journey Through Energy and Power

The basic currency of our fossil-fueled civilization, the C–C and C–H chemical bonds, contain about 0.7×10^{-18} J (0.7 aJ); the kinetic energy of a flying mosquito is about 150 nJ; striking a keyboard button with one finger requires 20 mJ; 1 kg of good quality coal contains about 30 MJ (about three times the daily food intake of a human being); the Hiroshima bomb released 84 TJ; a 1000 MW power plant operating for 7000 h per year produces 25 PJ, about one-tenth of the energy delivered by the largest H-bomb ever tested (240 PJ); a typical Caribbean hurricane carries 38 EJ; the global fossil fuel yearly consumption has reached 380 EJ [4, 7]; the solar radiation hitting the Earth annually amounts to 5 500 000 EJ. The short trip we have just made entails 43 orders of magnitudes along the energy scale.

It is also funny to travel through the scale of power and realize the rate at which different phenomena or entities consume or release energy. The minimum discernible signal at the antenna of a good FM radio receiver is 2.5 fW; the average consumption rate of a living cell is about 1 pW; the power of a laser in a CD-ROM drive is 1 mW; a burning candle develops 5 W of heat (mainly) and light energy; a washing machine working at 60 °C requires 800 W; the engine of a Formula 1 car is rated at 550 kW, the four engines of a B-747 airplane at 80 MW; a strong average thunderstorm releases energy at the rate of 100 GW, a large volcanic eruption at 100 TW; a Richter magnitude 8 earthquake is capable of delivering energy at the rate of 1.6 PW, spreading devastation hundreds of kilometers away from the epicenter through tsunamis.

These digits should probably arouse more respect for Mother Nature by human beings.

2.8
Some Energy Key Parameters

There exist a few parameters that are commonly used to compare energy sources, technologies, and performances [3]. Here we list some of the most significant.

- *Energy density* ($J\,l^{-1}$) is the amount of energy stored in a given system or region of space per unit volume and is commonly used to characterize fuels. *Specific energy* ($J\,kg^{-1}$) is the energy per unit mass and is largely utilized to describe the energy content of food. These parameters are essential whenever an energy supply has to be transported. For instance, airplanes could never fly overseas powered by natural gas whose energy density, at ambient conditions, is about 1000 times smaller than that of liquid kerosene. On the other hand, mountain climbers do not carry carrots in their backpacks but chocolate or power bars.

- *Energy concentration* ($J\,m^{-2}$) is the spatial density of energy resources, a critical determinant in assessing the infrastructure needed for exploitation [3]. Small hydrocarbon fields have an energy concentration of less than $1\,GJ\,m^{-2}$, whereas the most concentrated oil field in the world (al-Burkān, Kuwait) prorates to $1\,TJ\,m^{-2}$, about 10 times the Canadian oil sands in the Athabasca basin [3]. In biofuel crops this parameter provides the limit of harvesting possibilities. In all instances, energy concentration reminds us of the inexorable limits of spaceship Earth.

- *Energy efficiency* ($J\,J^{-1}$) is the ratio of the amount of energy output to the amount of energy input and is used to describe the performance of energy converters. Among the most efficient energy machines stand water turbines, that convert potential energy into electricity with efficiency of about 90%.

- *Power density* ($W\,m^{-2}$) is the power harnessed, generated or used over the land or sea surface. It can be used to describe energy extraction sites (e.g., oil fields), generating facilities (e.g., power stations), or consumption centers (e.g., buildings). This parameter reveals why it is impossible to power a steel mill with the solar radiation intercepted by its roof (Section 9.4).

- *Energy intensity* (J per unit of currency) is the amount of energy required to produce a given economic product or service. The ratio between TPES (total primary energy supply) and GDP (gross domestic product) is a common indicator of the energy intensity (EI) of an entire economy: high intensities indicate a high cost of converting energy into wealth. Hence, EI is often taken as an indicator of the energy efficiency of a nation's economy or, in simple terms, of the capability of doing more with less. Unfortunately, this parameter is strongly biased and comparisons of energy intensities in different countries have to be made with caution [52]. First of all, both parameters that define EI are inherently inaccurate. The GDP is not calculated in the same way in different countries and also does not capture black market transactions that, even in the richest industrialized countries, can be up to 25% of the real national economy [52]. TPES are calculated with conversion factors which are not internationally established [55] and their (in)accuracy also depends on the specific technology considered (what is the primary energy equivalent of hydroelectricity versus thermoelectricity?). Additionally, consumption at the national level is the result of a complex pattern of peculiarities such as natural (e.g., nations with colder climates consume more energy), structural (e.g., the presence of a heavy industrial system), technical (e.g., average commuting distance in the transport sector), historical (e.g., how the energy infrastructure has developed over time), and cultural (e.g., in some countries artificial cooling is abused) [52].

2.9
Energy Pervasiveness Versus Energy Illiteracy

In modern societies, we consume energy 24 hours a day, 7 days a week. Even when we sleep the fridge is working, the furnace is burning gas, the alarm clock is on. From the very first instant after waking up, our energy consumption rises suddenly: we take a bath, we eat breakfast, we commute to work. There we find a warm or cooled place, we switch on a PC, we send an e-mail to the opposite side of the planet. Its trip takes just a fraction of a second and there must be quite a bit of energy consumed (no idea where) to make this possible.

Energy is the most omnipresent but, at the same time, the most unnoticed entity that pervades our daily life. We continuously and inadvertently make use of it by pushing buttons, turning keys, buying tickets, opening taps. We are proud to be educated citizens in a "knowledge-based" society, in which we are literally bombed by a steady deluge of information. Unfortunately, the large majority of us know little about the very stuff the makes our modern life what it is: a relentless flow of energy. This makes all of us energy emperors with legions of virtual energy servants working for us continuously (Chapter 3).

Sure, we often complain that energy is expensive. But this is irrational hysteria. Who has ever realized that, even at $80 per barrel (bbl), a liter of crude oil is cheaper than a liter of Coca Cola? Never thought that a liter of refined gasoline in Europe is much cheaper than a liter of mineral water in a pizzeria? And, just to make the comparison even more astounding, the former has a tax burden of over 60%, the latter of less than 10%. How many people crying over the electric bill know the average price of a kWh of electricity, and how much work can be done with it? Moreover, who knows how many kWh of electricity are consumed annually by his/her household? And what about the cubic meters of natural gas burned to make our life more comfortable?

Most likely, we might cut our energy expenditures by reducing consumption, but this option is not practically viable, because we do not have any quantitative control of our consumption. Gas and electricity meters are often placed in the most neglected corners of our house and we are sort of scared by them. On the other hand, the monthly bill is somewhat obscure and rarely states with clarity how much energy was consumed.

Another issue that prevents us from being more conscious and effective with energy saving is our absolute misperception on consumption. Many people in affluent countries tend to associate (or even overlap) the concept of "energy" with that of "electricity", and believe that most of their consumption is electrical. This is profoundly wrong because, though the role of electricity continues to enlarge,

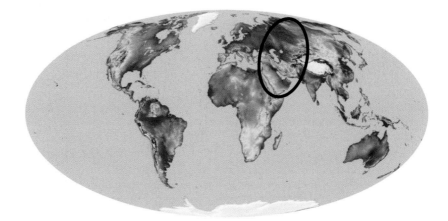

Figure 2.2 The so-called "energy strategic ellipse", an area stretching from the Arabian Peninsula to Western Siberia, where about 70% of the world's proven oil and gas reserves are concentrated.

even in the most affluent countries it represents only 20–30% of final energy uses. In fact, contrary to common perception, over 50% of energy consumption in modern houses is spent in the most trivial fashion: warming up water for heating, washing, and cooking. Then the second entry of our personal energy budget is transportation.

Energy is not only crucial in our personal daily life, but is also a key political and strategic issue. The so-called strategic energy ellipse (Figure 2.2) is not by chance the hottest geopolitical region in the world, which concentrates all the major international conflicts and tensions. Indeed, an increasingly larger part of the present and future world's energy supply originates in this area and its control is vital for maintaining the high level of prosperity reached in some limited areas of the world [57]. The only remaining military superpower, the US, in the last 20 years has spent trillions of dollars, thousands of human lives, and stupendous amounts of energy (equivalent to the consumption of entire nations) [52] to keep its influence on this area (Chapter 7).

Energy is most important for the present and the future of mankind. Energy illiteracy is an enemy that we have to fight obstinately if we want to have a chance to transform the present energy system.

2.10
Key Numbers: an Abacus for Energy Literacy

As pointed out in Section 2.7, dealing with energy numbers can trigger vertigo, due to the vast broadness of scale involved. Here we present, with a few comments in the captions, Figures 2.3–2.8 and Tables 2.3–2.4, with key energy data that may be useful landmarks for not losing the way in the immensity of this

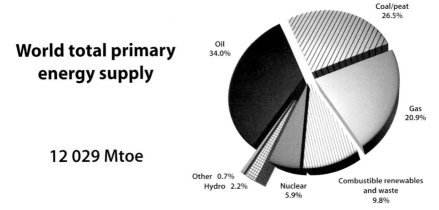

World total primary energy supply

12 029 Mtoe

Figure 2.3 World total primary energy supply by source in 2007. The large share of combustible renewables is mostly due to (estimated) consumption of non-marketed biomass in developed countries [4].

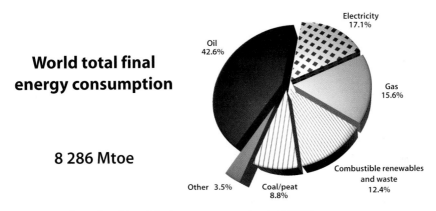

World total final energy consumption

8 286 Mtoe

Figure 2.4 Share of world total final energy consumption in 2007 [4].

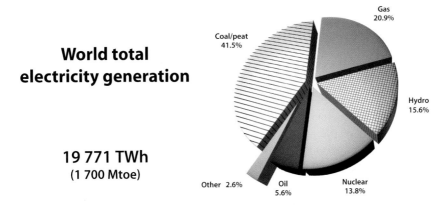

World total electricity generation

19 771 TWh
(1 700 Mtoe)

Figure 2.5 World total electricity generation by source in 2007 [4].

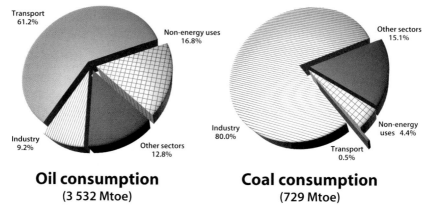

Oil consumption
(3 532 Mtoe)

Coal consumption
(729 Mtoe)

Figure 2.6   Shares of world oil and coal consumption in 2007 [4].

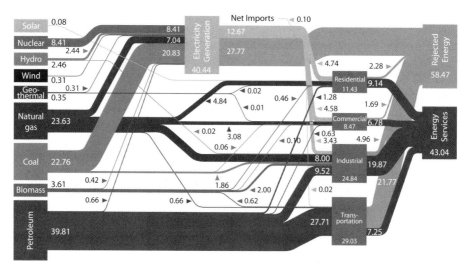

Figure 2.7   The complex pattern of energy transformation in the US. The primary energy input is on the left, the overall final result is on the right and shows that, out of the primary 101.5 quads of energy consumed in 2007, only about 43 quads were used to provide energy services and over 58 quads were lost (i.e., not utilized to provide energy services) because of inefficiencies in energy production, distribution, and use [58].

multidisciplinary and multifaceted problem. We propose them as an abacus to everyone who wants to take up the challenge of increasing energy knowledge and awareness. Further specific data on oil, gas, and coal are tabulated and illustrated in Chapters 4–6. Another useful source of information is the Appendix "Did you know that ... ?" at the end of the book.

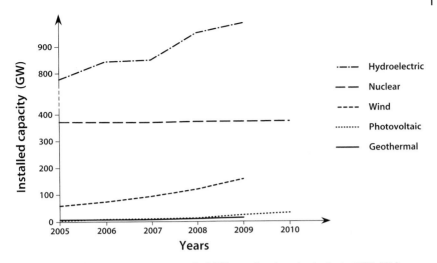

Figure 2.8 Global installed capacity trend of different electric technologies in 2005–2010.

Table 2.3 Population, primary energy and electricity consumption, and CO_2 emissions from the top 10 world energy consumers [7, 59]. For comparison, data for the whole world and Ethiopia are also reported. Data are for 2009 unless otherwise noted.

Country	Population[1]	Primary energy		Electrical consumption		CO_2 emissionn	
		Total (Mtoe)	Pro capite (Mtoe/pers)	Total[2] (TWh)	Pro capite (kWh/pers)	Total (Mton)	Pro capite (ton/pers)
U.S.	307 006 550	2 182	7.11	3 923.8	12 780.9	5 832.8	19.0
China	1 323 591 583	2 177	1.64	2 835.0	2 141.9	6 533.5	4.9
Russia	140 041 247	635.3	4.54	840.4	6 613.2	1 729.4	12.3
India	1 156 897 766	468.9	0.41	568.0	491.0	1 494.9	1.3
Japan	127 078 679	463.9	3.65	1 007.1	7 191.2	1 214.2	9.6
Canada	33 487 208	319.2	9.53	536.1	16 009.1	573.5	17.1
Germany	82 329 758	289.8	3.52	547.3	6 647.7	828.8	10.1
France	64 420 073	241.9	3.76	447.2	6 941.9	415,3	6.4
South Korea	48 508 972	237.5	4.90	386.2	7 961.4	542.1	11.2
Brazil	198 739 269	225.7	1.14	403.0	2 027.8	428.2	2.2
Ethiopia	85 237 338	2.7	0.03	3.13	36.7	5.97	0.1
World	6 776 917 465	11 164	1.65	17 109.7	2 524.7	30 377.3	4.5

1) in 2009, according to US Census Bureau, http://www.census.gov.
2) in 2007, latest data available from Energy Information in August 2010, http://www.eia.doe.gov.

Table 2.4 Key global energy data.

Solar energy input on Earth's surface	5 500 000 EJ
World consumption of primary energy	470 EJ
World average power demand	15 TW
World oil annual consumption	31 Gbbl (3.9 GToe)
World gas annual consumption	3 Tm^3 (2.7 GToe)
World coal annual consumption	7 Gt (3.3 GToe)
World installed electricity capacity	4.5 TW
Annual CO_2 emission by fossil fuel combustion	30 Gt

3
Energy in History

"Historia magistra vitae."

Cicero

3.1
Historia Magistra Vitae

History is the teacher of life, and life is nourished by energy. A critical examination of the periods in the history in which major changes in energy use have occurred can be helpful to understand better our present relationship with energy as well as to guide our decisions and choices.

From simple cellular creatures to large complex mammals, the very nature of life consists indeed in energy absorption and conversion for growth, self-preservation, and procreation. The cycle of life has continued for millions of years and, throughout eons, the right combination of ecosystems and geological conditions caused the formation of fossil fuel deposits in some limited areas of the planet. About 2 million years ago humans started to manufacture tools for hunting. Then, perhaps 500000 years ago or much earlier, man discovered the use of fire, a force that could be utilized both to create and to destroy [60]. It provided warmth and light, but could also transform the living into dead. Since then, man was able to affect the environment by selectively creating and destroying what was needed for his survival. In those early times, however, the ratio of food and wood to humans was very high, so man's impact on energy resources was insignificant. Then, when man learned to use fire for craft, like melting metals and hardening clay, his role in causing environmental changes started to increase. This role has magnified exponentially over time, and our modern societies' consumption patterns indeed have their roots in the utilization of tools and fire by early mankind.

About 18000 years ago, with the domestication of animals, man was able to capture a new substantial source of energy. Domesticated animals like sheep, goats, pigs, and cows provided not only a reliable and predictable source of food, but also mobile stores of energy for nomad populations. Around 10000 years ago, some groups of nomads began to settle in fertile river valleys, where they could

Energy for a Sustainable World: From the Oil Age to a Sun-Powered Future. Nicola Armaroli and Vincenzo Balzani
© 2011 WILEY-VCH Verlag GmbH & Co. KGaA, Weinheim
ISBN: 978-3-527-32540-5

develop more reliable sources of food and energy. Permanent settlements and the population growth led early human civilization to use ever greater quantities of trees, soil, freshwater, and animals to satisfy their rising demand for energy and food. Around 8000 years ago, oxen were used to pull carts, wagons, and simple wooden plows. The more powerful horses were employed only 2000 years later.

Agricultural societies were successful for long time periods, but eventually collapsed for several reasons, primarily because of shortfalls of food, the most important energy resource. The first record of such a collapse concerns the Sumerian civilization of Mesopotamia. The Sumerian society developed innovative and vast irrigation systems that for many generations increased crop yields, but forcing agriculture to cope with increase of population strained the fertile soil, which is not a renewable resource [61]. Progressive erosion of the soil caused a decline in crop yields, thereby playing an important role in the collapse of the society [60].

The history of civilization shows, indeed, that progress occurs when abundant energy is available, whereas decline takes place when excess energy is lacking [62]. Of course, other factors may also be involved [63], but most of them are related to energy by positive or negative feedbacks.

The fall of the Roman Empire, for example, was a complex phenomenon where different negative factors reinforced each other. A society, and particularly an empire, is a complex entity and, hence, its collapse must be related to complexity [64]. Complexity gives a benefit, but it is also a cost that must be paid for by energy availability. As the Roman Empire continued to expand, complexity (e.g., of bureaucratic structures, civilian construction, and long-range military ventures) enlarged, with ever increasing energy costs. Emblems of that time's complexity are suggestions concerning original but improbable ways to use energy, like a warship powered by oxen (Figure 3.1 [65]). By the end of the Roman Empire, Italy and most of the Mediterranean territories had been stripped of forest cover, and in several places soil was severely degraded.

Collapse of other ancient and modern societies, caused by a variety of factors directly or indirectly connected to shortfalls of energy resources, have followed similar patterns [63]. Often, collapse was accelerated by wars originated by competing demands on scarce energy resources. The lesson that too large resource absorption can lead to collapse of a society is clear from history and should be taken into serious consideration by our society, that is based on exaggerated energy consumption.

3.2
Animal Power

Many stationary labor tasks, such as the pumping of water and the milling of grain, began to be mechanized in antiquity. Energy conversion into mechanical movements capable of facilitating land cultivation and transport over long distances, however, came only in the 19th century or even more recently.

Figure 3.1 Ox-driven paddle-wheel warship described in the anonymous 4th–5th century military treatise *De Rebus Bellicis*: "In its hull, oxen, yoked in pairs to capstans, turns wheels attached to the sides of the ship; paddles, projecting above the circumference or curved surface of the wheels, beating the water with their strokes like oar-blades as the wheels revolve, work with an amazing and ingenious effect, their action producing rapid motion."

Old agricultural societies required heavy human labor, particularly to plow, collect, and transport crops. While people can work constantly at a rate of 50–80 W, the sustained power of working animals ranges from 300 W for smaller oxen to 700–800 W for good horses. Healthy animals could carry on at this rate (with short rests) for hours. The work of a strong horse was equivalent to one day's work of at least 10 strong men, but the horse's consumption of 4 kg of oats per day had to be obtained by the cultivation of an extension of land that would have provided wheat for six adults [3].

The emergence of cities marked the beginning of sedentary societies that produced enough surplus of food energy to allow a portion of the population to engage in activities other than crop cultivation and animal feeding and husbandry. Cities were the primary drivers of increasing social and economic complexity. For millennia only a small fraction of the total population was urbanized but, at the end of the first century AD, Rome had more than half a million people. Whereas sea transport had been diffuse since the beginning of civilization, land transport evolved very slowly and was based, in different countries, on oxen, horses, camels, elephants, and yaks. Ancient Romans know very well the importance of transportation and therefore constructed an excellent network of good roads that are still at the basis of some land transportation in European countries. Long-distance horse-drawn transport of goods was replaced by rail only after 1840, but draft horses remained indispensable on farms and also for distributing goods and moving people within cities until the first decade of the twentieth century.

In 1910, there were 24.2 million horses and mules on US farms. It has been estimated that these animals required approximately 50 Mt of feed, obtained from about 35 Mha, or 22% of the nation's harvested area in that year. Another 15% of the area was needed to feed horses and mules in industries and cities. On US farms, horse numbers peaked at 26.7 million in 1918, when there were already 85 000 tractors and 89 000 trucks. In China, in the early 1950s the agriculture was still powered by more than 50 million oxen, horses, water buffalo, and donkeys [3].

About 50 000 horses were required to keep Victorian London's public transport running. According to one writer of the time, these horses deposited 1000 tons of manure on the roads every day and, in 1894, it was estimated that in 50 years every street in London would be buried under 2.7 m of manure [66]. In 1903 London had 3623 horse-buses, each one requiring 12 horses per day. There were also 11 000 horse-powered cabs and countless carts and other vehicles, all working constantly to deliver the goods needed by the then largest city in the world.

The great horse-manure crisis vanished when millions of horses were replaced by millions of motor vehicles that eventually led to the climate and pollution crisis we are experiencing nowadays.

3.3
Human Slaves and Energy Slaves

Besides using animal power, all the great civilizations of the past made extensive use of slaves. It has been estimated that it took 100 000 slaves over 20 years to build the Great Pyramid in Egypt; in China, the Great Wall's main portion took 100 years to build, while innumerable slaves died during the construction. Today slavery is formally outlawed in many countries and condemned by the Universal Declaration of Human Rights. Nevertheless, the practice continues in various forms around the world [67].

In the past, rich people had many slaves at their service for a variety of tasks but, even today, each of us has a number of unseen energy slaves at his/her service [68] (see box).

3.4
Waterwheels and Windwheels

The process of mechanization extended gradually over many centuries. Waterwheels and windwheels were the first simple mechanical machines [69]. Waterwheels were used for crop irrigation, grinding grain, supplying drinking water to villages, and later to drive sawmills, pumps, forge bellows, and tilt-hammers and to power textile mills. As early as the first century, the horizontal waterwheel, which is very inefficient in transferring the power of the water current to the milling mechanism, was being replaced by waterwheels of vertical design [70]. The

Our Virtual Energy Slaves

The concept of "energy slave", introduced by Richard Buckminster Fuller in 1944, can be used to understand how lucky we are to live in a world energized by intrinsically cheap and abundant resources, much more powerful than human or animal power. A healthy person can generate a power of 800 W for a short time (e.g., running up on a staircase), but during a continuous day of work he/she cannot develop more than 80 W. Now let us see how this amount of power compares with that we employ in our normal, everyday life.

When we look at a football game on TV, we mobilize a power of 120 W for 1.5 hours. If we could not plug our TV set to the electricity distribution system, we would need one and a half human slaves pedaling for us all the time of the football game to produce the electricity needed (Figure 3.2).

For operating a washing machine (800 W) by human energy, about 10 slaves would be needed. Electrical heating of a small room (2.5 kW) corresponds to the work of over 30 human slaves. A small European green cutter (3.5 kW) consumes in one hour the energy produced by an entire working day of more than four human slaves. A medium-sized car (engine power 80 kW) running on a highway consumes an amount of energy equivalent to that produced by the work of 1000 human slaves. Not even a Roman Emperor had so many slaves available to draw his cart. Nowadays, we can assemble and immediately use such a great number of energy slaves simply by turning the key of our car. A Boeing 747-400 needs 80 MW to take off, which means that as many as 1 000 000 human slaves should be employed to raise it by human muscular work.

But who powers our energy slaves? Mostly, fossil fuels. The power of a big coal electric plant (800 MW) corresponds to that of about 10 million human slaves. The 100 GW of electric power installed in Italy corresponds to the power of 1.3 billion human slaves. On average, each modern Italian citizen has as many as 55 energy slaves at service 24/7, all year long. In the US, each citizen is served by not less than 135 energy slaves.

On growing, we have been taught to command these unseen and maybe even unappreciated energy slaves. They are always ready to assemble and serve, if we only have money to pay their incredibly low salaries. One liter of gasoline costs about €0.5 (before tax) and has an energy content of 12.9 kWh. The energy produced by a full day's work of a human slave is 800 Wh. Therefore, the energy produced by one liter of gasoline corresponds to that produced by about 16 human slaves whose one-day salary should be as low as €0.03 to compete with the cost of the gasoline-based energy slaves.

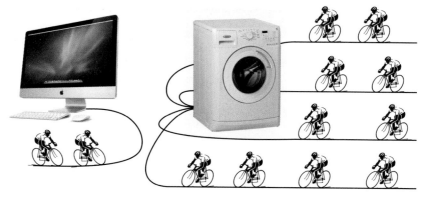

Figure 3.2 The electric power needed to operate a modern wide screen all-in-one desktop computer corresponds to that produced by two persons. For operating a washing machines, the power of 10 people would be needed.

efficiencies of wheel operation improved greatly once iron began to replace wooden hubs and shafts.

The invention of windwheels presumably goes back Heron of Alexandria (ca 10–70 AD), who apparently used it to operate an organ. The first wind-powered grinding grain machines were reported in Persia as early as the 7th century. These remained substantially unchanged until the 12th century, when the horizontal-axis windmill made its appearance in France and England, where the concept was probably brought by Crusaders coming back from the Middle East regions. In northwestern Europe, windmills were used starting from the last quarter of the 12th century. The Dutch refined the technology substantially and used windmills extensively for grain grinding and seawater drainage, thus shaping their country [71]. Dutch settlers brought the windmill to America in the mid-18th century, which constituted an icon of rural America until the early 20th century. The current use of windmills to generate electricity is described in Chapter 13.

3.5
From Wood to Coal

Until the end of the 19th century, much of the energy used by mankind was obtained by wood, that was also increasingly used, century after century, for housing and shipbuilding and any kind of useful device. In 1881, the 31 million US inhabitants still obtained 90% of their energy from wood.

The increase in population and, as a consequence, in wood consumption led to progressive deforestation in several European countries. Therefore, another type of energy source, coal, began to be used (Chapter 6). Coal had long been known because in some places coal seams outcrop at the surface and it was locally used since antiquity for heating, pottery making, and, particularly in China, for metallurgy. At the end of the 16th century, coal demand started to increase, mainly in Britain. Coal was commonly used in households as well as in a variety of small industries (bricks, tiles, glass, soap, sugar, etc.). The amount of coal output increased from 2 Mt in 1650 to 10 Mt in 1700 and 16 Mt in 1815. The development of coal consumption required various problems concerning mining, transport, and industrial uses to be solved [3]. Coal has a major advantage over wood, that is, about 50% higher energy density, but it is more difficult to collect. Furthermore, emissions generated by burning coal produce not only carbon dioxide, not feared at that time, but also a variety of ashes that gave London a deserved reputation as being unfit for human habitation.

At the beginning, mining was completely based on human work, including a large number of women and children. Workers were constantly endangered by cave-ins from unsupported ceilings and high methane levels that caused recurrent, and often deadly, explosions. Before 1800 a great deal of coal was left in place as extraction was primitive and coal was abundant. Large-scale coal mining developed in the following decades. Rising coal demand led also to deeper mines, up to 300 m by 1830.

Coal fueled the transition from traditional artisan economies to modern mass manufacturing. The use of coal surpassed that of wood and crop residues around 1870 in several European countries, around 1880 in the US and 20–30 years later in Russia and Japan. Besides providing coke, heat, and power for stationary and mobile steam engines, coal was also extensively employed to produce town gas, used as a source of urban light. The first public street lighting with gas took place in London on January 28, 1807. In 1812, the first gas company in the world was founded in London. Less than 2 years later, on December 31, 1813, Westminster Bridge was lit by gas. Gas lamps dominated urban lighting, indoor and outdoor, until the 1880s, when they began to be replaced by incandescent lights. From the 18th to the 19th century, the availability of abundant and easy energy caused major changes in agriculture, manufacturing, mining, and transport. These changes had a profound effect on the socioeconomic and cultural conditions throughout Europe, North America, and eventually other parts of the world.

Coal's share in the aggregate global use of commercial energy reached 95% in 1900. Then, its relative importance declined because of the increasing use of oil, but its absolute production grew roughly sevenfold between 1900 and 1989. Coal has now three principal markets: electricity generation, coke production, and cement production (Chapter 6).

3.6
Steam-Powered Machines

Rudimentary steam engines, the first machines designed to convert the chemical energy of fuels into mechanical energy, were designed in Britain in the first decade of the 18th century. It took several decades before arriving at the first efficient steam engine, patented by James Watt in 1769. The typical power of Watt's steam engine was about 20 kW, but five times more powerful engines soon became available. In the first decade of the 19th century the original Watt's engine was modified by introducing much smaller, high-pressure boilers which allowed the development of steam-powered transportation on land (tractors and railways) and water (steamboats). The first public railway, from Liverpool to Manchester, opened in 1830, and by 1850 the fastest locomotives traveled at over 100 km h^{-1}. On water, the concept of paddle-wheel river boats was successfully extended to larger ocean-crossing ships. Coal-powered steamships played an important historical role because they carried most of the 50 million emigrants who left Europe from 1850 to 1914 and, in the same period, they powered Europe's colonial expansion.

Despite their success, Watt's steam engines had two drawbacks: weight and limited power capacity. In 1884 steam turbines were invented, in which the expansion of the steam takes place in multiple stages, resulting in a much higher efficiency. The first model of the steam turbine was connected to a dynamo that generated only 7.5 kW of electricity, but powers several orders of magnitude higher were obtained in a short time. Large steam turbines are currently used in power stations to produce about 80% of the world's electricity. They are used also in nuclear-powered ships and submarines where steam is generated by a nuclear reactor.

By the end of the 19th century there were also some steam-driven cars and trucks in London, Paris, and New York; they disappeared from the market with the advent of vehicles powered by internal combustion engines. Steam turbines were also extensively employed in ships but then oil-powered diesel engines prevailed and in 1911 the Royal Navy began its conversion from coal-powered to oil-powered ships [72]. This conversion may be taken as the indication of the shift from the coal to the oil era.

3.7
Road Vehicles

Early automobile designs used a variety of engines including those powered by steam, but they were too heavy for small vehicles. The history of road transportation began to change in 1876, when the German engineer Nikolaus Otto built the four stroke internal combustion engine that, however, was still fairly heavy and relatively inefficient. Nine years later Karl Benz and Gottlieb Daimler in Germany succeeded in creating lighter and more efficient Otto-derived gasoline-powered engines and in putting them on wheeled carriages. Shortly afterwards, in 1892, Rudolf Diesel introduced the diesel engine. The first modern configuration for a road vehicle appeared in 1891, when the French engineer Emile Levassor put the engine in front of the driver instead of under the seats and used a rear-wheel drive layout [73]. This design, known as the "Système Panhard," quickly became the standard for all cars because it gave better balance and improved steering. Many technical innovations followed in a few years, but cars, produced in small workshops, were very expensive and often unreliable, so that their number was low.

Things changed in 1908, when Henry Ford in the US introduced his Model T, the first car produced on assembly lines, with completely interchangeable parts, deliberately built and priced to create a new mass market. The manufacturing arrangement of the assembly line relied on the subdivision of labor into a large number of simple repetitive tasks. In 1913, when the assembly line was implemented, the car could be put together in 98 minutes. The huge productivity gain offered by the assembly line was undeniable and by 1914 Ford could reward his workers with doubled salaries and shorten working time periods. In that year, an assembly line worker could buy a Model T with 4 months' pay. Introduced at $850 in 1908, after World War I the Model T was sold at $265. When its production ended in 1927, in total, more than 15 million Model Ts had been manufactured, more than any other model up to 1972, when Model T production was surpassed by the Volkswagen Beetle. The manufacturing arrangement based on assembly lines and subdivision of labor was strongly criticized because it is physically demanding, requires high levels of concentration, and can be excruciatingly boring. However, it was copied by many types of industries.

3.8
Aircraft

3.8.1
Conventional Engines

Internal combustion engines made also possible the construction of the first airplanes. After several years of studies and experiments, on December 17, 1903, on a beach in North Carolina the Wright brothers made the first controlled, powered, and sustained heavier-than-air human flight. Their airplane was launched off a monorail track and was powered by a 12 hp engine. The first flight by Orville Wright covered 37 m in 12 s.

In the following years airplanes developed much faster than cars. Louis Bleriot flew from France across the English Channel on July 25, 1909. By 1914, plans were in place to cross the Atlantic via Iceland. War interrupted these designs, but fostered the development of more powerful engines and more reliable structures that were used for fighter and bomber planes. The first air raid was made on December 24, 1914, by a single German plane that dropped one bomb in Dover, England. The heaviest raid of the war was that made on London, June 13, 1917, when 160 persons were killed and 432 injured. Several thousand airplanes of at least 70 different types were used during the war [74].

After the end of World War I, there was a continuous succession of important aviation events [75]. Regular airmail service commenced in the US and in some European countries. On February 8, 1919, the first commercial flight between France and the UK transported 11 paying passengers from Paris to London. In 1922 there was the first commercial night flight between the two cities. By 1925, a number national airlines had begun regular services on various routes. In the late 1920s several speed records were made by Italian and German aircraft and the first trans-Atlantic crossings were accomplished. On November 28, 1938, a Lufthansa plane made the airline's first non-stop flight from Berlin to Tokyo, covering the distance of 14 228 km in 46 h. In 1939 the first flight of a turbojet plane was accomplished, marking the beginning of a new era that progressively led to complete abandonment of the reciprocal engine in aircraft.

On September 3, 1939, World War II broke out. Planes played a fundamental (and horrific) role in the war, particularly with bombing raids on industries and populated towns. The Allies alone dropped 2 700 000 tons of bombs on Germany in 1 440 000 bombing raids, losing 79 265 American and 79 281 British men as well as 18 000 American and 22 000 British planes. At the beginning of August 1945, 801 planes were used in a single bombing raid on Japan [74]. On August 6 and 9, 1945, atomic bombs were dropped on Hiroshima and Nagasaki (Chapter 8); on August 14 Japan surrendered.

3.8.2
Jet Engines

Jet aircraft had no real impact during WWII, but after the war both military and civilian airplanes began to be equipped with jet propulsion. For fighters, jet propulsion roughly doubled the speed of air combat, while bombers became much larger, and with greatly increased ranges. By the 1950s the jet engine was almost universal in combat aircraft, and by the 1960s all large civilian aircraft were also jet powered.

The progress made by aviation in less than one century can be evidenced by a few comparisons:

- **Speed:** The speed of the first flight by Wright brothers, on December 17, 1903, was 11 km h^{-1} (37 m in 12 s). On February 7, 1996, a Concorde supersonic jet made the fastest Atlantic crossing of a commercial jet plane, taking just 2 h 52 min 59 s. The maximum speed was Mach 2.04 and the cruising altitude 18 300 m. Despite outside temperatures of −55 °C, Concorde's skin heated up to 127 °C at the nose, and 91–98 °C on the fuselage and wings because of friction heating, with an estimated length expansion of about 20 cm during flight.

- **Range:** On March 10, 1919, Australia's Prime Minister announced a £10 000 reward to the first aviator who could fly from Great-Britain to Australia in less than 30 days. On November 9, 2005, a Boeing 777-200 LR completed the world's longest non-stop passenger flight, traveling 21 602 km eastwards from Hong Kong to London, in 22 h 22 min.

- **Number of passengers:** In 1919, the first commercial Boeing plane, the Boeing Model 6 (length 9.53 m, wingspan 15.32 m), had two passenger seats. Today, the largest passenger airliner in the world, the Airbus A380 (length 73 m, wingspan 79.8 m), is a double-deck, wide-body construction capable of carrying 555 passengers. It can fly non-stop from Boston to Hong Kong, 15 200 km.

3.9
Electricity

3.9.1
Early Development

In the history of electrical energy, at least a few breakthroughs must be highlighted. In 1800 the Italian physicist Alessandro Volta invented the electric cell. In 1831 Michael Faraday discovered electromagnetic induction, which allows conversion of mechanical energy into electrical energy and vice versa. In 1879 Thomas Edison patented the incandescent carbon filament lamp. He is remembered primarily as an inventor, but he was also an effective entrepreneur: in 1882 he opened the world's first steam-powered generating station to provide electricity for street lighting by direct current (DC) transmission. In 1888 Nikola Tesla invented the first practicable alternating current (AC) motor along with the polyphase power transmission system later implemented by George Westinghouse. In a few years electricity literally changed the way of life of citizens in the affluent countries providing artificial lights capable of extending the length of the day. Conversion of industrial power from steam to electricity and the adoption of an increasing variety of household energy converters proceeded rapidly in the following years, until we reached the present-day pervasive dependence on electricity (Chapter 13).

One of the most outstanding achievements of the use of electricity in the last century has been the development of Information and Communication Technology (ICT).

3.9.2
From Wayfarers to ICT

Communication of information has always played a fundamental role in the progress of mankind, and the amount of accessible information has always depended on the quality and quantity of energy available. Until 1800, most information was communicated by wayfarers and only in special cases was it conveyed by an organized system capable of transmitting written documents.

The first well documented postal service, reserved to government correspondence, was organized in Rome at the time of Augustus Caesar (62 BC–14 AD). The service was provided by two-wheeled carts pulled by oxen or horses. Starting from 1290, a horse-based postal service was operating between Italian city states and in 1505 the Holy Roman Emperor Maximilian I established a postal system reaching Rome, Naples, Spain, Germany, and France by couriers. Since the mid-19th century national postal systems have generally been established as government monopolies. The development of transportation progressively led to a faster and more reliable postal service. However, even not long ago, communication of information over long distances was rather slow.

Since the advent of e-mail and Internet in the mid-1990s, the transfer of information can be nearly instantaneous and the volume of exchanged information has increased exponentially year after year. The term ICT is now used to indicate all technologies, essentially based on electricity, that are employed for the communication of information. ICT tools can be used to find, explore, analyze, exchange, and present information, giving users quick access to ideas and experiences from a wide range of people, communities, and cultures. It should also be pointed out that ICTs have exacerbated pre-existing disparities between people of the world who can or cannot afford to use the electric energy, to know the English language, and to learn the latest technologies. This gap is known as the digital divide.

A main role in ICT is played by computers. ENIAC (Electronic Numerical Integrator and Computer) was the first general-purpose electronic computer, designed in 1943 to calculate artillery firing tables for the US Army's Ballistic Research Laboratory. ENIAC was a huge, very complex, modular, digital computer, without any moving parts. It contained 17 468 vacuum tubes, 7200 crystal diodes, 1500 relays, 70 000 resistors, 10 000 capacitors, and around 5 million hand-soldered joints. It weighed 30 t, occupied an area of 63 m^2, consumed 150 kW of power and had an average error-free running period of only 5.6 h.

Since the 1960s, vacuum tubes in computers have been replaced by transistors, which are smaller, faster, cheaper to produce, require less power, and are more reliable. In the 1970s, integrated circuit technology and the subsequent creation of microprocessors further decreased the size and cost and increased the speed and reliability of computers. With the evolution of the Internet, personal computers are becoming as common as televisions and telephones in the household. It has been estimated that currently there are more than one billion computers around the world. In the last few years semiconductor manufacturing has pushed miniaturization much below the micrometer level [76], while the possibility of using molecules as computation elements (chemical computer) is being actively investigated [77].

In spite of the relatively low power consumption of modern desktop computers (around 150 W) and of other electronic devices, ICT guzzles an enormous amount of electricity because of the very large number of appliances and the extended time of usage. A recent German study has shown that ICT already accounts for 10% of total electricity consumption and is set to increase by more than 20% by 2020 [78]. Another study has shown that server farms and telecommunication infrastructure are responsible for roughly 3% of worldwide electricity consumption [79]. If the present growth trend of 16% per year continues, as the increase in Internet traffic and the number of mobile phone subscribers suggest, this consumption will rise by a factor of 30 in only 23 years: the current level of world electricity consumption. The doubling of consumption predicted by the WEC for 2050 [80] would thus be reached 20 years earlier, due to server farms and telecommunication infrastructure alone.

Currently, ICT systems are responsible for the same amount of CO_2 emissions as global air travel. If the growth of ICT energy consumption continues at the present pace, it might endanger ambitious plans to reduce CO_2 emissions and tackle climate change. Moreover, a lot of energy is also consumed in the construction of the appliances. It has been estimated that 2700 kWh of energy investment

Figure 3.3 Hidden energy cost: making a computer requires an amount of energy equivalent to that contained in 250 liters of oil.

is needed to make a desktop PC (Figure 3.3). This means that a computer, before being switched on the first time, has already consumed about three-quarters of all the energy needed during its life cycle [81].

Sizeable amounts of the electricity used by ICT are consumed when equipment is switched off or is not performing its main function. A European, Japanese, Australian, or North American home often contains 20 devices constantly drawing standby power. Today, these standby losses are of the order of 50% of the electricity used by ICT, representing a huge opportunity for change and improvement [82]. The International Energy Agency estimates that standby mode could be causing a full 1% of the world's greenhouse gas emissions. Increasing the energy efficiency of ICT systems is clearly a major R&D challenge for the decades to come.

Part Two
Fossil Fuels

Energy for a Sustainable World: From the Oil Age to a Sun-Powered Future. Nicola Armaroli and
Vincenzo Balzani
© 2011 WILEY-VCH Verlag GmbH & Co. KGaA, Weinheim
ISBN: 978-3-527-32540-5

4
Oil

> "My father rode a camel. I drive a car. My son flies
> a jet-plane. His son will ride a camel."
>
> Saudi proverb

4.1
What is Oil

Most geologists and geochemists believe that crude oil was formed by heat and
pressure transformation of marine biomass in sedimentary marine or lacustrine
basins [83]. The transformation process is very complex and takes geological time.
The commercially exploitable oils were formed up to 300 million years ago. In the
first stage of the transformation, microbial aerobic degradation returns a significant part of the sediment carbon to the atmosphere as carbon dioxide. Subsequent
anaerobic fermentation by bacteria releases methane and hydrogen sulfide. Eventual burial of organic matter in anoxic muds and progressive sinking at depths
down to a few kilometers lead to oil and gas, with the ratio of gas to oil formation
increasing with increase in temperature. At depths exceeding 5 km all carbon–
carbon bonds are broken over geological time, with formation of methane. The
rate of underground chemical transformations is affected by pressure, presence
of water, heat-tolerant bacteria, and surrounding minerals, with transition metals
playing the role of catalysts. The formations from which oil and gas are extracted
are generally different from the source rocks in which they were originally formed.
Once liberated from the source rock, they can migrate upwards to form shallow
oil or gas fields (called "reservoirs") or even appear as surface oil seeps.

The biogenic origin of hydrocarbons is supported by several types of experimental evidence [84]; however, there is an alternative theory first put forward by
Russian and Ukrainian scientists, who suggested that oil has an abiogenic origin
and originates from reactions of hydrogen with carbon under the high temperatures and pressures encountered in the Earth's mantle [85, 86]. Other studies,
however, indicate that the abiotic theory is invalid [87].

Energy for a Sustainable World: From the Oil Age to a Sun-Powered Future. Nicola Armaroli and
Vincenzo Balzani
© 2011 WILEY-VCH Verlag GmbH & Co. KGaA, Weinheim
ISBN: 978-3-527-32540-5

The fraction of the original phytomass carbon remaining in the extracted oil is less than 0.5% (10% for coal); it can be estimated that the annual global consumption of fossil fuels burns organic carbon that would require about 400 years of current annual phytomass net primary productivity [88]. This shows that the current fossil fuel consumption rate is unsustainable, since it is based on an essentially nonrenewable energy store.

Depending on their chemical composition, crude oils range from light mobile liquids of reddish brown color to highly viscous black materials. Their main components are: alkanes (C_nH_{2n+2}, commonly termed paraffins); cycloalkanes (C_nH_{2n}, cycloparaffins), and arenes (C_nH_{2n-6}, aromatics). The lightest alkanes, methane (CH_4) and ethane (C_2H_6), are gases at ambient conditions. Propane (C_3H_8) and butane (C_4H_{10}) are also gases but they are easily compressible to liquids [liquid petroleum gas, (LPG)]. Compounds with 5–16 carbon atom chains are liquids, and those with longer chains are solids. Cycloalkanes are the most important components of crude oils along with aromatics such as benzene (C_6H_6), its derivatives (toluene, ethylbenzene, and xylenes), and some polycyclic compounds (naphthalene, anthracene, phenanthrene). Carbon accounts for 83–87% of crude oils by weight and hydrogen for 11–15%. The most important impurity is sulfur, derived from proteins of organic matter: oil with a low sulfur content (<0.5%) is called "sweet" crude, and that with a high sulfur content (>2%) is called "sour" crude. Other elements contained in oils as impurities are N and O, as well as traces of various metals (Al, Cu, Cr, Pb, V). The energy content of oil is around 42–44 MJ kg^{-1}.

Differences in oil composition result in specific densities that may range from 0.74 to 1.04 g ml^{-1} (in most cases between 0.8 and 0.9 g ml^{-1}). Hence the standard volume measure used in the oil industry, namely a barrel (bbl, roughly 159 liters) has no single mass equivalent.

The specific density of oil is commonly measured in terms of degrees of American Petroleum Institute gravity, °API [89]. Crude oils with densities above 31.1 °API are classified as light, whereas those with densities lower than 22.3 °API are classified as heavy. Many of the world's important crude oils are medium or moderately light. The higher the °API the higher is the price, since light oils lead to larger amounts of the most valuable final products, namely gasoline and kerosene.

4.2
Oil History, Exploration, Drilling, Production

4.2.1
History

Some varieties of crude oils have long been known and used by several civilizations around the world. Natural seepage of oil was used in the Middle East and the Americas for a variety of medicinal, lighting, and other purposes. The modern oil industry dates from the middle of the 19th century when oil wells were drilled in

Romania and then, starting from 1859, in the US. The first oil well, drilled by Edwin Drake near Titusville in Pennsylvania, was around 20 m deep and produced 10 barrels of oil per day [90].

4.2.2
Exploration

Early explorers were guided only by surface signs of oil's presence, but in the second decade of the last century oil exploration began to be based on techniques used for geophysical investigations. Particularly successful was reflection seismology, a technique based on clocking the time needed for sound energy waves (produced by dynamite or special devices) sent underground to return to the surface after reflection from the interfaces formed by different types of rock formations. In the second half of the last century a progressive development of the seismographic technique, based on electronic data recording, computation, processing, and visualization, led to three-dimensional (3D) and time-lapse (4D) seismic surveys. These techniques are invaluable for locating new production wells, monitoring movement of hydrocarbons in a reservoir, and identifying areas that have been bypassed by existing wells and are worth returning to. Oil exploration has been pursued worldwide for decades and, nowadays, only a few partially unexplored regions still exist such as West Africa, Northern China, Eastern Siberia, areas off of the east coast of Latin America, and Arctic and Antarctic regions. It should be pointed out that, despite the impressive progress in surveying techniques, the risk of drilling dry holes and wasting huge amounts of money cannot be completely eliminated.

4.2.3
Drilling

At the beginning, drilling was carried out by percussion (cable-tool), powered by small steam engines. A major technical advance was the introduction of rotary drills in 1895, a technique which became dominant worldwide only after 1950. The entire assembly of structures and machines used in rotary drilling is called a rig that consists of several interconnected components.

During the early decades of oil exploration the drillers did not have any reliable means to monitor the progress of their operations. Nowadays it is possible to obtain and record valuable information on a number of physical variables in or around the well while drilling. The rate of penetration depends not only on the quality but also on the weight of the bit; modern bits are covered by a layer of fine-grained synthetic diamonds [91]. Rock cuttings produced by an advancing drilling bit are removed from the borehole by a circulating mixture of fluid called drilling mud. As the well deepens, a hollow steel pipe, or casing, is lowered into the hole. The casing prevents the hole from crumbling, seals off undesirable fluids, and prevents oil from contaminating freshwater sands near the surface. Better drills have led to a steady increase in drilling depths. Production from wells deeper than 5000 m is common but, beyond this, problems often arise because of high temperature, high pressure, and the presence of corrosive environments.

The first drilling from wharves was done in California as early as 1897 and the first oil well drilled out of sight of land was built in 1947. In the following years offshore drilling rapidly progressed up to the construction of drill ships capable of working in waters up to 3000 m deep and semisubmersible rigs. Nowadays almost one-third of oil production is supplied by offshore wells drilled by ca. 650 marine drilling rigs. One of the most explored region is the Gulf of Mexico, but now offshore West Africa has a greater potential [92]. The offshore production platform Statfjord B of 800 000 t, placed in the North Sea in 1982 and expected to remain in production until 2019, is the heaviest object ever moved by humans. The cost of offshore oil from deep water of the Gulf of Mexico was estimated to be \$58 b^{-1} in 2007, compared with an average of \$25 b^{-1} for onshore US and \$13 b^{-1} for the Middle East. The 2010 catastrophic oil spill off the coasts of Louisiana has risen much concern about offshore drilling [93, 94]. Early drillers sank vertical wells. It was later realized that much larger volumes of hydrocarbon-bearing strata can be reached by directional and near horizontal drilling from a single drilling site (Figure 4.1). The world's longest extended reach well, BP's Wytch Farm M 11 in England, is only 1605 m deep but taps reserves more than 10 km from the surface wellsite.

Big oil wells initially flow spontaneously. As production continues, the pressure of the reservoir decreases and pumps are used to lift oil to the surface. Even for

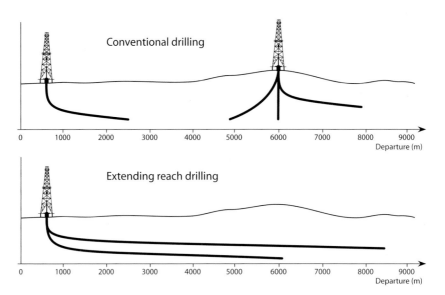

Figure 4.1 Comparison between conventional directional drilling (a) and extending reach drilling (b). The second solution eliminates the need for a second drilling pad.

the lightest oils, primary recovery (i.e., production from natural flow and pumping) yields a minor fraction of the fuel originally present in the parental rocks. In most oil fields, the amount released is around 25–35%, but these rates can be increased by 10–50% by secondary recovery methods, for example, by injecting into the well water, steam, or surfactants. After secondary recovery, about half of the oil is still in the underground reservoir. A long-standing hope is recovering the second half. Unfortunately, after the water flood, the remaining oil is in isolated droplets in the reservoir rocks, and even faster water flow will not recover additional oil.

The largest conventional field in the world, Al-Ghawar in Saudi Arabia (280×30 km), since 1965 has been producing oil with the assistance of peripheral water flooding obtained by delivering huge volumes of seawater (seven million barrels per day) from the Gulf. In 2003 water saturation of extracted oil was already 33%. Not much more is known about Ghawar because the oil company and Saudi government closely guard production details.

The US has been explored for oil more heavily than any place on Earth. In Texas alone, nearly one million wells have been drilled. As of the end of 2008, there were close to 850 000 producing oil wells worldwide, 500 000 of which were in the US [92]. In 2006, the average daily extraction was about 100 bbl per well per day worldwide, but the average production in the US was only 10 bbl per well per day, with 30% of the wells producing less than 1 bbl per day. In Saudi Arabia (about 1600 wells) and Iran the average daily production was 5700 and 3400 bbl per well, respectively.

4.2.4
Production

Production of oil was about 300 t in 1850, 4 Mt in 1880, 22.5 Mt in 1900, 100 Mt in 1920, 1.052 Gt in 1960, and 3.23 Gt in 1979. Until 1975 the US was the largest producer, but it lost its primacy to the USSR in 1975 and in 1977 the second position to Saudi Arabia, which was the largest producer between 1992 to 2008. In 2009 Russia took the lead. In recent years, the consumption trend has been as follows (in Mbbl per day): 63.1 in 1980, 66.7 in 1990, 76.4 in 2000, 83.5 in 2005, 85.6 in 2007, and 84.1 in 2009. According to the EIA, the total production should rise to 94.7 in 2020, 99.9 in 2025 and 105.4 in 2030 [95].

Table 4.1 [7] shows the top 10 producers, consumers, and reserve owners. The US is the third producer and by far the largest consumer. Five of the top 10 consumers (Japan, India, Germany, South Korea, and Brazil) do not appear in the list of the top 10 producers. Taken together, eight non-OPEC nations, the US, Mexico, Brazil, Russia, China, India, Norway, and the UK, have combined reserves below 12% of the world total, yet they produce 38% of the world's oil. Viewed from a per capita standpoint (excluding small Persian Gulf countries) the largest consumers are Saudi Arabia, Canada and the US with 38, 24, and 22 bbl per person per year, respectively. In general, developed countries use more than 10 bbl/person per year, whereas China and India use 2.4 and one bbl/person, respectively.

Table 4.1 Top 10 countries in terms of oil production, consumption, and reserves [7].

Production		Consumption		Reserves	
Country	**kbbl/day**	**Country**	**kbbl/day**	**Country**	**Gbbl**
Russia	10032 (12.9%)	U.S.	18686 (21.7%)	Saudi Arabia	264.6 (19.8%)
Saudi Arabia	9713 (12.0%)	China	8625 (10.4%)	Venezuela	172.3 (12.9%)
U.S.	7196 (8.5%)	Japan	4396 (5.1%)	Iran	137.6 (10.3%)
Iran	4216 (5,3%)	India	3183 (3.8%)	Iraq	115.0 (8.6%)
China	3790 (4.9%)	Russia	2695 (3.2%)	Kuwait	101.5 (7.6%)
Canada	3212 (4.1%)	Saudi Arabia	2614 (3.1%)	U.A.E.	97.8 (7.3%)
Mexico	2979 (3.9%)	Germany	2422 (2.9%)	Russia	74.2 (5.6%)
U.A.E.	2599 (3.2%)	Brazil	2405 (2.7%)	Libya	44.3 (3.3%)
Iraq	2482 (3.2%)	South Korea	2327 (2.7%)	Kazakhstan	39.8 (3.0%)
Kuwait	2481 (3.2%)	Canada	2195 (2.5%)	Nigeria	37.2 (2.8%)

Table 4.2 Top 10 countries or regions as oil importers and exporters [7].

Importers		Exporters	
Country/ Region	**kbbl/day**	**Country/Region**	**kbbl/day**
Europe	10308	Middle East	16510
U.S.	8893	Former Soviet Union	6868
Other Asia Pacific	4590	West Africa	4263
China	4086	S. & Cen. America	2588
Japan	3545	North Africa	2232
India	2928	Canada	1938
Singapore	930	Mexico	1282
Canada	785	Other Asia Pacific	807
S. & Cen. America	504	Europe	464
East & S. Africa	439	East & S. Africa	297

Table 4.2 [7] shows the top importers and exporters. The US imports about 48% of its oil consumption, Japan 100%, China 47%, and Europe 82%. Forty-five countries export crude oil and more than 140 import oil or refined oil products. This reflects the peculiar property of oil, which is a liquid not an ore, being concentrated by natural geological processes. Oil fields cover less that 0.1% of the continents and continental shelves [96].

The average oil production per capita and per year peaked in 1980 at 5.26 bbl, declined to 4.44 bbl in 1993, then increased to 4.58 bbl in 2005, and dropped again to 4.54 bbl in 2006 and 4.30 bbl in 2009.

4.3
Oil Transportation

4.3.1
Pipelines

Enormous amounts of oil are transported between different countries over long distances. Pipelines are the least expensive choice for moving oil on land [83]. Oil transport by pipelines is reliable, safe, clean, environmentally acceptable, and highly efficient (a 1 m diameter pipe can carry 50 Mt of oil per year) [83]. Steel oil pipelines have inner diameter from 10 to 120 cm and are buried at a typical depth of about 1–2 m, but the Trans Alaska pipeline (1280 km) that crosses permafrost territory had to be built on elevated supports (Figure 4.2) above the ground and heated to 50 °C to keep the oil fluid.

Oil is pushed through pipelines by centrifugal pumps displaced at a distance of 30 to 160 km, and usually flows at speed of about 5–12 km h^{-1}. The pipelines are inspected and cleaned using inspection gauges called pigs. These devices are launched from pig-launcher stations and travel through the pipeline to be received at any other station downstream, cleaning wax deposits and material that may have accumulated inside the line. In the US, there are about 90 000 km of oil trunk pipelines, mostly of 20–60 cm diameter. Norway, Siberia, Caspian and Middle East oil fields are connected to European nations by an extended network of pipelines and new pipelines connecting Siberia fields with Pacific Ocean (China and Japan) are under construction or planned [97]. Some pipelines are used to transport two

Figure 4.2 Trans-Alaska pipeline (photograph by Luca Galuzzi, www.galuzzi.it).

or more different products (e.g., gasoline and diesel) in sequence according to advanced schedules.

4.3.2
Tankers

Although pipelines can be built under the sea, it is economically and technically more convenient to use ships. The first tanker ships appeared in the period 1870–1880 and substantial technical development occurred until WW II, when tankers played an important role in providing energy supply in war scenes [98].

After WWII the size of tankers started to grow [92, 99]. Until 1965, the capacity of the vast majority of oil tankers was 60 000 t or less and oil from the Middle East to the US was shipped passing through the Suez Canal and the Mediterranean sea. In 1967, Egypt closed the Suez Canal after the six-day war and the problem of fueling the US and Western Europe with cheap oil was solved by very large crude carriers (VLCCs) and ultra-large crude carriers (ULCCs) with up to 350 000 and 500 000 dead weight tons (dwt), respectively, shipping around the tip of Africa.

The world's largest supertanker ever built was the *Seawise Giant* in 1979. This ship was built with a capacity of 565 000 dwt, a length of 459 m, 31 541 m^2 of deck, and a draft of 25 m, which prevented its navigation not only through the Suez Canal, but also in the English Channel. The ship was damaged during the Iran–Iraq war, renamed *Knock Nevis*, and converted into a permanently moored storage tanker. Supertankers stopped growing not because of technical problems or excessive cost, but because of operational considerations, since they can dock only at a limited number of deepwater ports. Alternatively, they have to transfer their oil to special offshore terminals connected to onshore storage sites. Furthermore, supertankers must follow restricted routes in channels, require very long distances (up to 3–4 km) and times (up to 15–20 min) to stop, and have a turning diameter up to 2 km. Currently, outdated tankers are being retired faster than new ones are being built, which is taken as an indication that oil production has reached a peak [96].

Along the oil transit shipping routes there are two important checkpoints: the Strait of Hormuz and the Strait of Malacca. The Strait of Hormuz (Figure 4.3) is the only sea passage to the open ocean for the oil produced in the Persian Gulf. To traverse the Strait, ships pass through the territorial waters of Iran and Oman under the transit passage provisions of the United Nations Convention on the Law of the Sea. The Strait at its narrowest is 55 km wide. Ships moving through the Strait follow a scheme which separates inbound from outbound traffic to reduce the risk of collision. The traffic lanes in some parts are as narrow as 3 km. About one-third of the world's oil supply, including three-quarters of all Japan's oil, passes through the Strait, making it one of most important strategic sites of the world and a place of high tension between the US and Iran [100].

The Strait of Malacca is an 805 km stretch of water between Peninsular Malaysia and the Indonesia island of Sumatra, its narrowest passage being 2.8 km wide. It is the main shipping channel between the Indian Ocean and the Pacific Ocean,

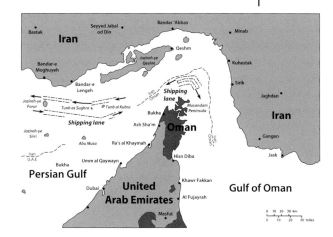

Figure 4.3 A key geopolitical hot spot: the Strait of Hormuz, the only sea passage to the open ocean for Persian Gulf oil.

linking major Asian economies. About one-quarter of all oil carried by sea passes through the Strait, mainly from Persian Gulf suppliers to China, Japan and South Korea.

Other important checkpoints for oil shipment are the Turkish Straits, linking the Black Sea to the Mediterranean, the Strait of Bab-el-Mandeb, connecting the Red Sea to the Gulf of Aden, and the Suez Canal, connecting the Mediterranean and the Red Sea.

4.4
Oil Refining

Crude oil must be processed, or refined, for conversion into commercially usable products: transportation fuels, fertilizers, petrochemicals, lubricants, wax, asphalt, and many other products. An updated list of oil refineries, including location, crude oil daily processing capacity, and the size of each process unit in the refinery, can be found in the *Oil and Gas Journal*. The world's largest refinery, with a capacity of 940 000 barrels per day (bpd), is in the Paraguaná Peninsula in Venezuela; three of the world's 10 largest refineries, including the second one (840 000 bpd) are in South Korea. At least 10 refineries in the world can process more than 500 000 bpd.

Oil refineries are typically large, sprawling industrial complexes with extensive piping running throughout, carrying streams of fluids between large chemical processing units [101]. Many sequential and feedback operations take place. The process starts with desalting that removes suspended solids, inorganic salts, and water-soluble metals. Then oil is distilled in crude distillation towers where the various products are separated according to different boiling points: gases (methane and ethane), liquefied petroleum gases (propane and butane), light naphtha (compounds with 5–7 carbon atoms), heavy naphtha (6–10 carbons), kerosene (10–14 carbons), and diesel oil (14–18 carbons), leaving a residue of solid material. Catalytic cracking and a variety of other processes (reforming, isomerization, alkylation, hydrocracking, hydrotreating) allow the refiners to remove, as hydrogen sulfide and ammonia, most of the sulfur and nitrogen present in liquids and to adjust to some extent the proportions of final products to meet specific demand.

Oil refining is the most energy-intensive manufacturing industry where energy consumption represents 35–50% of operating costs. Of course, refineries produce enormous amounts of carbon dioxide and in some cases they are responsible for very dangerous releases of gaseous and liquid hydrocarbons into the environment (Section 7.8.1).

4.5
Oil Storage

Oil production and oil delivery can be interrupted for several technical and political reasons: hurricanes as often occur in the Gulf of Mexico, oil spills from pipelines, strikes or turmoil, embargoes due to political controversies, local wars in the regions where oil is produced or transported across, and political decisions of the OPEC countries. For all these reasons, large volumes of crude oil are stored in all the delivery and consumption places. Huge storage tanks are present at export oil terminals to feed pipelines and/or fill shipping tankers. Refineries must also have on-site capacity for storing crude oil for at least 2 weeks of operation. Most important, large amounts of oil are stored by oil-importing nations as a strategic reserve.

According to the EIA [102], approximately 4.2 billion barrels of oil are held in strategic reserves in OECD countries, of which 1.4 billion are government controlled; each country has typically reserves equivalent to about 90 days of imports. The US Strategic Petroleum Reserve, 726 million barrels at the end of 2009 [103], is the largest and was used at the beginning of the 1990 Gulf War and on the occasions of damage caused by hurricanes to offshore oil platforms, terminals, pipelines, and refineries concentrated in the Gulf of Mexico. Several experts believe that the current US reserve is too small. At current US consumption levels of 19 million barrels per day (bpd) and imports of 9 million bpd, it corresponds to 38 days of total consumption and 80 days of import protection. Experts believe that the reserve should guarantee oil for at least 90 days of interruption in imports, or even five times larger [92]. Such an improved reserve would correspond to the total annual production of Saudi Arabia. It should also be noted that currently the maximum total withdrawal capability from the US strategic petroleum reserves is only 4.4 million bpd. Large and properly managed petroleum reserves could somewhat reduce the volatility of the oil market, temper OPEC's cartel power, and allow an extended and stable transition period for the development of alternative transportation energy resources.

Recently, other non-IEA countries have begun to create their own strategic petroleum reserves. Furthermore, there are several specific agreements, for example, between South Korea and Japan, and among Germany, France and Italy, to share their oil reserves in case of an emergency. As the world's second-largest oil consumer, China now relies on imports for about half of its oil needs. It imported 203.5 million tons of crude oil in 2009, and by 2020 it is expected to import 60% of its oil. Currently, its strategic crude oil reserve capacity covers only 21 days of its economy's needs. If everything goes as planned, China will increase its strategic crude oil reserve capacity to 90 days by 2020 (700 million barrels) [104].

4.6
Unconventional Oil

The liquid which shoots out of the ground when the drill reaches a reservoir is only part of the oil resource, known as conventional oil. There are other types of oil known as unconventional oils [83, 92, 96], which include medium heavy oil (mobile, 25–18°API), extra heavy oil (somewhat mobile, 20–7°API), tar sands (or oil sands) and bitumen (non-mobile, 12–7°API), and fine-grained sedimentary rock known as oil shale that may contain significant amounts of kerogen (a solid mixture of organic chemical compounds). Some of these oils exist in vast amounts, greater than the remaining conventional oil, but they are not easy to convert into usable crude oil [96]. However, as higher prices justify expensive secondary recovery from conventional oil fields, mining some unconventional sources is also becoming price effective. Exploitation of unconventional sources causes severe damage to the environment [105].

Tar sands exists in large quantities, especially in Canada (Alberta) and Venezuela (Orinoco Belt), whereas the US owns vast resources of oil shale, which in principle might cover its current consumption for 270 years. The Canadian Athabasca tar sand deposits contains 300 billion barrels of economically recoverable oil [106] and make Canada's total oil reserves the second largest in the world, after Saudi Arabia's. In this area hydrocarbons are recovered to the extent of approximately 75% by surface mining, which destroys the environment (Figure 4.4), whereas the remainder lies deep and has to be produced by strip mining. Extraction of oil from tar sands needs vast amounts of water and huge amounts of energy to boil it, plenty of space for the waste [107], and leaves persistent contamination to local ecosystems [108]. As an average, to obtain a single barrel of oil from Alberta oil sands requires digging out 4 tons of material, producing up to three times more greenhouse gas pollution than conventional oil extraction and 3–5 barrels of waste water. The process recovers up to 60% of the oil, but about one unit of energy as natural gas for steam generation is needed for every five units of heavy bitumen. In 2009, output from Canada tar sands was about 1.49 million bpd and it is estimated that it could reach 3.3 million bpd by 2020.

The Orinoco Belt in Venezuela consists of large deposits of extra heavy crude oil known as the Orinoco oil sands. The Orinoco Belt is perhaps the largest oil

Figure 4.4 The tar sands mine site at Syncrude's Mildred Lake plant, Canada.

sand deposit in the world, with about 1.2 trillion barrels, of which 270 billion barrels are thought to be economically recoverable. A commercial product from this area is Orimulsiòn, a boiler fuel (coal or gas substitute) which is a mixture of 70% natural bitumen, 30% water, and a small amount of an additive that stabilizes the emulsion. Venezuela has recently gone through a nationalization process of the oil extraction operations that were previously conducted by private companies under operating services agreements. The nationalization process was completed on May 1, 2007, and, 1 year later, Venezuela and China formed a joint venture to produce oil in Venezuela's Orinoco Belt to supply a new 400 000 bpd refinery to be constructed in China.

Globally, significant unconventional oil resources are not limited to Canada and Venezuela. Russia, for example, is estimated to have similarly large reserves of extra-heavy oil and the potential US resource base is estimated at 40 billion barrels. In 2008 the Italian oil company ENI announced an agreement with the Republic of Congo to exploit tar sands in a large area (1790 km^2) which is estimated to hold reserves of between 500 million and 2.5 billion barrels.

In the US, the oil shale resources total 2 trillion barrels. As much as 750 billion barrels has a richness of 25 gal t^{-1} or greater and could be produced with adaptation of existing technologies [109]. An oil shale industry could be initiated shortly, with an aggressive goal of 2 billion bpd by 2020. Ultimate capacity could reach 10 billion bpd, a value comparable to the long-term prospects for Alberta's tar sands.

Indeed, with further technology development and depending on the price of conventional oil in the future, some of these additional unconventional resources might eventually become economical, but not before having solved important technical and environmental problems [96, 105]. First, most of it needs to be dug out in strip mining rather than drilled. Once dug out, it needs to be heated to

450–500 °C, and enriched with hydrogen via steam before the resulting oil is separated. The remaining sludge has increased in volume by 30% through the process and needs to be disposed of. Oil shale production creates more than four times as much greenhouse gases as conventional oil production, it uses vast quantities of water, and wastes around 40% of its initial energy in production. Oil shale has one-eighth the energy of conventional oil on an energy-per-ton basis.

With potential recovery of about 3 trillion barrels globally, there is about as much oil in oil shale as the total global conventional oil endowment estimated in the USGS 2000 Assessment [92].

4.7
Petrochemicals

The processing of crude oil opened up the route to petrochemicals, because cracking produces, besides fuels, also unsaturated hydrocarbons containing one or more C=C bonds, in particular ethylene, propylene, butylenes, and butadiene. These olefins can be readily used and further transformed by chemical reactions. They are produced in very large quantities and constitute the basic building blocks for obtaining synthetic polymers and a variety of other artificial products which have made our lives better, safer, longer, and more comfortable [110].

Plastic packaging of food allows for better conservation and protection against contamination. Plastic bottles and containers for all kinds of beverages, mostly made from poly(ethylene terephthalate) (PET), are safer and lighter than glass bottles. Waste-disposal garbage bags are made of plastics. In the bathroom, shampoo and shower gels are composed of synthetic soap formulations and their unbreakable bottles are made out of polyethylene, polypropylene or poly(vinyl chloride) (PVC). A large part of our clothes are based on synthetic fibers such as polyesters, polyacrylics, or rayon. PVC is the material of choice to replace wood for windows, doors, and furniture. DVDs and CDs are all made using polymers, as was the case for the old vinyl records and videotapes. When we sit in our motor car, we are literally surrounded by materials derived from hydrocarbons. Even structural steel and aluminum frames of cars are increasingly replaced, at least in part, by new generations of composite plastics so as to reduce weight and achieve better fuel efficiency. Tires are made from artificial rubber and the roads are asphalted, that is, covered by heaviest fractions of the crude oil refining.

Petrochemicals have made a very important contribution to the progress of healthcare. Disposable syringes, blood bags, artificial heart valves, and artificial joints and limbs are just a few of commonly used plastic medical products. Petrochemicals are also the basis of synthetic pharmaceutical products used to prevent and heal a great variety of health problems. Last but not least, petrochemicals are essential for modern agriculture. Oil is the basic feedstock (with natural gas) for the production of fertilizers and pesticides, which are indispensable to avoid crop losses from diseases and insects. In total, about 6% of crude oil is used today to produce petrochemicals [111] and, in view of the great benefits given by the petrochemical use of oil and natural gas, a question arises: is it wise to burn these precious natural resources as fuels?

4.8
Oil as a Fuel

4.8.1
World Picture

As countries develop, industry, rapid urbanization, and higher living standards drive up energy use, most often of oil. Oil is used as a fuel for heating, generation of electricity, industrial uses, and predominantly as transportation fuels (Figure 4.5). Detailed data on the oil consumption by various countries are available [7, 102]. In 2009, oil consumption was subdivided as follows: North America, 26.4%; South and Central America, 6.6%; Europe and Eurasia, 23.5%; Middle East, 8.7%; Africa, 3.7%; Asia Pacific, 31.1%. Top world oil consumers are listed in Table 4.1 [7].

World crude oil demand grew by an average of 1.76% per year from 1994 to 2007, with a high of 3.4% in 2003–2004. The growth in 2007 was 1.3%, whereas in 2008 there was a 0.6% decrease, followed by a much larger drop (−1.4%) in 2009 because of the global economic crisis. In 2008, a large decrease occurred in US consumption (−6.4%), consolidated in 2009 (−4.9%) and in several European countries. Large growths, however, were were observed in China and India with +6.7% and 3.7% in 2009, respectively. In the long term, world demand for oil is projected to increase by 1.4% per year until 2030, which means 37% over 2006 levels [102]. In the same period, the average growth per year is expected to be 3.8% for China and 2.4% for India. However, these mid- to long-term projections have

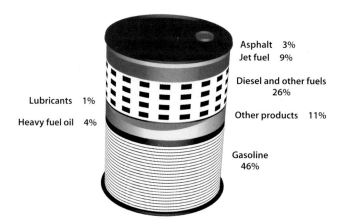

Asphalt   3%
Jet fuel   9%
Diesel and other fuels 26%
Lubricants   1%
Other products   11%
Heavy fuel oil   4%
Gasoline 46%

Figure 4.5   Products made in the US from a typical barrel of oil.

to be treated with great caution: there is an embarrassing record of wrong forecasting of energy consumption trends [52].

Transportation has seen the largest growth in demand in recent decades, mainly because of an increasing number of vehicles for personal use. Power generation and industrial energy uses are increasingly switched to other sources, but at present transportation demand can only be satisfied by an increase in oil consumption. Air transportation would be literally impossible without liquid fuel, which makes it possible to travel between any cities in the world in less than 24 hours.

A significant factor in increasing oil demand has been human population growth and, in recent years, production has often outpaced this growth. The world population increased by 6.2% in 2000–2005 whereas, in the same period, global oil production increased by 8.2% [7]. The exploding numbers of oil-fueled road vehicles has so far drawn much less attention than the population increase. In 1939 the world's roughly 2.3 billion inhabitants shared a total of around 47 million motor vehicles. Today's 6.8 billion human being have around 900 million motor vehicles to fuel. In several developing countries, the rate of increase in vehicles is more than 10 times larger than the rate of population growth.

In 2009, a total of 61.7 million new road motor vehicles were produced worldwide [112]. Compared to the previous year, there was a decrease of 34.3% in the US, 17.6% in Italy, and 20.3% in France, and an increase of 48.3% in China, and 12.9% in India.

4.8.2
US and Developed Countries

Although demand growth is highest in the developing world, the US is by far the world's largest consumer of oil with 21.7% of the total 2009 consumption (Table 4.1). As much as 85% of the US energy consumption comes from fossil fuels, 40% from oil [102]. In 2009 the US produced about 7.2 million bpd of oil and consumed about 18.7 million bpd, with an imported share of 48%. In the US, transportation accounts for almost 70% of the petroleum used for fuel and US vehicles consume about 6% of the global primary energy supply. In 2009, US drivers used 522 billion liters of gasoline at a cost of about $360 billion.

The automotive industry crisis of 2008–2009 in the US and more generally in developed countries occurred mainly as a result of the global financial crisis and the related "credit crunch." A role was also played, however, by changes in consumer buying habits and increasing competition from the public transport sector. Other contributing factors were pricing pressures on raw materials and substantially more expensive automobile fuels, which, in Canada and the US, caused customers to turn away from large vehicles such as SUVs. In certain countries, particularly in the US, the industry also suffered from relatively cheap imports available from Japan and Europe. In November 2008, the Big Three US manufacturers (General Motors, Ford, and Chrysler) indicated that unless additional funding could be obtained over the short to medium term, there would be real dangers of bankruptcy. Likewise several European and Japanese vehicle manufacturers received direct or indirect financial support from their governments. Indeed, in advanced countries the number of cars is reaching a plateau since it will be difficult to convince people to own more than one car each. However, there are opportunities for car makers, particularly in the US, to replace large, heavy, high-consumption vehicles with smaller, lighter, and more efficient ones.

The US way of transportation and more generally the US way of life are taken as a model by most developed and underdeveloped countries. Discussing America's addiction to oil (Section 4.9) may help to understand what is currently happening in other countries and what could be the fate of spaceship Earth.

4.8.3
China and India

China and India occupy the second and fourth places in the list of top consuming nations (Table 4.1). China has doubled oil consumption from 1999 to 2009 and India's oil imports are expected to double from 2009 and 2030. In 2007 the oil and gas producer PetroChina became the world's first $1 trillion company, with a capitalization twice that of the closest rival, ExxonMobil.

In 1975 only 139 800 automobiles were produced annually in China; then production started to rise steeply and reached 5.7 million in 2005 and about 13 million in 2009 [113]. Since November 2009 China has been the world's largest automobile producer and consumer [114]. In spite of this growth, the number of cars per 1000 people in China is 36, compared to 840 for the US and about 650 for several Western Europe countries [115]. In other words, in terms of vehicle ownership, China is where the US was in 1916 and India (fewer than 20 vehicles per 1000 people) is further behind. It seems unlikely that China and India will rapidly climb the motor vehicle ownership trajectory laid by the US and followed by some Western European nations. In fact, even if Chinese and Indian incomes doubled every 7 years (10% annual growth) from the current levels, most people in these countries will not be able to afford to buy even a Nano Tata car ($2000) for decades.

At present, China consumes about 2.4 barrels of oil per capita annually, about 9 times less than the consumption per capita in the US. At the level of US consumption (22 barrels per person), today it would consume 29.4 billion barrels of oil, a quantity comparable to the current annual world oil consumption of about 31 billion barrels [7]. It should also be noted that a strong increase in the number of cars in the largest Chinese and Indian metropolitan areas would cause a substantial degradation of local air quality, although in 2010 China imposed the Euro 4 emission standards nationwide. It is easy to predict that in coming years oil imports by China and India will grow strongly because of their large population and this will most likely result in a stress on global oil demand. The expectation of a growing future dependence on oil imports has brought China to acquire interests in exploration and production in places like Kazakhstan, Russia, Venezuela, Iran, Saudi Arabia, Canada, and particularly Africa [116]. Despite these efforts, it has become increasingly dependent on Middle East oil [92].

China encourages the development of clean and fuel-efficient vehicles in an effort to sustain continued growth of the country's automobile industry. By the end of 2007, the average fuel consumption for all types of vehicles was required to decrease by 10%. Furthermore, the proportion of vehicles burning alternative fuels such as compressed natural gas (CNG) and LPG will be increased and the research and development of electric and hybrid vehicles will be intensified (Section 13.7.1).

The picture represented above shows that the current transportation model, almost exclusively based on fossil fuels, is unsustainable. Vehicle fuel economy is a most important target, but a real revolution is needed in the transportation sector.

4.9
America's Addiction to Oil

The incredibly high oil consumption of the US has its roots in the first half of the last century, when the country dominated the expanding world oil market. Since then, cheap gasoline was considered by many Americans to be as much a birthright as the right to carry a gun. In the late 1940s, there was a first, unanticipated negative signal: transition from oil-exporting to oil-importing nation [117]. In 1950, about 8% of US consumption was covered by net imported oil.

In 1956, President Dwight Eisenhower launched the interstate highway system which resulted in the development of commercial trucking, family vacations, and daily commutes. During the 1960s, US energy consumption increased by 51%, compared to 36% during the previous decade. The US imported a fixed quota of oil, but it was still the world's major oil producer and, because of its surge capacity, was not yet dependent on foreign oil. In 1960, the Organization of the Petroleum Exporting Countries (OPEC) was founded and began to play a major role in the oil market [118].

In 1970, US oil production, after more than a century of steady increase, reached its peak (Section 4.10). At that time there was an explosive demand growth, with more fuel needed for new and larger cars. The fuel efficiency was as low as 13.5 mpg (mile per gallon, 17.4 liters per 100 km), and the increasing popularity of suburban living required longer commutes. At the same time, more energy was needed for larger houses and offices whereas rising environmental concerns forced a switch from coal to other fuels.

Under such circumstances, and also in the hope that nuclear power could supply 25% of electric energy by 1980 and 50% by 2000, in April 1973 president Richard Nixon ended restrictions on foreign oil imports and appealed for voluntary energy conservation; in August net oil imports reached 36% of consumption (Figure 4.6). On October 15, 1973, the Arab members of OPEC proclaimed an oil embargo in response to the US decision to re-supply the Israeli military during the Yom Kippur war. In November, President Nixon called for new energy-saving actions and launched Project Independence, pledging that technology would end reliance

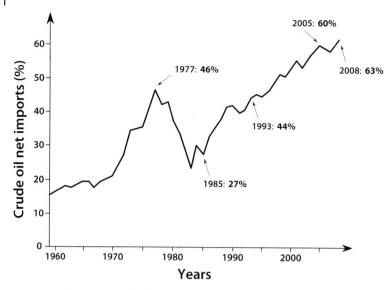

Figure 4.6 Oil net imports by the US (1958–2008) [119].

on foreign oil within 7 years. In the meantime, OPEC increased the price of oil from $3 to $12 per barrel [119].

Despite President Nixon's attempts to enforce conservation measures, the fuel crisis in the US continued to escalate. Several States lowered speed limits and mandated motorist to limit their purchase of gasoline to odd or even days, on the basis of the number on their license plates. The embargo ended in March 1974 and in June the oil shortage was over, because of the increase in oil imports. In 1975 President Gerald Ford was forced to admit "Americans are no longer in full control of their own nation destiny, when that destiny depends on uncertain foreign fuel at high prices fixed by others." The embargo stimulated the passage of the Federal Energy Policy and Conservation Act, which set the Corporate Average Fuel Economy (CAFE) standards that led to substantial efficiency gains in automobile and light truck fleets over the next decade [120]. The increase in fuel economy in the first years of the standards came through both increases in fuel efficiency and significant decreases in the weight of new vehicles.

In his first public speech in 1977, President Jimmy Carter demonstrated that he had clear ideas about the US energy crisis, saying that "the amount of energy being wasted which could be saved is greater than the total energy that we are importing from other countries" (see also Section 10.2). He called for sacrifices for the common good and said that the energy challenge should be considered "the moral equivalent of a war" [121]. He created the Department of Energy (DOE) and proposed an impressive energy program composed by more than 100 actions, which included an increase in gasoline taxes, taxes on automobiles failing to meet mileage efficiency standards, and tax credits for installing solar equipment, insulating houses, and using alcohol-blended fuels.

Unfortunately, his plan found strong opposition in the Senate and was approved only in part; in 1977 US oil imports increased to 46% of consumption. In 1978 US oil production rose due to the Prudhoe Bay oil field in Alaska, and the share of foreign oil dropped to 43%. The collapse of oil production in Iran at the end of 1978 caused a second energy crisis in the US and reinvigorated efforts to find new solutions. President Carter called for the bold goal of generating 20% of the nation's energy from renewable energy by the end of the century. A tax of $227 billion on the oil industry made possible generous incentives to homeowners and businesses who switched to renewable energy. The revenues from the taxes also made possible extended aid to poor families affected by higher energy prices. During the early 1980s the federal efficiency standards adopted by the Carter administration had their greatest impact and, together with continuous Alaskan oil production and displacement of oil by coal and nuclear power to obtain electricity, made major contributions to lower the share of oil import. Hence Presidents Ford and Carter succeeded in reducing net oil imports, which decreased from a share of 47% in 1977 to 28% in 1982, with the OPEC consumption dropping from 34% to 17% (Figure 4.6).

In the following years, President Ronald Reagan relied more on the free market and less on government regulations to deal with energy [119]. In his first 10 days in office, he announced decontrol of crude oil and gasoline, whose price increased immediately. He tried without success to abolish the Department of Energy and halved the energy part of the budget, including conservation and alternative fuels, while doubling expenditures for the nuclear weapons program. He ordered an end of restrictions on temperature control in federal buildings, and recommended deep cuts in the research and development budget for solar energy. Residential solar tax credits expired at the end of 1985 and were not renewed. In his second term, President Reagan's attention shifted to defense, space, and health and the budget for solar research and development was less than one-quarter of its peak under President Carter. In the meantime, the speed limit of 55 mph was elevated to 65 mph on certain roads.

In summer 1986, while the Iraq–Iran war was continuing, Saudi Arabia decided to gain market share by increasing oil production by 4 million bpd, more than doubling the production rate of the year before. This caused a sudden drop in oil price from $25 per barrel in 1985 to $10. Such a low price killed any residual attention on alternative fuels, but also struck the domestic oil industry, whose production began to decrease, reaching in 1988 its lowest level since 1977. Oil imports increased dramatically, with a share jumping from 27% in 1985 to 33% in 1986. The car and truck fuel economy began to decrease because the low gasoline price encouraged people to use larger vehicles like pick-ups, vans, and SUVs; in a few years, such vehicles gained nearly half of the US car market. Furthermore, there was a steadily increase in the average distance traveled per year.

In 1990 Iraq invaded Kuwait. President George H. W. Bush declared that the Kuwait invasion could become a major threat to US economic independence and urged the major oil-producing nations to increase output to protect the world economy. On January 16, 1991, when he ordered military attacks on Iraq, the release of oil from the strategic petroleum reserve was announced. On February 27 the war was over and the price of gasoline soon returned to its pre-war level.

President Clinton was not very active on energy issues. There was an overabundance of oil on the market and consumers were quite satisfied about the price of gasoline, as low as about $1 per gallon. President Clinton mandated to increase the efficiency of refrigerators and air conditioners, but the average mileage of new vehicles continued to decrease because more drivers moved to SUVs, vans, and trucks and the administration relied only on voluntary cooperation from US automobile makers to improve vehicle efficiency. Meanwhile the threat of climate change was opening up another area of concern about oil consumption (Chapter 7).

In an attempt to increase the price of oil, in 1998 the OPEC nations decided to cut production. On September 11, 2001, there was the terrorist attack in New York and in March 2003 the US decided to wage war against Iraq. The official reason for this decision was that Iraq aided, trained, and harbored terrorists and possessed and concealed some of the most lethal weapons ever devised, but it was supposed from the beginning, and it is now clear, that the unstated reason was a closer control of the oil fields of the Persian Gulf region, which supplied a large part of US imported oil. Even when OPEC began to increase output in 2003, the cost of oil continued to rise because of the increasing need for oil by emerging economies, particularly China. In 2005, the US dependence on foreign oil reached a share of 60%. The reason why the demand for oil imports has not declined in response to higher prices is that a society, in the absence of actions by the government, has a limited ability to adjust its energy consumption habits in the short term.

No action to repair such an alarming situation was taken by the government until 2007 when Energy Independence and Security Act [122] found bipartisan support in both houses of Congress and was signed by President George W. Bush into law in December. It was indeed a very strong Act that did take a major step towards US energy independence. It was focused on automobile fuel economy, development of biofuels, energy efficiency in public building lighting, funding of research and development of renewable energy technologies, new railroad infrastructure, and modernization of the electricity grid. After more than two decades of inaction, the approval of this law with a very large majority both in the Senate and the House of Representatives and with full agreement of the White House made it clear that theUS was turning the page on energy policy. In the meantime, increases in oil imports and oil prices raised the US trade deficit to $439 billion in 2008 [123], then it decreased to about $195 in 2009 as a consequence of the economic crisis.

The economic downturn offered the opportunity to intervene drastically in energy efficiency and CO_2 production. The choice of President Obama to appoint Steven Chu, Nobel Prize Winner in Physics and supporter of the use of renewable energies, anticipated other steps in the right direction in 2009. As stated by Richard Ernst, "Who else, if not the scientists, is responsible for setting guidelines for defining progress and for protecting the interests of future generations?" [124].

President Obama was aware that America had lost almost 30 years with wrong decisions on energy policy. Therefore, the $787 billion American Recovery and Reinvestment Plan (February 2009) contained many energy-related issues such as energy efficiency, renewable energy, house insulation, electric cars, and public transportation, including the construction of high-speed railroads along the Eastern and Western coasts. Even more important, 5000 km of a new electric grid were planned to exploit the electricity produced by photovoltaic panels.

There is still a long way to go for the US towards independence from foreign oil and to reduce carbon emissions to reasonable low values. According to the America's Energy Future report [58], energy-efficiency technology for the transportation, building, and industrial sectors exists today, or is expected to be developed in the normal course of business, that could save about 30% of the energy used in the US economy by 2030. The report notes, however, that there are formidable barriers to improving energy efficiency and suggests that particular attention should be paid to buildings, infrastructures, and other long-living assets. The need to reduce carbon dioxide emission could be an opportunity to shift part of transportation from private to public vehicles and from road to rail. Hopefully, it will also spur Americans to moderate insane and sometimes ridiculous attitudes such as setting air conditioning systems at unreasonably low temperatures.

4.10
Oil Price

Oil is a colossal business. Of the top five world companies in 2008, oil corporations accounted for 61% of revenues. Prediction of oil price in the near future is a difficult enterprise, as shown by the random shifts observed in the last 35 years [7, 130], which are the result of several independent factors including demand, national and international policy, fears of shortage, embargoes, wars, discovery of new reserves, and speculation. It is reasonable to assume that in the long run the price of oil will increase, until it will become possible to use a more convenient resource.

The price of oil reached an all-time record in July 2008 at over $147 per barrel. The effect of the oil price on economy is known as a price shock. European countries have high taxes on fuels, hence price shocks could potentially be mitigated by temporarily lowering the fiscal burden. This method is less effective in countries with much lower gas taxes, such as the US, where jumps in oil price are promptly transferred to the gasoline pump. The price of oil is substantially controlled by OPEC, which possesses about 70% of the current oil reserves. Assuming a price of $60 per barrel, the value of the 944 billion barrels of the OPEC reserves is worth $56 trillion, more than the current annual gross world product [92]. OPEC nations obtain maximum profits from oil extraction, compared to many other oil-producing nations, because its members are generally the lowest-cost producers for geographic and geological reasons [131]. OPEC exercises control of the price of oil by placing quotas on the production of its members. If just six OPEC

Oil Wars

The largest external cost of oil is that caused by oil wars. As to December 31, 2009, the US expense for the US-led Persian Gulf wars amounted to $710 billion [125]. According to the Congressional Budget Office, by 2017 the total for the three oil wars (two in Iraq and one in Afghanistan) could amount to $2.4 trillion including interest [126], or nearly $8000 per person in the US. The Iraq war cost over $400000 per soldier, compared to less than $100000 (in 2007 dollars) in WWII. Joseph Stiglitz, Nobel Prize Winner in Economic Sciences in 2001, has stated that the total cost of the Iraq War for the US economy will be 3 trillion dollars in a moderate scenario based on conservative assumptions [127]: "The war in Iraq will have an impact on US spending for a long time and many people will rightly ask what could have been accomplished with such an enormous amount of money concerning health, education, science, research and public infrastructure repairs badly needed in the US, for example, bridges, highways, water systems, dams and power grids."

But before the financial aspects, one should consider the loss of lives, the injuries, and the destruction caused by the oil wars. As August 22, 2010, at the end of war operations, 4415 American soldiers died in Iraq; each of these "priceless" lives cost $500000 to the Pentagon budget, the amount paid out to survivors in death benefits and life insurance. Along with them, a much larger number of "no cost" Iraqi citizens succumbed [128].

There is another connection between war and oil, because any means used in war is powered by large amounts of oil derivatives. Military fuel consumption for aircraft, ships, ground vehicles, and facilities makes the US Department of Defense (DOD) the single largest consumer of oil in the world. In 3 weeks of combat in Iraq, the Army burned 40 million gallons of fuel, an amount equivalent to the gasoline consumed by all Allied armies combined during the four years of World War I [129]. In 2006, the US Air Force consumed around 2.6 billion gallons of jet fuel, which is the same amount of fuel that US airplanes consumed during WWII between December 1941 and August 1945.

Armored vehicles have very low fuel efficiency. For instance, the Abrams tank can travel less than 4 l/km and Bradley fighting 1.2 l/km of fuel. The modern American "GI" is indeed the most energy-consuming soldier ever seen on war fields. The famous Third Army of General Patton during WWII had about 400000 men and used about 400000 gallons of gasoline per day, that is, 1 gallon per day per soldier. This amount went up to 9 gallons per day per deployed soldier in the Vietnam War and to 15 gallons nowadays.

countries (Saudi Arabia, Kuwait, United Arab Emirates, Venezuela, Nigeria, Iran) were to increase their annual production from 1% to 2% of their reserves, it would equal the combined production of the US, UK, Russia, Norway, Brazil and China. As oil prices fell to $35 per barrel in December 2008 because of the global recession, OPEC decided on production cuts which in a few months raised the cost to around $80 per barrel. OPEC believes that a fair price of oil is around $75 per barrel since lower prices would lead to a failure of industry to invest in oil development and a consequent supply crunch in the future.

It should be pointed out that the market price has little to do with the real cost of oil (Chapter 7). The military cost to the US government to keep a steady oil flow of oil is estimated to be around $50 billion [92, 132]. Given that the US imports about 640 Mbpy from the Middle East [7], the standing armed force cost alone amounts to about $78 per barrel. That hidden cost, not paid by American motor vehicle users but by taxpayers, is higher than the average market cost of a barrel of oil in 2009.

4.11
Oil Peak and Reserves

4.11.1
A Non-renewable Resource

Some people believe that a continuous supply of oil is part of Nature, an idea supported by several classical economists who deny the idea of depletion of resources. But oil is a finite good, unevenly distributed among the world's nations, not renewable in our timescale, with no easy substitution in sight.

Oil depletion is an issue of the utmost importance [3, 58, 83, 88, 92, 96, 111, 119, 133–135]. We often encounter statements like "With present rate of consumption oil will last 48 years". Such a number is obtained by dividing the reported reserves (e.g., 1200 Gbbl) by annual consumption (e.g., 25 Gbbl). Such statements are misleading since they lead people to believe that oil supply is secured for some decades. But oil production will not stay flat for 48 years and then suddenly drop to zero. In fact, oil production will rise to a peak and then begin to decline. If new reserves are discovered, they just push the peak ahead. Any sign of declining production may reflect in a price increase and, perhaps, in various political actions, including wars. Therefore, when the last drop of oil will be extracted and consumed is a matter of no practical interest: the most important event regarding our future reliance on oil is when production will reach a peak. In principle, the concept of peak production can be applied to any non-renewable resource, for example, metals.

The status of oil depletion and, in particular, the date of the oil peak are subjected to extensive investigations. The results, however, are disappointing because several intrinsic technical, economic, and political problems make it difficult to predict the production and consumption trends.

4.11.2
Oil Reserves

It is rather difficult to estimate the quantity of oil that can be extracted from an oil field. Only a fraction of oil can in fact be recovered from a geological reservoir and only this amount can be classified as a *reserve*. All reserve estimates involve uncertainty, depending on the amount of reliable geological and engineering data available and, even more so, on the interpretation of those data. Accordingly, oil reserves must be categorized as *proved* or *unproved* [136]. Proved reserves are those claimed to have a reasonable certainty (normally at least 90% confidence) of being recoverable with existing technologies and under existing economic and political conditions. Unproved reserves have a probability of being recovered to the extent of 50% or less.

There are more sophisticated systems of definitions that incorporate oil quality, geological factors, continuous evolution of technology, oil request from the market, economic feasibility, as well as forecast on some of these factors [136]. Experience shows that initial estimates of the size of newly discovered oil fields are usually too low, so that successive estimates of the ultimate recovery of fields tend to increase.

Data on the oil reserves are reported annually by EIA [102] and BP [7]. Proved reserves (in Gbbl) grew from 1006 in 1989 to 1086 in 1999 and 1333 in 2009 [7]; Table 4.1 shows the official data for proved reserves by country. However, it is difficult to know if such data are fully reliable because oil producers have to report the size of their proved reserves by taking into account oil that can be produced economically, based on oil prices at the end of each year. Accordingly, when the price increases the amount of oil that can be produced economically also increases. Correct estimation of overall proved reserves is made even more difficult because data are provided by oil companies and countries, using practices not internationally standardized and, not rarely, manipulated for economic and political purposes. In recent years, big oil companies such as Royal Dutch Shell and BP had to announce a reduction of proved reserves, leading to calls for an industry-wide standardization of accounting practices [92].

Concerning countries, there are doubts about the reliability of official OPEC reserve estimates, which are not provided with any form of audit or verification that meet external reporting standards. The dramatic increases in reported reserves by practically all Persian Gulf countries in the last two decades [92] is likely to be related to the fact that production quotas of the OPEC countries are partly based on the level of their reserves. These countries gave unconvincing or no details about the reasons for the upgrade, denying accusations of political factors.

4.11.3
Oil Peak

Although some believe that oil production will never end due to inexorable technological advances and increases in oil price, it should be pointed out that there is an insurmountable physical limit: the energy needed to produce the oil itself, defined by the EROI (Section 2.4). When the energy needed to obtain a barrel of oil is higher than the energy obtainable by the barrel, it makes no sense to produce that oil. This concept further underpins the notion that a peak will be reached.

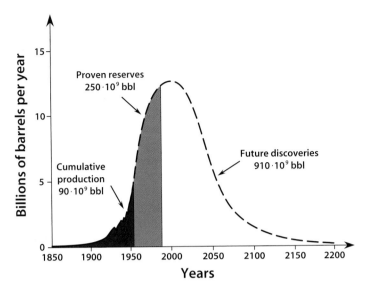

Figure 4.7 Hubbert's 1956 prediction of ultimate world oil production based upon initial reserves of 1250 billion barrels.

The first scientist to discuss the oil peak was M. King Hubbert [137]. In 1956 he pointed out that the rate of oil production of any reserve over time would resemble a bell-shaped curve based on the limits of exploitability and market pressures and predicted that oil in the US would peak between 1966 and 1971. This prediction proved correct. He also predicted a peak around 1990 for the world production based on initial reserves of 1250 billion barrels (Figure 4.7), later updated to 1995–2000, but this prediction did not turn out to be accurate. The empirical Hubbert model was analyzed from a theoretical viewpoint by several authors who proposed modified versions of his original logistic model, to allow for real-world factors such as new reserve growth data. The central features of the Hubbert curve (that production stops rising, flattens, and then declines) remain unchanged, albeit with different profiles.

Current predictions of the timing of peak oil include the possibilities that it has recently occurred, that it will occur shortly, or that a plateau of oil production will sustain supply for up to 100 years. The Association for the Study of Peak Oil and Gas (ASPO) is the most influential organization supporting the "peak oil" theory [138]. ASPO, founded by Colin Campbell in 2000, is a network of scientists affiliated with a wide array of global institutions and universities. In 2008 ASPO predicted the oil peak (including non-conventional sources) for 2010. Some experts and associations do believe that the oil peak has already occurred; one significantly nominated Thanksgiving Day, November 24, 2005, as World Oil Peak Day [96]. Optimistic estimations [139–141] push the peak 20 or more years in the future, and believe that it will simply be a challenging opportunity rather than a tragedy [142].

CERA, a consulting company specialized in advising governments and private companies on energy markets [143], forecasts that there will be growth of production capacity through 2030, with no peak evident. It also predicts that there will be no unique picture of the course of future of supply, since oil depletion is a complex, multicomponent system.

There are several reasons which seem to indicate that the oil peak should not be far away [92, 134, 135, 144, 145], but there are also valid counter-arguments to imminent global oil depression that have been discussed recently in much detail [92]. A decline in production does not necessarily indicate scarcity; as happens with other non-renewable resources, it is often caused by a decrease in demand. Furthermore, between 1986 and 2003, there has been a significant lull in spending on oil exploration by major energy companies. The Middle East, Russia, and Africa contain three-quarters of the world's oil reserves and yet account for only one-seventh of exploratory drilling. Global supplies of unconventional oil greatly exceed the estimated global conventional oil endowment, even though some authors argue that unconventional resources cannot mitigate the peak of conventional oil and that total production could peak within 5 years [146].

The disagreement about the date of the peak is accompanied by disagreement about its consequences on the economy and, ultimately, the life of mankind. Most of the people believing in an early peak also believe that the peak is a turning point in human history [96, 133–135, 147–149] and perhaps, coupled with a steep rise in global population and per capita oil demand, it will lead to massive unemployment, homelessness, breadlines, and a catastrophic end of industrial civilization [150]. Others believe that there is no reason to see the transition to the post-oil era as a period of unmanageable difficulties or total economic and social catastrophes [92, 140–142]. They point out that past energy transitions have always been among the most important stimuli of technical advances, promoting resource substitution. Perhaps, one of the wisest statements on oil peak is: "Peaking is one of these fuzzy events that you only know clearly when you see it through a rear view mirror, and by then an alternate resolution is generally too late" [149].

A recent thorough report notes that despite much popular attention, the growing debate on peak oil has had relatively little influence on energy and climate policy [16, 151]. On the technical side, the report points out that anticipating a forthcoming peak is far from straightforward because oil supply is determined by a complex and interdependent mix of "above-ground" and "below-ground" factors. Methods of forecasting future oil supply vary widely in terms of their theoretical basis, their inclusion of different variables and their level of aggregation and complexity. Each approach has its strengths and weaknesses. The report [151], which focuses on conventional oil, synthesizes more than 500 publications and compares 14 forecasts of world production, some of which are shown in Figure 4.8. Many forecasting methods are considered overly pessimistic, but forecasts that delay the peak beyond 2030 are believed to require assumptions that are at best optimistic and at worst implausible. Large resources may be available, but are unlikely to be accessed quickly and may make little difference to the timing of the global peak. There is a significant risk of a peak before 2020, which is not distant in view of the long

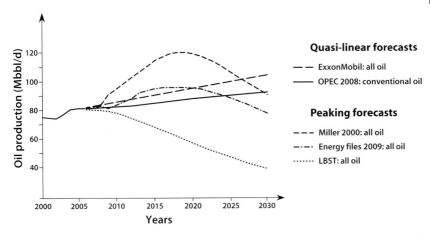

Figure 4.8 Some examples of peaking and quasi-linear forecasts of oil production as reported in the 2009 report by UKERC [151].

times needed to develop alternatives. The conclusion of the report is that without sufficient investment in demand reduction and substitute sources of energy, a decline in the production of conventional oil could have a major impact on the global economy also because there are uncertainties over the extent to which the market may be relied upon to signal oil depletion in a sufficiently timely fashion [151].

It is indeed time to think about an energy transition from oil, and more generally from fossil fuels, to other energy resources. Such a transition must occur as soon as possible owing to concerns about resource availability and even more for environmental and climate reasons (Chapter 7). Hopefully, the fossil fuel age will end when mankind becomes sufficiently clever to understand that it can safely rely only on the plentiful and inexhaustible resource provide by the Sun, which accompanies our spaceship Earth along its aimless trip in the Universe (Chapters 9–10).

5
Natural Gas

"First, we have to find a common vocabulary for energy security.
For Americans it is a geopolitical question.
For the Europeans, right now, it is very much focused
on the dependence on imported natural gas."

Daniel Yergin

5.1
What is Natural Gas and Where It Comes From

Natural gas was named as such since it occurs *naturally*, that is, not manufactured from coal like the so-called *town gas*, that was extensively utilized much earlier than natural gas. Most of the gas and oil extracted so far was formed between 10 and 270 million years ago. In between this huge time span occurred the "dinosaur" periods of the Cretaceous, Jurassic, and Triassic. Natural gas normally occurs at greater depths than oil, unless it has moved upward in later times. Normally, the deeper the deposits, the greater is the fraction of methane, the simplest hydrocarbon, and the ultimate step of degradation of the organic matter.

Natural gas consists primarily of methane, but also includes ethane and propane and basically all the isomers of butane and pentane plus non-hydrocarbon molecules such as carbon dioxide, nitrogen, and hydrogen sulfide. Hydrogen sulfide can make natural gas highly toxic and has to be carefully removed during natural gas treatment. Natural gas can also be formed through the transformation of organic matter by microorganisms, methanogens, that break down organic matter chemically to produce methane. They are found in areas near the surface of the Earth that are devoid of oxygen. The methane thus produced usually escapes into the atmosphere but, sometimes, can be trapped and possibly recovered. Notably, methanogens also live in the intestines of most animals, including humans. In this regard, the production of methane by livestock is a relevant contributor to the release of greenhouse gases into the atmosphere, along with rice cultivation [152].

A third way by which some believe that natural gas might be formed is through abiogenic processes [153], likewise oil (Section 4.1).

Energy for a Sustainable World: From the Oil Age to a Sun-Powered Future. Nicola Armaroli and Vincenzo Balzani
© 2011 WILEY-VCH Verlag GmbH & Co. KGaA, Weinheim
ISBN: 978-3-527-32540-5

5.2
Gas Properties and Definitions

Natural gas is a rather elusive substance, being colorless and odorless. Its density is approximately half that of air, hence it will rise and disperse when released into the atmosphere. This is an excellent property, in light of the fact that natural gas is highly flammable and potentially explosive in confined spaces. Every year hundreds of people are killed or injured by gas explosions in coal mines, factories, and residential buildings worldwide. Natural gas that is delivered to end-users is non-toxic on inhalation but possesses a 21-fold higher global warming potential compared to CO_2 in a 100 year period (Section 7.2.3). The energy density of natural gas at standard temperature and pressure conditions (20 °C, 1 atm) is 35 kJ m^{-3}, that is, 10 000 times less than that of oil, which occurs as a liquid. Nonetheless, its high caloric value and relatively clean burning make it an excellent fuel for heating and cooking even though, at odds with common perception, nowadays its industrial exploitation dwarfs residential usage (Figure 5.1). Upon combustion, the CO_2 emission factor (2.35 t$_{CO_2}$ toe^{-1}) is approximately 26% lower than that of oil and 41% lower than that of coal.

The gas found above and/or dissolved in oil deposits (a frequent case) is called *associated* and it may come out with oil. Sometimes it is reinjected back underground the keep the pressure high and facilitate oil extraction. *Nonassociated* gas is that in reservoirs containing only gaseous hydrocarbons. Finally, *stranded* gas refers to associated or nonassociated gas that, being not connected to a pipeline due to the remote location, is not marketable. Stranded associated gas produced in oil fields is sometimes burned (flared) or even vented into the atmosphere. This practice, which has represented a gigantic waste of natural resources for decades, is being reduced, also due to the ever-increasing value of natural gas as an energy and feedstock commodity [154].

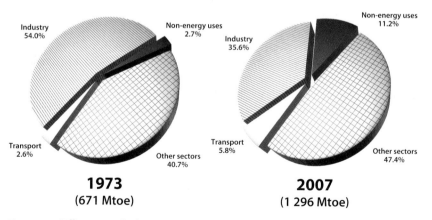

1973
(671 Mtoe)

2007
(1 296 Mtoe)

Figure 5.1 Differences in final use gas consumption between 1973 and 2007 worldwide [4]. Other sectors include agriculture, commercial and public services, residential.

5.3
Brief Historical Notes on Gas Exploitation

Natural gas has been known for millennia [155]. The first intentionally drilled well to extract natural gas was made in Fredonia (New York) in 1821, 38 years before the first oil well. For most of the 1800s, natural gas was essentially used as a fuel for lamps, particularly in city streets, since there was no pipeline infrastructure to bring it to individual homes. In 1885, Robert Bunsen invented a burner that mixed air with natural gas, showing the great potential of gas for cooking and warming.

Only very few pipelines were built until the end of WWII. Afterwards, the US began building its pipeline network to distribute and exploit the abundant gas resources that were found in its territory. Until the late 1960s the US was contemporarily the first consumer, the first producer and the first holder of gas reserves in the world then, in the following four decades, it had only the consumer record.

After the discoveries of vast gas reserves in Russia, the Middle East and the North Sea in the 1960s–1970s, the role of natural gas in the world energy portfolio has steadily increased in the last two decades: nowadays it provides 21% of the world's primary energy supply [4]; in 1990 it provided 16%. The gas world scenario and markets are still evolving with the rise of the liquefied natural gas (LNG) market, characterized by a relatively more flexible distribution infrastructure than the traditional pipeline markets (Section 5.5). Additionally, in the last few years, advances in drilling techniques have made possible the exploitation of vast unconventional gas reserves in the US (Section 5.9), where the extraction of shale gas has re-established the country as the world's top gas producer in 2009, ahead of Russia [7, 156]. If this trend were to be followed in other areas of the world, a sudden global oversupply of natural gas might occur, leading to a potential revolution in the geopolitics of hydrocarbons.

5.4
Gas Production, Consumption, and Reserves

In Table 5.1 are reported the top 10 countries in terms of production, consumption, and reserves in 2009. Russia is the biggest exporter and the second producer of natural gas in the world [7]. Notably, over 70% of its huge production is used domestically, the remainder being addressed to East and West European markets, where Ukraine, Belarus, Germany, and Italy are the largest importers. The US is now back as the largest world producer, but its huge internal production does not match the ever-increasing demand; it imports, mainly from Canada via pipeline, about 15% of its supply [7, 157]. Canada, in its turn, still maintains a good export capacity but internal demand tends to rise also as a consequence of the production of oil from tar sands, which requires huge amounts of gas (Section 4.6).

Gas reserves are extremely concentrated. Nearly 55% of the world's conventional natural gas reserves are located in only three countries (Russia, Iran and Qatar) [7], inside about 20 giant and supergiant fields [158]. Its concentration makes

Table 5.1 Top 10 countries in terms of gas production, consumption, and reserves [7].

Production		Consumption		Reserves	
Country	Gm³/year	Country	Gm³/year	Country	Tm³
U.S.	593.4 (20.1%)	U.S.	646.6 (22.2%)	Russia	44.38 (23.7%)
Russia	527.5 (17.6%)	Russia	389.7 (13.2%)	Iran	29.61 (15.8%)
Canada	161.4 (5.4%)	Iran	131.7 (4.5%)	Qatar	25.37 (13.5%)
Iran	131.2 (4.4%)	Canada	94.7 (3.2%)	Turkmenistan	8.10 (4.3%)
Norway	103.5 (3.5%)	China	88.7 (3.0%)	Saudi Arabia	7.92 (4.2%)
Qatar	89.3 (3.0%)	Japan	87.4 (3.0%)	U.S.	6.93 (3.7%)
China	85.2 (2.8%)	Great Britain	86.5 (2.9%)	U.A.E.	6.43 (3.4%)
Algeria	81.4 (2.7%)	Germany	78.0 (2.6%)	Venezuela	5.67 (3.2%)
Saudi Arabia	77.5 (2.6%)	Saudi Arabia	77.5 (2.6%)	Nigeria	5.25 (2.8%)
Indonesia	71.9 (2.4%)	Italy	71.6 (2.4%)	Algeria	4.50 (2.4%)

natural gas the most geopolitically hot commodity in the current world scenario [159], also because its main distribution infrastructure (gas pipelines) is "rigid" and crosses continents and countries [160]. The big game of new gas pipeline projects that radiate from northern Siberia and the Middle East is ongoing. For several decades only Western European countries competed for this energy resource. Then the number of gas-thirsty consumers increased elsewhere: Eastern Europe, the Middle East, China, India. Several gas pipeline projects (and related diplomatic conflicts) are being developed to establish new links such as the off-shore North-European Gas Pipeline project that will connect Russia and Germany directly through the Baltic Sea. This off-shore route will be more expensive than an on-shore connection but will avoid disputes with Eastern European countries about transit rights [161]. Gas is vital for the Western European economy. Less than 2% of gas reserves are located in the EU [7], but the European continent happens to be surrounded by some of the major gas production basins in the world: Russia, the North Sea, North Africa and the Middle East [161, 162]. The access to vast gas reserves has been one of the main factors leading to the collapse of the coal industry and the impasse of the nuclear sector in the Old Continent in recent decades. In Figure 5.2 is depicted the gas flow by pipeline to Europe.

Notably, an increasing share of electricity is produced in Europe by gas power plants, with The Netherlands (>60%) and Italy (50%) largely reliant on this primary source to feed the electrical grid. Germany gets almost half of its imported gas from Russia [157, 163]. The UK is steadily increasing its gas dependence while its once large domestic production peaked in 2000 and is dramatically decreasing (−45% since the peak) [7]. France is less dependent on gas, due to its large fleet of nuclear power plants. The Dutch government has stated that peak gas production in the country occurred in 2007–2008 and it will have to become a net importer of natural gas by 2025; to preserve its resource base, an upper limit for the produc-

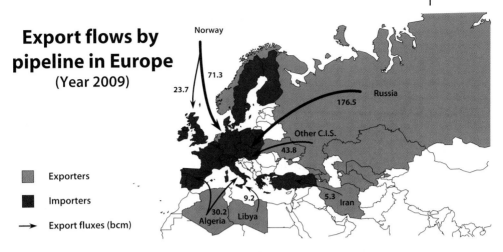

Export flows by pipeline in Europe (Year 2009)

Norway

71.3

23.7

Russia

176.5

Other C.I.S.

43.8

Exporters

Importers

Export fluxes (bcm)

9.2

30.2
Algeria   Libya

5.3   Iran

Figure 5.2 Export flows to Europe of natural gas by pipeline in 2009 [7].

tion of natural gas has been set by law. The only country left in Europe with gas production still on the rise is Norway, now the fifth largest world producer [7]. Its production peak is expected to occur between 2015 and 2020 [164]. It is easy to predict that until 2030 European countries will be increasingly dependent on Russian gas supplies, although it is worth emphasizing that the three Siberian supergiant fields that now accounts for about 60% of Russian production are already in production decline [161, 165]. In 2009 the Russian production decreased by 12% [7]. The IEA estimates that Gazprom, the Russian and world's largest gas company, will have to invest an average of $17 billion per year until 2030 in exploration and production to keep up with increasing domestic and external demand.

The most obvious strategy for Europe to limit its dependence on Russian gas supply is to expand imports by shipment, as will be discussed in the next section.

5.5
Liquefied Natural Gas (LNG)

LNG takes up about 1/600th the volume of natural gas at ambient pressure, which means that a lot of it can be transported in a given volume such as a ship. The liquefaction process is carried out in special facilities, where the gas is cooled repeatedly until it reaches a temperature of $-162\,°C$, when it becomes liquid at atmospheric pressure. Under these conditions it can be shipped in special vessels equipped with four or five separated spherical containers. At destination, the LNG is reconverted back to the gas state in a special regasification facility and funneled to the existing gas pipeline grid. The whole process is schematized in Figure 5.3.

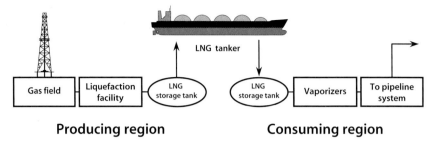

Producing region   **Consuming region**

Figure 5.3   The LNG chain.

For distances longer than 3000–5000 km the transport of natural gas in liquefied form may become more economic compared to pipeline transport. The largest LNG tankers in service can transport up to 140 000 m^3 of LNG. There are about 300 LNG ship in service worldwide and several dozen are under construction. There are now about 65 marine liquefaction terminals and 180 regasification facilities under construction or planned around the world [166]. The major LNG supplier for the US is Trinidad and Tobago (6 Gm^3 in 2009) [7]. By far the largest LNG regasification capacity in the world has been developed by Japan since the 1970s in the aftermath of the first oil crisis [155]. The global LNG market is expected to increase substantially relative to the pipeline market by 2020, covering 36% of the projected 1500 Gm^3 demand [158].

The so-called LNG train, that is, the whole system that connects production fields to the final consumers, is extremely energy intensive. It has been calculated that on the way from Qatar, the largest LNG exporter, to the east coast of the US (7000 miles, about 11 000 km), not less than 15% of the initial shipment is consumed to power the tanker and maintain cryogenic conditions [155]. Additionally, the deployment of a large LNG infrastructure requires huge and concerted investment from producing and consuming countries as well as from energy companies. Indeed, as in the case of the nuclear industry (Chapter 8), it is not straightforward to fund industrial activities which take many years to be finalized and which may attract fierce public opposition over safety concerns and undergo long-lasting legal conflicts. Nowadays the LNG industry has very high safety standards. Nonetheless, potential hazards are associated with LNG handling, particularly accidental spills that may lead to explosion, formation of vapor clouds, or direct contact with the cryogenic material. A vapor cloud formed by an LNG spill could drift downwind into populated areas and burn, in the presence of accidental ignition, if the concentration of natural gas is between 5 and 15% in air. The most severe accident occurred in 1944 in Cleveland, OH, when a steel tank failed, spilling liquid gas onto the city streets and killing 128 people in a tragic fire [166]. Given their huge energy density, LNG facilities and ships are attractive targets for terrorist attacks. This somewhat complicated context makes the social acceptability of LNG infrastructures limited, especially in affluent countries where environmental regulations are stricter.

5.6
Natural Gas Processing

Natural gas produced at the wellhead normally contains contaminants and natural gas liquids (NGL), hence it must be processed ("cleaned") before it can be safely delivered to the high-pressure, long-distance pipelines that transport the product to the final consumers [167]. This process is less complex than oil refining but equally important, albeit largely ignored by the general public; a general scheme of the natural gas industry is reported in Figure 5.4. Notably, in the aftermath of Hurricanes Katrina and Rita in 2005, several gas-processing facilities along the US Gulf Coast were severely damaged and had to stop gas delivery to pipelines.

Natural gas processing starts at the wellhead. The conversion into pipeline-quality dry natural gas involves several steps to remove oil (in associated gas), NGLs (see below), water, and a variety of other gases such as helium, nitrogen, hydrogen sulfide, carbon dioxide, and others. Additionally, it is sometimes necessary to install scrubbers and heaters at or near the wellhead, that serve primarily to remove sand and other large-particle impurities. The heaters are used to keep the temperature of the gas high enough to avoid the formation of hydrates with the water vapor content of the gas stream. These natural gas hydrates (Section 5.9) are crystalline ice-like solids or semi-solids that can hamper the safe flow of gas through the pipeline valves. Once NGLs have been collectively removed from the natural gas stream, they must be separated into their base components to be useful. The process used to accomplish this task is called fractionation, which is based on the different boiling points of the various hydrocarbons in the NGL stream. Final products obtained include ethane, liquid petroleum gas (LPG) (propane, butanes, propane–butane mixtures, ethane–propane mixtures), isopentane, and small quantities of finished products, such as motor gasoline, special naphthas, jet fuel, kerosene, and distillate fuel oil.

Some deposits of natural gas contain significant amounts of carbon dioxide and hydrogen sulfide; owing to the rotten smell caused by its sulfur content, it is

The natural gas industry

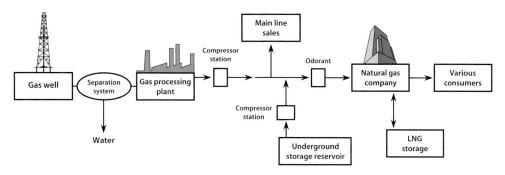

Figure 5.4 The natural gas industry.

commonly called "sour gas" [168]. In order to be utilized, it has to be cleaned to remove the sulfur compounds, which are extremely harmful to breathe and corrosive to pipelines. The process for removing hydrogen sulfide from sour gas is commonly referred to as "sweetening". The sulfide extracted from the natural gas streams is a valuable market product by which, upon treatment, elemental sulfur is obtained. Sulfur production from gas processing plants accounts for about 15% of the total production of this element in the US.

5.7
Transport, Storage, and Distribution

5.7.1
Transport

Once the natural gas has been processed, it becomes of "pipeline quality" and can be conveyed and marketed to end-users. Starting from the 1950s a huge gas infrastructure has been built around the world and now the most affluent countries are crossed by wide pipeline networks operating at different pressures termed *high* (250–20 atm for long-distance transport), *medium* (10–20 atm, to feed large industrial utilities) and *low* (about 2 atm for domestic utilities). There are essentially three major types of pipelines along the transportation route: the gathering system, the long-distance high-capacity pipelines, and the terminal distribution network near end-users. The longest pipeline in the world (about 6500 km) transports natural gas from the supergiant fields in northwestern Siberia to European Russia and all the way to Western Europe, where it bifurcates to feed Germany and France northbound and Italy southbound.

Long undersea gas pipelines are a relatively recent achievement (Figure 5.5). A long (520 km) and deep (up to 1127 m) undersea connection was established in 2004 across the Mediterranean Sea, which conveys about 10 Gm^3 of natural gas from Libya to Italy. The most challenging deep-water transmission system to date is the Blue Stream project, which crosses the Black Sea and connects Russia to Turkey, avoiding third countries. It operates at a pressure of about 255 atm and achieves depths of 2150 m where the water pressure is 214 atm. At such great depths, care must be taken to avoid external stresses on the pipe which might cause implosion. This record depth will soon be overtaken by a new project that, in 2012, will connect Algeria to central Italy, going as deep as 2885 m. To date, the longest undersea pipeline in the world is Longeled, which connects Norway to the UK below the North Sea through a distance of 1166 km with a capacity of 25.5 Gm^3 per year. Among the major gas pipeline projects under scrutiny it is worth mentioning the Trans-Saharian (4128 km from Nigeria to Algeria, an opportunity to diversify European imports from Africa), the Alaskan Natural Gas Pipeline (2760 km, from Alaskan North Slope to the US Midwest), the Iran–Pakistan–India gas pipeline (2775 km, to supply the Indian subcontinent), and the Turkmenistan–China gas pipeline (1818 km, under construction), just to mention a few [169].

Figure 5.5 Saipem 7000, a 198 m long multipurpose crane vessel for placing off-shore hydrocarbon production platforms and laying pipelines in deep waters. It can lift 14 000 t up to 42 m (world record); its tower is 135 m high, and the overall weight of the complete pipelaying facilities is 4500 t; it can host an 800 people crew. It laid the Blue Stream pipeline between Russia and Turkey up to the record depth of 2150 m in the Black Sea. For more details, see www.saipem.eni.it. Photograph courtesy of SAIPEM.

The US, due to its geographic position, is substantially outside the major transnational gas projects: it can only plan to strengthen its connection with Canada. On the other hand, new large gas reserves have recently been discovered in South America, particularly in Bolivia, which might be exported as LNG to the US West Coast. Indeed, this has turned out to be a very delicate political case, which sparked heavy tensions in the South American continent, such as the so-called Bolivia gas conflict, which ended with the nationalization of the Bolivian hydrocarbon industry in May 2006. The import of South American gas by the US might decrease if the boom of domestic unconventional gas production continues (Section 5.9).

The natural gas chain is an extremely energy-rewarding business. Exploration claims on average less than 1% of the discovered fuel. The steady flow of gas through high-capacity pipelines over long distances is guaranteed by pumping stations which are located every 100–150 km; the energy embodied in a steel pipeline is less than 0.1% of the transported fuel during the pipeline lifetime [3]. It is estimated that in Canada and the US the pumping work claims around 3% of the transported natural gas [3].

Leakages from the world gas pipeline infrastructure are subjected to continuous monitoring due to obvious economic reasons and to the high global warming potential of the gigantic amount of methane that is currently extracted and handled (about 3000 billion m^3 in 2009 [7]). Leakages from Russian pipelines were high in the Soviet era, that is, 6–9% of the transported total [170]. A more recent assessment has found a less worrying rate of 1–2.5%, comparable to leakages estimated from US gas pipelines [171].

5.7.2
Storage

Traditionally, natural gas has been a winter fuel. Its seasonality has moderated in recent years due to the soaring utilization of gas to produce electricity, that has progressively shifted its yearly peak demand from winter to summer, due to the expansion of air conditioning in residential buildings.

Considering that gas demand varies also on a daily basis, it is not difficult to understand that an oscillating gas demand cannot be controlled by managing the gas flow in high-pressure pipelines. This makes storage facilities an essential part of the modern gas chain. Stored natural gas plays a vital role in ensuring that (i) any excess supply delivered during the summer months is available to meet the increased demand in the winter months and (ii) some buffer natural gas capacity is available in case of unforeseen supply disruptions due to accidents, conflicts, or political disputes. This latter scenario has materialized in recent years for Western Europe, which sometimes experienced a reduced supply of Siberian gas imports due to disputes between Russia and Ukraine, through which the Russian pipelines pass on going westward. Moreover, it has to be emphasized that, in an increasingly deregulated gas market, natural gas storage is also used for commercial reasons: it is stored when prices are lower, and sold when prices are higher.

Natural gas is typically stored underground, in large natural storage reservoirs. There are three main types of underground storage: (i) depleted gas reservoirs; (ii) aquifers, which are underground porous rock formations that act as natural water reservoirs and can sometimes be reconditioned and filled with natural gas; and (iii) salt caverns, which are caverns in underground salt formations which are created through the injection of fresh water and the removal of the core salt in solution.

In underground storage facilities, as in any other hydrocarbon reservoir, there is a certain amount of gas that can never be extracted (*physically unrecoverable gas*). In addition, underground facilities contain a volume of gas that must remain stored to provide the required pressurization to extract the gaseous fuel (*cushion gas*). The volume of natural gas in the storage reservoir that can be extracted during normal operations is termed *working gas*.

5.7.3
Distribution

This is the final step in delivering natural gas to end-users. While some large industrial, commercial, and electric generation customers receive natural gas directly from high-capacity pipelines at high and medium pressure, most other users receive natural gas from a local distribution company at lower pressures. In order to detect gas leakages easily, local distributors add a tiny amount of an odorant, typically a mercaptan (thiol), which is the source of the familiar rotten egg smell in natural gas.

5.8
Gas Uses: Energy and Feedstock

Natural gas is an extremely versatile commodity that, like oil, is exploited in a variety of applications in both the energy and the industrial sectors. For the use of natural gas in the production of electricity, see Chapter 13.

5.8.1
Energy Use

Nowadays gas is virtually the only primary energy source that is directly available in the residential buildings of most affluent countries. At a quite affordable price it provides space heating, hot water, and cooking. These services are much more practical and more energy efficient compared to the electrical counterparts. The most technically advanced gas furnaces, which exploit also the heat of exhausts, can now achieve efficiencies in excess of 95%. It is conceivable that, in the future, natural gas fuel cells and microturbines will provide to residential consumers the capacity to generate electricity on site, expanding the energy services offered by natural gas for residential needs.

Since the beginning of the automotive era in the late 19th century, compressed natural gas (CNG) was considered a primary fuel for transportation. However, although many companies have manufactured cars and even trucks running with CNG since then, this sector has never taken off due to a number of disadvantages that hampered mass-production, in particular limited driving range, trunk space, higher initial cost, and lack of an extended refueling infrastructure. Existing gasoline-powered vehicles may be retrofitted to allow the use of CNG, which is typically stored in steel containers at ambient temperature and pressures above 200 atm. Today CNG is growing worldwide as a transportation fuel although, probably, it will never compete successfully with liquid fuels. Pakistan holds the biggest fleet of CNG vehicles (2.3 million), followed by Argentina and Brazil [172]. European car makers are increasingly offering CNG cars, also equipped with a small backup gasoline tank. They perform better in terms of air pollution and GHG footprint than classical gasoline or diesel cars and thus often benefit from government subsidies. Italy has the by far largest fleet of CNG cars in Europe (over 600 000), thanks to an extended network of refueling stations. In the US there are just 110 000 CNG vehicles [172].

CNG vehicles are increasingly used in big metropolitan areas for large fleets that drive many kilometers a day, for example, taxicabs, transit and school buses, airport shuttles, garbage trucks, and delivery vehicles. These fleets alleviate urban pollution and help to meet the ever-increasing stringent air quality regulations.

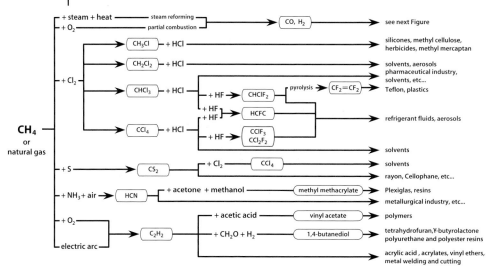

Figure 5.6 Main uses for methane in the chemical industry.

5.8.2
Natural Gas as a Feedstock

Natural gas is utilized to make chloromethanes, carbon sulfide, hydrogen cyanide, and acetylene which allow the production of solvents, synthetic textiles, herbicides, refrigerants, and so on (Figure 5.6). Methane and the higher hydrocarbons that are separated during natural gas processing (Section 5.6) are also a feedstock to prepare ethylene and propylene, the starting materials to make a myriad of molecules and polymers on a large scale [173].

The main use of methane as a chemical feedstock is in the preparation of synthesis gas (*syngas*) via steam reforming:

$$CH_4 + H_2O \rightarrow CO + 3H_2 \tag{5.1}$$

This mixture of carbon monoxide and water is the gateway to a large variety of products and intermediates such as pure hydrogen, ammonia, methanol, aldehydes and, by means of the Fischer–Tropsch process, even liquid hydrocarbons such as naphtha, kerosene, and liquid fuels [174]. The main uses for syngas are schematized in Figure 5.7. Ammonia is one of the most widely produced inorganic chemicals, with an annual world output exceeding 130 Mt. It is the starting chemical to make a wealth of products such as fibers, resins, detergents, explosives, insecticides, and, most notably, fertilizers. Indeed, the Haber–Bosch process for the synthesis of ammonia, and hence fertilizers, enabled the so-called green revolution by boosting agricultural productivity; ultimately, it triggered the world population explosion that started in the 20th century and is still ongoing [175]. Although

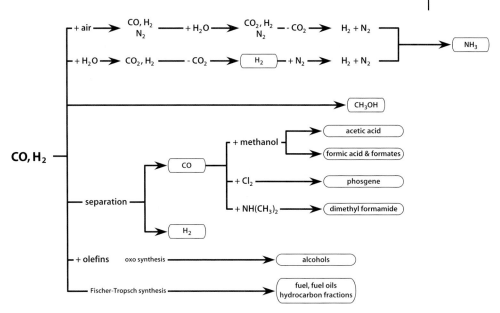

Figure 5.7 Main uses for syngas in the chemical industry.

totally unperceived by the general public, the food bonanza that we have enjoyed in the last 50 years is largely based on the plentiful availability of natural gas to make nitrogen fertilizers. In other words, natural gas is a hidden but essential ingredient of our daily food supply but, since there is no free lunch, the massive introduction of anthropogenic nitrogen in the biosphere implies unpredictable consequences on the delicate equilibria of the nitrogen [176, 177] and even of the carbon cycle [178, 179].

More than 50% of the cost of fertilizers comes from natural gas itself and soaring feedstock prices can jeopardize the ammonia and fertilizer industry, as happened in the US during the frequent price jumps that occurred after 2000, following the so-called US gas crisis [180] induced by severe shortages in the natural gas supply and exacerbated by catastrophic hurricanes in the gas-producing area of the Gulf of Mexico.

5.9
Unconventional Gas

As in the case of oil (Chapter 4), our planet possesses vast and largely untapped resources of unconventional gas, namely gas that is more difficult and less economically attractive to exploit because the technology to collect it is not fully developed or still more expensive compared to conventional extraction technologies [181]. Admittedly, the meaning of "unconventional" is quite loose if one thinks that almost 50% of the current US domestic natural gas production already comes from reserves classified as "unconventional," to be further increased in the future [95].

There are several classes of unconventional gas that we briefly review:

- **Deep and geopressured natural gas:** The so-called deep gas occurs typically at around 5000 m (15 000 ft), that is, somewhat deeper underground than conventional gas deposits. Technical obstacles for deep drilling include high temperatures and pressures as well as acid and corrosive gases like H_2S. Enormous gas resources are stored (geopressurized) down to 15 km in deep aquifers [52]. It is hard to predict if they will ever be exploited economically with a net energy gain.

- **Tight gas:** This gas is stuck underground in extremely hard rock or in unusually impermeable and non-porous sandstone or limestone formations. Its extraction at acceptable rates of gas flow is difficult and costly and implies deep fracturing and acidizing. Currently tight gas covers 32% of the domestic US gas production [95], thanks to recent advances in drilling techniques [182].

- **Shale gas:** This occurs in reservoirs predominantly composed of shale, a very soft rock primarily composed of silt and clay formed from the mud of shallow seas that existed about 350 million years ago. Although only 10% of this unconventional resource is recoverable, it is estimated that there are 21 Tm^3 of technically recoverable shale gas in the US, representing a large and growing share of the total available fuel resource. Currently shale gas covers 11% of the US domestic natural gas production and is on the rise [95, 183]. There is currently great interest in the development of shale gas in China as well as in some European countries such as Poland, Germany, the UK and Austria [184]. The European potential is vast but perspectives for development are less optimistic than in the US for a variety of different reasons, not least the stronger environmental sentiment in the densely populated Old Continent. In fact shale gas exploitation is more land demanding and drilling techniques may contaminate underground water reservoirs [185]. On the other hand, the exploitation of domestic gas resources may loosen the dependence of Europe on Russia.

- **Coalbed methane (CBM):** Many coal mines contain natural gas which is adsorbed into the solid matrix of the coal itself or of the surrounding rock. Coalbed gas is essentially methane with up to a few percent of carbon dioxide and trace amounts of ethane, nitrogen, and a few other gases but virtually no hydrogen sulfide. Due to its composition it is odorless and may constitute a major threat for coal miners because, by combining with the air coming from above ground, it makes an explosive mixture termed firedump (or grisou) that has caused innumerable fatalities in coal mines worldwide. Today coalbed methane has become a viable unconventional form of natural gas, but its margin of expansion is predicted to be limited [95]. Australia is evaluating the possibility of building up an infrastructure to produce and export its vast CBM gas resources and become the only exporter of unconventional gas, a resource which is normally developed only for domestic markets [186].

- **Gas hydrates:** Hydrates, in general terms, are compounds that contain water molecules; the most classical examples are hydrated inorganic salts. A special family of hydrates is represented by water clathrates, that is, three-dimensional cages of water molecules that are able to trap other molecules physically inside their nanocavities. The chemical bonds that make up the hydrate structures are exactly the same as those of ice, namely hydrogen bonding, but in hydrates there are further van der Waals interactions between water and the guest molecules that stabilize the crystalline water phase and raise the melting temperature of the material. The first naturally occurring methane hydrates were found in permafrost sediments of the Arctic by Soviet geologists in 1965.

  The upper-depth limit for the existence of methane hydrates is about 150 m below the surface in continental polar regions where surface temperatures are below 0 °C. In oceanic sediments they occur where bottom-water temperatures approach 0 °C and the water depth exceeds 300 m; in warmer oceans they may occur down to 2000 m and are obviously more difficult to recover. Fully saturated gas hydrates have one CH_4 molecule for every 5.75 molecules of water, which means that 1 m^3 of hydrate can contain as much as 164 m^3 of methane. It is estimated that the global mass of organic carbon locked in gas hydrates can be as high as 10 Tt, thus exceeding the element's total in oil, gas and coal reserves combined [187]. More conservative estimates indicate 1.6–2.0 Tt [188].

  China, Japan, and India, big energy consumers with limited domestic energy resources, place great expectation on the future exploitation of their own methane gas hydrate resources and actively promote research projects in this field. However, the exploitation of methane hydrates as an abundant source of hydrocarbons remains a big challenge. Thus far, the only method that has been successful in recovering methane economically from gas hydrates is depressurization, which is applicable only to hydrates that exist in polar regions beneath permafrost. Other methods under study include heating by injection of steam or hot water, injection of antifreeze, or, most notably, injection of carbon dioxide [189]. Although successful extraction of methane from hydrates has been demonstrated on a small scale, the first commercial production might begin not before 10–15 years and hydrates will probably not provide a significant contribution to the world gas supply for the next 30 years [190, 191].

Gas Hydrates and Climate

Carbon dioxide gives rise to water clathrates more stable than those involving methane. Therefore, it has been suggested that injection of CO_2 in water containing methane gas hydrate resources could yield a sort of underground exchange between the two gases. This strategy would combine, in a synergistic approach, two antithetic processes in terms of climate effects: hydrocarbon exploitation and carbon sequestration [192].

Gas hydrates represent a very important field of research in marine geology and climate change science. In fact the role of these widely occurring materials is considered to be key in the stability of the sea floor and in the overall balance of the global carbon cycle, due the climate-warming potential of methane [188, 193]. The destabilization of methane hydrate deposits as a consequence of increased ocean temperature or pressure, following rapid sea-level variations, might have devastating effects on the stability of the climate [194]. It has been estimated that almost 1 Tt of carbon might be released in response to a 3 °C warming of the planet, yielding an additional 0.4–0.5 °C of hydrate-driven warming [188].

Some have argued that the sudden release of methane from gas hydrates, triggered by some natural catastrophic events such as comet impacts or massive undersea volcano eruptions [195], prompted abrupt climate changes in the past, in turn causing mass extinctions. This theory is popularly known as the "clathrate gun hypothesis", meaning that external triggering events may destabilize ocean gas clathrates inducing global climate change. However, recent paleoclimatic studies have shown that marine clathrates were stable during past abrupt warming episodes [196], suggesting a more prudent approach in establishing a straightforward link between Earth climate trends and release of biogenic methane from the ocean.

6
Coal

"Coal is everything to us. Without coal, our factories will become idle,
our locomotives will rust in the shed, our houses uninhabitable.
Our rivers will forget the paddlewheel,
and we shall again be separated by days from France,
by months from the United States."

Editorial, *The London Times* (1866).

6.1
What is Coal

The general term "coal" indicates several classes of carbonaceous fossil materials
which differ in their carbon and energy contents, therefore some prefer to talk
about "coals," which is probably more accurate. Here we will stick to the singular,
for the sake of simplicity.

Coal is a carbon-based sedimentary rock consisting of decomposed organic
materials, various minerals and water; it can be found on the Earth's surface and
down to thousands of meters underground. Coal was generated by the massive
accumulation of dead land-based plants, mainly trees, throughout millions of
years. As a result of climatic variations through geological times enormous swampy
forests sank below water and their debris was rapidly covered by large quantities
of mud and sand, initially generating peat. This premature burial sheltered the
debris from the air, prevented quick rotting and favored anoxic decomposition of
the organic matter. Under these conditions they were slowly transformed by the
action of increasing temperature and pressure, converting peat to coal (and, occa-
sionally, coal to diamond).

The chemical elements found in coal are the same as found in vegetation
and in the soil that contributed to its formation. The largest coal resources
were generated in the so-called carboniferous period (360–286 million years
ago), but coal was accumulated also in the Jurassic, Cretaceous, and Paleocene
periods up to 55 million years ago. The youngest and poorest quality lignites
and peats were formed starting from 1.8 million years ago. The large variety
of coals makes classification difficult and there is basically no uniform

Energy for a Sustainable World: From the Oil Age to a Sun-Powered Future. Nicola Armaroli and
Vincenzo Balzani
© 2011 WILEY-VCH Verlag GmbH & Co. KGaA, Weinheim
ISBN: 978-3-527-32540-5

international standard of definition. Below only the main general categories are given.

- *Anthracite* is hard, black and slow burning coal of the highest rank; it contains over 93% carbon, and a low percentage of volatile matter, moisture, and incombustible minerals (ash). It exhibits the highest heating value of any type of coal, typically around 33 MJ kg^{-1}, represents less then 1% of known coal reserves and approximately 8% of current world coal production, almost entirely concentrated in China [197]. It is mainly used for residential and commercial space heating.

- *Bituminous coal* is a black coal containing a tarlike substance called bitumen. The carbon content of bituminous coal is in the range 83–92%. It constitutes about 50% of current world production, again with China as top producer [197]. It is a general-purpose fuel with a heating value of 27–33 MJ kg^{-1} used to generate steam electric power as well as power and heat in manufacturing plants. It is also widely used to produce coke for iron- and steel-making.

- *Subituminous coal* is a dark brown to black coal which is waxy and relatively soft. It has a carbon content of 78–83% and represents about 30% of the current world coal output. It exhibits a heating value of 19–27 MJ kg^{-1} and is used in electricity production and cement manufacturing and is the most suitable choice for conversion to liquid and gaseous fuels.

- *Lignite* or brown coal is the lowest rank coal material containing only 25–35% of carbon. It is crumbly, has high moisture content and is characterized by a heating value of 8–20 MJ kg^{-1}, that is, often lower than air-dry wood (~15 MJ kg^{-1}). It is used mainly for electricity generation and, to a minor extent, to make synthetic natural gas and fertilizers. It represents about 15% of the extracted coal worldwide [197].

- *Peat* is not strictly considered coal but a form of biomass. Areas with a naturally accumulated peat layer at the surface (peatlands) cover about 2% of global land (~3 million km^2), mainly in boreal and tropical areas. Being some thousand years old, peat is nominally neither a fossil fuel nor a renewable biomass, but it has to be considered a fossil fuel on the scale of human civilization. The conversion of peatland into agricultural land has been extensively pursued in southeastern Asia in the last few decades. This is mainly accomplished by setting ablaze vast areas of peatland. Indonesia is currently the world's third largest carbon emitter, when deforestation and land use changes are included, mainly due to the destruction of its ancient peat swamp forests, particularly in Borneo [198, 199].

6.2
Coal Extraction

Modern coal extraction involves both surface and underground mining, with about 40% and 60% global share, respectively. Open-cast coal mining recovers a greater proportion of the coal deposit (around 90%) than underground methods, as more layers of coal seams may be exploited. Large open-cast mines can cover an area of many square kilometers and use very large pieces of equipment. In some cases, for instance in the Appalachian region, the exposure of coal deposits is obtained by progressive removal of mountain tops [200], a practice that is highly controversial for the related environmental impacts which involve mountain deforestation, use of explosives, and production of large amounts of excess rock and soil that eventually have to be disposed of [201]. In the US, almost 70% of current coal output comes from surface mines [52] and it has been estimated that CO_2 emissions from coal extracted in this way are up to 17% higher than for underground feedstock, when related forest and soil disturbances are factored in [202].

Modern underground mining in developed nations involves several technologies and has very little to do with the methods of the nineteenth century. The most utilized technology is the so-called long-walled mining which makes use of huge rotating drums that move mechanically back and forth across a wide coal seam. The loosened coal falls on to a pan line that takes the coal to a conveyor belt for removal from the work area; the recovery rate can be as high as 90%. Despite these advances, it is estimated that every year several thousand workers, mainly in China, where underground coal mining is still prevalent [203], die in accidents caused by leaks of poisonous gases such as hydrogen sulfide, explosions of coalbed methane, collapsing of mine slopes, flooding, malfunctioning of mining equipment, and so on [204].

Current world annual production of coal is around 6.6 billion tons [197]. To feed human civilization through the extraordinary achievements and horrific tragedies of the twentieth century, it has been estimated that 250 billion tons of coals were burned along with 125 billion tons of oil and 60 000 billion cubic meters of natural gas [52]. Interestingly, in Western Europe, the leading extracting area for centuries, the coal industry has laid moribund for several decades. In 1920 there were 1.25 million miners in the United Kingdom, now they are reduced to fewer than 10 000 [52].

6.3
Coal Transportation and Industrial Uses

Coal is a bulky commodity and is more expensive and less convenient to transport over long distances than liquid and gaseous fossil fuels: in overseas shipping the cost of transportation may account for up to 70% of the delivered cost of coal. Although around 900 Mt of coal were traded internationally in 2008 (90% via ship) [197], production for local or domestic use is dominant. More than 60% of coal used for power generation is consumed within 50 km of the mining site. Quite often the power station is next to the mine, just because it is easier, cleaner, and cheaper to dispatch large amounts of electricity than moving huge quantities of a solid material. Transportation is made not only by road, rail or ship but also by slurry pipelines.

The main use of coal is in electricity generation (Chapter 13). With an average share of 41.5% globally, coal and peat are by far the principal fuels for electricity production, with gas ranking second at 21% [4]. The next two main markets for coal in affluent countries remain the production of metallurgical coke and cement manufacturing. Coke is produced by baking low-ash, low-sulfur bituminous coal at temperatures above 1000 °C in the absence of oxygen. It is widely used to make pig iron (carbon content 3.5–4.5%) that is eventually transformed into steel (carbon content 0.2–2.1%). Nowadays steel is one the most recycled materials in the world (80% in the US, about 40% globally).

A commercial coal market that has seen a steady growth in recent years is the production of cement, mainly driven by the Chinese economic boom: over 2.7 Gt of cement were consumed globally in 2008, half of which was in China [205]. Coal is used as an energy source to reach temperatures as high as 1500 °C and partially melt the basic raw mixture of limestone, silica, iron oxide, and alumina that makes up cement. Large amounts of energy are required for this industrial activity: around 0.5 kg of powder coal is needed for every kilogram of final product. Other minor uses of coal include the production of activated carbon filters for water purification, carbon fibers as reinforcement materials, and silicon metals which are used to produce silicones and silanes, in turn employed to make lubricants, water repellents, resins, cosmetics, hair shampoos, and toothpastes.

6.4
Coal Gasification

Coal is widely used to make syngas, a gas mixture of carbon monoxide and hydrogen which is the starting point to make liquid fuels [206, 207] and a large variety of chemicals, as discussed in Chapter 5. This process, called Fischer–Tropsch, involves a variety of competing chemical reactions and leads to a series of products and undesired byproducts. The most important reactions are those leading to the formation of alkanes, which can be described by the following general equation:

$$n\text{CO} + (2n+1)\text{H}_2 \rightarrow \text{C}_n\text{H}_{(2n+2)} + n\text{H}_2\text{O} \tag{6.1}$$

Process conditions and catalyst composition are thoroughly tuned to favor higher order reactions ($n > 1$) and minimize methane formation. Most of the Fischer–Tropsch alkanes tend to be straight-chained, although some branched alkanes are also formed. Historically, this process has been extensively utilized as an *extrema ratio* in case of lack of oil supply: petroleum-poor but coal-rich Germany and South Africa made liquid fuels by domestic coal during the Nazi and apartheid regimes to respond oil embargoes imposed by democratic nations.

Clearly, the biggest problem with synthetic gas from coal is the production of massive amounts of CO_2. The single largest point source of carbon into the atmosphere of the whole planet is the SASOL coal-fed synfuel facility in Secunda, South Africa, which emits 20 million tons of CO_2 per year, about the equivalent of ten 800 MW natural gas combined-cycle power stations. A recent report for the

National Academy of Sciences analyzed the potential for production of liquid transportation fuels from coal and biomass, even combined, in the US [208].

Underground coal gasification (UCG) is a further possibility for exploiting coal resources in the future. This process involves the injection of steam and air or oxygen into an underground seam of coal and makes syngas *in situ*. Feasibility studies and small scale demonstrators suggest that UCG could potentially increase world reserves by as much as 600 Gt, almost doubling current estimates [158]. Extensive deployment of this technique might pose concern about underground water contamination [209].

6.5
Coal Production, Consumption, and Reserves

Table 6.1 shows the top 10 countries in terms of coal production, consumption, and reserves [7]. The first important fact to highlight is that not a single country of the Middle East is included among the top reserve holders, making this commodity attractive to diversify primary energy imports from industrialized countries. Nonetheless, the three main coal reserve holders store over 62% of the total amount, a share even higher than the three top kings of oil and gas (Tables 4.1 and 5.1).

China is by far the biggest coal producer with a share of almost 46%. This huge output is now almost completely used for domestic consumption, which is notable for a country that used to be a relevant coal exporter in the late twentieth century [210]. The US ranks as the second world producer and fifth world exporter but its coal production is increasingly utilized to cover a soaring internal demand [211]. Interestingly, in the period 1980–2010 the production of coal in the US increased

Table 6.1 Top 10 countries in terms of production, consumption, and reserves of coals (including all types of coal combined) [7].

Production		Consumption		Reserves	
Country	Mtoe/year	Country	Mtoe/year	Country	Mton
China	1552.9 (45.6%)	China	1537.4 (46.9%)	U.S.	238308 (28.9%)
U.S.	539.9 (15.8%)	U.S.	498.0 (15.2%)	Russia	157010 (19.0%)
Australia	228.0 (6.7%)	India	245.8 (7.5%)	China	114500 (13.9%)
India	211.5 (6.2%)	Japan	108.8 (3.3%)	Australia	76200 (9.2%)
Indonesia	155.3 (4.6%)	South Africa	99.4 (3.0%)	India	58600 (7.1%)
South Africa	140.9 (4.1%)	Russia	82.9 (2.5%)	Ukraine	33873 (4.1%)
Russia	140.7 (4.1%)	Germany	71.0 (2.2%)	Kazakhstan	31300 (3.8%)
Poland	56.4 (1.7%)	South Korea	68.6 (2.1%)	South Africa	30408 (3.7%)
Kazakhstan	51.8 (1.5%)	Poland	53.9 (1.6%)	Poland	7502 (0.9%)
Colombia	46.9 (1.4%)	Australia	50.8 (1.6%)	Brazil	7059 (0.9%)

by 40%, but in China by 400% [210]. A current major coal producer is also India, with a sixfold increased output in the last three decades, in an effort to keep up with booming internal demand which can be satisfied only by continuously increasing imports [210]. A key country in the world coal market is Australia, which is by far the largest exporter with a share of about 25% globally [210]; in 2008 it exported 115 Mt of steam coal and 137 Mt of coke mainly to Japan, South Korea, Taiwan, and India. The whole proved recoverable reserves of all coals combined are now estimated at about 0.85 Tt [158], that is, about 130 years at current consumption levels, although some recent debate was sparked by some scientists who, by applying a Hubbert's approach to coal production, forecast a production peak around 2020 followed by a 30 year long plateau and an inexorable decline. These projections, however, are highly questionable [212].

Environmental consequences of coal combustion (Chapter 7) will likely continue to be much more important in determining the extent of coal exploitation, rather than concerns about its physical availability.

6.6
Carbon Capture and Sequestration (CCS)

Given the large availability of coal, its primary role in the global production of electricity, and the relentless increase of world electricity demand, several strategies are being scrutinized to reduce the amount of CO_2 dumped in the atmosphere by coal burning. Geologic sequestration at depths below 800–1000 m is one of the most promising strategies [213, 214].

About 60% of global CO_2 produced by human activities comes from more than 8000 large industrial point sources spread over five continents [215]. They are either fossil fuel power stations or industrial facilities making cement, steel, metals, synthetic oil, refined oil products, and so on. The largest contribution comes from the fleet of about 2300 large coal power plants which total about 1800 GW of installed capacity and alone emit about one-third [216, 217] of the 30 Gt of CO_2 currently injected into the atmosphere by our industrial civilization [218]. These large and relatively few point sources are the primary targets in the effort to curb greenhouse gas emissions.

In view of the increasing demand for electricity expected in the next two decades (and in the absence of unexpected events of some sort that may alter forecast expansion trends), it has been estimated that 1400 GW of new coal power plant will have to be put on-line by 2030 [219]. China alone is currently opening the equivalent of a 1000 MW new coal power plant per week, a capacity comparable to that of the entire UK each year [220]. The US produces about 1.5 Gt of CO_2 from coal burning power plants per year. If only 60% of this amount were to be captured, compressed, and liquefied for geologic sequestration, its daily volume would about equal the total current US oil consumption of 20 million barrels per day [220]. At the global level, according to the Intergovernmental Panel on Climate Change, humanity must prevent itself from emitting (or must soak up) 650 Gt of carbon by 2100 to keep the concentration of CO_2 under 450 ppm and prevent catastrophic warming [15].

Carbon capture and sequestration is still in its infancy and it is not guaranteed that it will ever become possible on a large industrial scale due to technical and economic hurdles [221–223]. As of 2010, there are only 12 CCS large facilities that run at an experimental level or are under construction around the world [224]. When fully operating, in 2016, they will sequester about 13 Mt of CO_2 per year [224], the equivalent of about three large coal power plants. The ambition of CCS, however, is enormous: to fit all coal and gas power stations by mid-century and reduce world CO_2 emissions from the electricity sector by 20% relative to the current level, while doubling electricity supply.

CCS will cut power plant output substantially due to its intrinsic energy consumption [225] and will necessarily lead to a price increase of electricity from coal. This cost penalty cannot be forecast precisely at the present stage (estimates range between 10 and 60%), also because it is not yet clear which CCS technology will ultimately prevail among the following three under study [223] (Figure 6.1):

- *Postcombustion capture* implies chemical separation of CO_2 from the flue gas stream. A solvent, typically monoethanolamine [226], absorbs CO_2 and is then regenerated in heating columns at 150 °C [227]. The main advantage of this technique is that, in principle, it can be retrofitted to existing power stations and can be more readily deployed. Drawbacks include large energy consumption, vast space requirements (the equipment is sometimes as large as the power station itself) and the use of huge amounts of solvents with related environmental risks.

- *Oxyfuel combustion* is less developed than amine-related approaches, but holds promise because it could be used to capture not only CO_2 but also other pollutants. This method modifies combustion conditions by burning coal in an oxygen-enriched environment by injecting pure oxygen diluted with CO_2 and water.

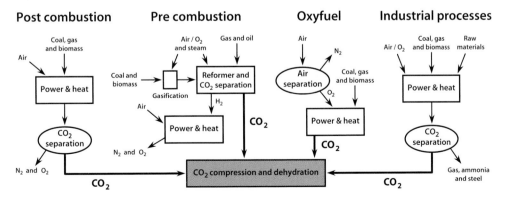

Figure 6.1 Overview of CO_2 capture systems.

- *Precombustion capture* is probably the most promising approach. The fossil fuel is first converted into syngas and then sent to a water gas shift reactor, where most of the CO reacts with water to produce CO_2 and hydrogen. At that point, CO_2 is under pressure and can be removed with a physical solvent-based system in a relatively facile way, but still needing substantial implementation [223].

Sequestration presents complex technical challenges also for the construction of wide transportation pipeline systems and for the assessment of safe underground reservoirs [214]. Transporting CO_2 on the gigaton scale through an effective world-wide distribution network indeed appears to be an almost impossible economic, technical, and diplomatic effort.

Another major challenge for CCS is the availability and reliability of underground geologic storage sites (see Box). In the meantime other CO_2 sequestration schemes are being investigated. One is the so-called ocean storage, which is essentially the use of the ocean as a sort of CO_2 dump (Figure 6.2) However, the ecological impact of ocean storage is an issue of great concern since acidification of ocean surface waters due to the increase of atmospheric CO_2 concentration is already affecting marine ecosystems (Section 7.3.4). Underwater gas reservoirs occur naturally such as Africa's Lake Kivu. They are constantly monitored in an effort to prevent catastrophic consequences in case of sudden release, as already happened in the past [228].

Time is running short for the decarbonization of our energy system and CCS is the last opportunity for coal to demonstrate that we can still rely, for some

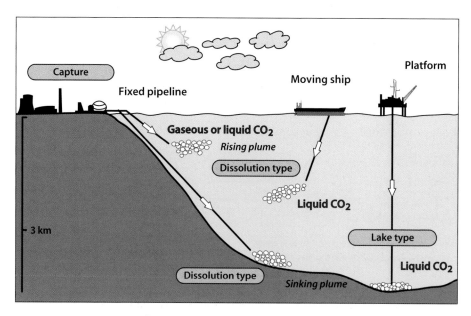

Figure 6.2 Strategies for ocean storage of CO_2.

CO₂ Underground Repositories

The underground repositories for CO_2 being scrutinized are depleted oil and gas fields, unmineable coal beds, and saline formations in sandstone (aquifers) or basalt formations; all of them can be located both onshore [231] and offshore [232] (Figure 6.3). The estimated carbon storage capacity over the entire planet is placed between 1 and 10 Tt [219, 231, 233], mainly consisting of saline aquifers. The lower, and probably more realistic, 1 Tt value would guarantee many decades of sequestration, hopefully sufficient to make CCS a viable bridging technology towards a carbon-free energy future.

At the present state of research it is impossible to assess the risk of CO_2 escape from geologic repositories through existing fractured rock, soil porosities, or natural catastrophic events such as earthquakes [214, 229]. At least seven potential routes of escape to the biosphere, both in the oceans and in the atmosphere, have been individuated and perennial monitoring will have to be implemented [214]. Sudden leaks would have immediate catastrophic effects in the proximity of the storage sites: the CO_2 is intrinsically non-toxic but at a 4% concentration in the atmosphere it becomes asphyxiant. Even minimal but constant leakages (well below 1% per year) would make pointless the whole endeavor; a delay in carbon release of decades or a few centuries would not affect ongoing warming trends substantially.

At depths below 800–1000 m CO_2 becomes supercritical and has a liquid-like density (about 500–800 kg m^{-3}) that provides the potential for efficient utilization of underground storage space and, in principle, improves storage security. However, the massive underground injection of CO_2 would represent an unprecedented type of interference with the lithosphere with potential side effects, such as the possibility of triggering "artificial" earthquakes [214]. Furthermore, little is known about the long-term fate of large quantities of CO_2 put into geologic storage.

---

decades, on an abundant fossil resource that will not alter irreversibly the Earth's climate stability [229]. In one decade we will find out whether CCS can be a revolutionary technology or just a costly ambitious plan poised to remain no more than a scientific curiosity and, at the very end, an enormous waste of economic resources [230].

6.7
Integrated Gasification Combined Cycle (IGCC)

The integrated gasification combined cycle (IGCC) is a coal-based technology for electric power generation which is gaining increasing attention because it might combine several highly desirable goals. It integrates coal gasification from the

Underground sequestration of carbon dioxide

Overview of geological storage options

1. *Depleted oil and gas reservoirs;*
2. *Use of CO₂ in enhanced oil and gas recovery;*
3. *Deep saline formations*
   - **a.** *offshore*
   - **b.** *onshore*
4. *Use of CO₂ in enhanced coal bed methane recovery.*

Oil or gas deposit

Stored carbon dioxide

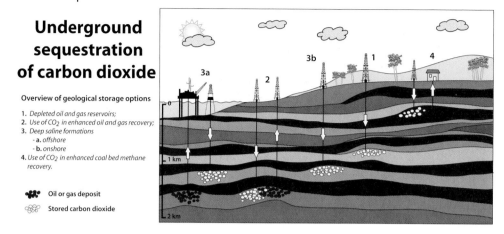

Figure 6.3 Geological options for carbon storage.

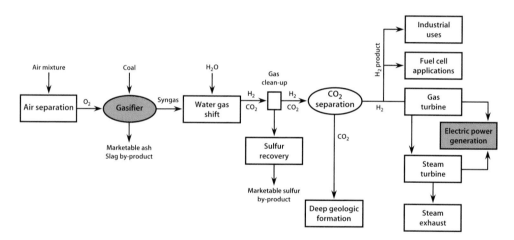

Figure 6.4 Block diagram of an advanced IGCC power plant with carbon sequestration.

chemical industry and combined-cycle power generation from the power industry. IGCC plants (Figure 6.4) turn coal into syngas, allowing the removal of impurities before combustion (see *Precombustion* in the previous section); this results in lower emissions of sulfur dioxide, particulates, and heavy metals. Additionally, excess heat from primary combustion and generation is passed to a steam cycle (similarly to a combined cycle gas turbine), boosting fuel efficiency of electricity generation to 50% or more, compared to the 30–35% of conventional pulverized coal plants. Other advantages of this technology include less water for cooling purposes and the removal of ash from feedstock in the form of a glassy, non-leachable slag. This can be more easily recycled, compared to bottom ash produced by conventional coal technologies, in the manufacture of cement or as asphalt filler.

In the future, if IGCC facilities can be retrofitted for carbon sequestration, environmental attractiveness could thus be enhanced but with an energy output loss of about 20% [214]. At present, IGCC technologies are still at the testing stage, with several economic and technical hurdles to be overcome before further development and commercialization [234]. In the US there are only two plants running, one under construction and one pending. It is expected that a decisive boost for this technology will come if stringent legislation on CO_2 emission were to be pursued, eliminating current economic uncertainties that make IGCC not cost-competitive compared to established power thermal technologies [235]. Since hydrogen can be used as fuel to produce electricity, the success of IGCC would constitute a bridge between today's fossil-fueled energy system and tomorrow's economy based on the use of hydrogen, not only for combustion but also for feeding fuel cells (Chapter 14).

7
Fossil Legacy

"Your grandchildren will likely find it incredible, or even sinful,
that you burned up a gallon of gasoline to fetch a pack of cigarettes."

Paul MacCready

7.1
The Energy Dark Sides

7.1.1
Inequalities and Externalities

There is no doubt that the massive exploitation of fossil fuel resources has been the single main factor causing the spectacular improvement of the quality of life of millions of people on Earth during recent decades. However, these lucky individuals and communities constitute a minority of the world global population.

Furthermore, we should not forget that the ever increasing exploitation of coal, oil, and gas is a gigantic conversion of resources placed in the lithosphere into waste products released into the biosphere, with various effects on the environment and human health. Such effects are typically not integrated into the pricing of a given energy source or technology and are termed external costs or *externalities*. These include immediate and short-term impacts related to the discovery, extraction, transportation, distribution, or burning of energy resources; for example, accidents in coal mines and oil spills during sea transportation or extraction. Such direct effects are complemented by others that are spread over time to future generations (e.g., the long-term storage of wastes) or over space to the entire planet, such as transboundary air pollution or alteration of the carbon cycle which triggers climate change. The latter consequences are the most ethically troublesome, because they affect also those who cannot enjoy the benefit of energy abundance, and even future generations.

Energy for a Sustainable World: From the Oil Age to a Sun-Powered Future. Nicola Armaroli and Vincenzo Balzani
© 2011 WILEY-VCH Verlag GmbH & Co. KGaA, Weinheim
ISBN: 978-3-527-32540-5

7.1.2
Monetizing Costs

The economic assessment of damage caused by energy systems is extremely complicated, questionable, and subject to continuous refining and updating [236, 237]. A recent assessment on the hidden price of modern energy systems estimates that fossil-fuel-based energy production, mostly from coal and oil, causes $120 billion worth of health and other non-climate-related damage in the US alone each year, that are not figured into the price of energy [238]. The study finds that nearly all of the $120 billion in damages is due to electricity generation ($63 billion) and transportation fuel production and use ($56 billion). Notably, this huge cost does not consider the impact of climate change, which is separately estimated to be between $7 billion and $700 billion.

Military costs definitely stand among the largest and most unaccounted externalities of our modern energy systems. Huge expenditures are made, even in times of peace, to guard and secure the overseas oil flow that keeps alive the world economy, and billions of dollars are disbursed as economic assistance of US governments to allied Arab regimes in the Middle East area. Even if the role of oil as *casus belli* may have been overplayed by some historians [50], it is hard to dismiss oil as the major cause of the wars waged in the Persian Gulf and nearby regions starting from the early 1990s, that had a huge economic and human toll. In principle, we should charge on our energy bill the costs of these war and reconstruction expenditures (see Box, Chapter 4).

7.1.3
Indirect Effects

A long list of other negative *indirect* side effects, sometimes subtle, can be further drawn, for example soil erosion [61], which is primarily the consequence of intensive agriculture, made possible by a huge mobilization of fossil energy. In turn, the wide availability of food, the enhanced consumption of meat and sweet beverages, and the adoption of more sedentary lifestyles have led to an epidemic of obesity, exposing many people, not only in affluent countries but also in some developing nations, to enhanced risk of diabetes, cardiovascular diseases, and cancer [239]. Other indirect externalities of fossil fuels are those connected to the huge and ever increasing fleet of transportation means [240]. They include accidents and fatalities (in 2009, 34 500 road fatalities in the EU), health effects caused by exposure to air pollutants and noise, climate impacts related to the emission of greenhouse gases, and soil and water contamination. The role of the transport sector in greenhouse forcing (see below) has been estimated as 15 and 30% of the total man-made CO_2 and O_3 contributions, respectively [241, 242].

Although a definitive accounting method is probably beyond our present and future ability, it is hard to question that there are no market prices so unrelated to "real" costs such as those of coal, oil, and gas.

7.2
Alteration of the Carbon Cycle by Fossil Fuel Combustion

7.2.1
Carbon Reservoirs and Fluxes

Carbon is the main building block of life: its concentration in living matter is about 100 times greater than that in the Earth's crust. On our planet, carbon is continuously exchanged among four main reservoirs: land, ocean, atmosphere, and lithosphere. This generates the so-called carbon cycle, a set of extensive and complex biogeochemical processes that are modeled by means of mathematical formulations that estimate the fluxes between reservoirs (Figure 7.1). In terms of time, the global carbon cycle can operate both on a geological scale (millions of years) and on a shorter biological/physical level (days to thousands of years). Obviously, we are mainly interested in the second one.

Figure 7.1 shows the amount of carbon stored in each reservoir [243]. By far the greatest pool is the deep ocean, which contains approximately 35 000 Gt of carbon, mostly as bicarbonate ions; a smaller amount (800 Gt) is localized on surface ocean. The atmosphere contains about 800 Gt of carbon as CO_2, a slightly higher value compared to storage in land vegetation, which embodies 600 Gt. The carbon stored in soil as organic detritus amounts to 2000 Gt, that is, more than three times the amount fixed in the plants that grow upon it. Organic carbon dissolved in

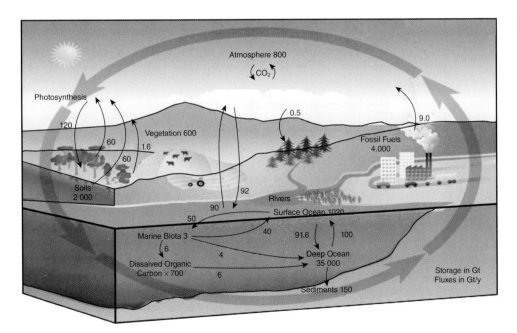

Figure 7.1 Schematic representation of the main reservoirs and fluxes of the complex global carbon cycle.

ocean is about 700 Gt, while that accumulated in fossil fuels (mainly coal deposits) is about 4000 Gt, of which about 1000 Gt are classified as proven reserves (about 670, 140, and 95 Gt of coal, oil, and gas, respectively) [244].

Yearly, 100 Gt of CO_2 are exchanged between ocean and air, whereas photosynthesis converts 120 Gt of carbon as CO_2 in terrestrial plants. These and other exchanges, mainly occurring in oceans and detailed in Figure 7.1, used to be substantially in balance. The equilibrium was disrupted by the massive exploitation of coal, oil, and gas started with the industrial revolution. Currently, about 9 Gt per year of carbon are taken out of the lithosphere and injected into the atmosphere [245]. A further 1.5 Gt come from land use changes such as deforestation [246] and intensive agriculture, that tend to reduce the carbon content of soil [247].

7.2.2
CO$_2$ Rise and Its Measurement

That fact that our planet had reached a substantial balance of the carbon fluxes among the four main carbon reservoirs has been inferred by the analysis of ice cores in Antarctica and other paleoclimatic probes (proxies), which allow one to "measure" indirectly the concentration of CO_2 in the atmospheric carbon pool throughout 800 000 years [248]. In this long period of time, it is found that the carbon dioxide concentration oscillated in a relatively narrow range between 170 and 290 ppm [243, 249] and, from the rise of our present civilization (nearly 6000 years ago) and the beginning of the fossil fuel era, it fluctuated even less (250–290 ppm). In the last several decades, in contrast, this value has started to deviate form this long-lasting range and the imbalance is unambiguously attributable to the burning of fossil fuels, for the beginning of the deviation is concomitant with the energy and industrial revolution.

CO_2 is colorless, odorless, and non-toxic, and its injection into the atmosphere is perhaps the largest and most unnoticed environmental human footprint on Nature. CO_2 is also a very stable molecule that, on average, can remain intact in the atmosphere for several decades, while the atmosphere is thoroughly mixed in about 1 year. Thus its concentration is substantially identical all over the planet.

Direct measurement of atmospheric CO_2 started in 1958 at the Mauna Loa Observatory at Hawaii [250], an ideal place to measure carbon dioxide levels with minimal influence from the continents' urban pollution or large forests. Five decades of measurements at this remote station have invariably shown that there is a seasonal oscillation of the CO_2 concentration measured: up during fall, winter, and early spring, then down during late spring and summer (Figure 7.2). During the warm northern season, terrestrial photosynthesis manages to absorb the excess of CO_2 injected into the atmosphere by human activities, but during the colder period the process is dramatically reduced. Despite oscillations, the yearly averages inexorably keep going up. From the first data registered at Mauna Loa in the late 1950s, the atmospheric concentration of CO_2 has increased by more than 20%, reaching 390 ppm in 2010.

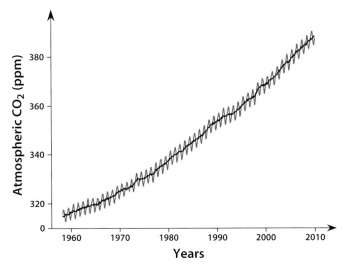

Figure 7.2 Historical record of CO_2 concentration at US Mauna Loa Observatory, Hawaii.

7.2.3
The Greenhouse Effect

While CO_2 is only a very small part of the atmosphere (0.039%), it plays a large role in the radiation balance of the planet. Such a balance is given by the interplay between the incoming shortwave solar radiation (UV–Vis) and the outgoing long-wave (IR) radiation that is delivered by the Earth's surface upon absorption of the incoming solar flux (see Figure 9.2). Our atmosphere is highly transparent to UV–Vis solar radiation but contains several polyatomic molecules [mainly H_2O, CO_2, CH_4, N_2O, O_3, chlorofluorocarbons (CFCs), and hydrofluorocarbons (HFCs)] whose modes of vibration match the frequencies of the outgoing infrared emission (i.e., heat), thus generating the natural greenhouse effect. This process is crucial to keep the average temperature of the Earth at the optimal life-sustaining value of 15 °C; without it, our planet would be a rather inhospitable place with an average temperature of −18 °C.

The abnormal accumulation of anthropogenic CO_2 and other greenhouse gases in the atmosphere has modified the Earth's heat balance, leading to a positive radiative forcing (i.e., heating), which, relative to the start of the industrial era (about 1750), is estimated to be $3 \, W \, m^{-2}$ over the planet's surface. It is primarily due to CO_2 ($1.66 \, W \, m^{-2}$), but the combined effect of CH_4, CFCs, HFCs, O_3, and N_2O is almost as large (Table 7.1) [15]. The greenhouse radiative forcing is partially counterbalanced by an anthropogenic negative forcing (i.e., cooling), mainly dictated by aerosol effects (Section 7.4). As a result, the net positive forcing responsible for ongoing global warming amounts to $1.6 \, W \, m^{-2}$, according to IPCC. The increase in atmospheric CO_2 concentration and the alteration of the carbon cycle leading to climate change have nothing to do with the occurrence of ice and warm ages

Table 7.1 Global warming potentials (GWPs) of some greenhouse gases (from IPCC report 2007). GWP represents how much a given mass of a chemical contributes to global warming over a given time period compared to the same mass of carbon dioxide.

Greenhouse gas	Chemical formula	Lifetime (years)	Radiative efficiency $(W\,m^{-2}\,ppb^{-1})$	Global warming potential	
				20 years	100 years
Carbon dioxide	CO_2	~100	1.4×10^{-5}	1	1
Methane	CH_4	12	3.7×10^{-4}	72	25
Nitrous oxide	N_2O	114	3.0×10^{-3}	289	298
CFCs	$C_nF_xCl_y$	45–1700	0.18–0.32	~5000–11 000	~5000–14 000
Halons	$C_nF_xCl_yBr_z$	16–65	~0.3	~3700–8 500	~1600–7 000
Carbon tetrachloride	CCl_4	26	0.13	2700	1400
Methyl bromide	CH_3Br	0.7	0.01	17	5
Methylchloroform	CH_3CCl_3	5	0.06	506	146
HCFCs	$C_nH_xF_yCl_z$	1.3–18	0.1–0.3	~270–5500	~80–2300
HFCs	$C_nH_xF_y$	1.4–270	0.1–0.4	~400–12 000	~120–15 000
Sulfur hexafluoride	SF_6	3200	0.52	16 300	22 800
Nitrogen trifluoride	NF_3	740	0.21	12 300	17 200
PFC-14	CF_4	50 000	0.10	5210	7390
PFC-116	C_2F_6	10 000	0.26	8630	12 200

over geological times. These phenomena depend upon astronomical factors that, evolving over the tens of millennia timescale, affect the amount of sunlight hitting the Earth and cause the long-term expansion and retreat of its ice cover [251].

Mentioning the above astronomical phenomena allows us to clarify and emphasize a key point. Dramatic climate change does not necessarily require a variation of the CO_2 level. But changes in the atmospheric CO_2 level always alter the Earth's climate, because it inexorably affects the overall heat balance [252].

7.3
Anthropogenic Climate Change

7.3.1
The Path to Present Understanding

In 1896, Svante Arrhenius first put forward the possibility of extensive human interference with the greenhouse effect. The first "modern" scientific paper addressing the issue and starting to raise some concern was published in 1957 [253]; one year later, CO_2 measurements started at Mauna Loa, certifying the possibility of rapid and severe global warming. Public attention, however, was attracted only in the 1980s. In 1988 the United Nations established the Intergovernmental Panel on Climate Change (IPCC), a scientific body aimed at evaluating the risk of climate change. Over the years the IPCC has released four reports, which are not based on original research but on a critical survey of peer-reviewed literature. Starting from 2010 the work of IPCC is independently reviewed by a committee of the Interacademy Council [254], and is probably going to be reorganized [255].

The latest IPCC report, released in 2007 [15], put the average increase of surface temperature at 2.0–4.5 °C by 2100, with a best estimate of about 3.0 °C. The evidence of a human footprint on climate grows bigger [256–258], whereas attempts to attribute the observed warming trends to natural factors, such as variations of the Sun's energy output, or to biased temperature measurements due to the so-called urban island effect, are increasingly discredited [259–261]. Nonetheless, since 2009 the debate has been stirred by a controversy over leaked e-mail messages written by climate scientists and an acknowledged mistake from the IPCC about the retreat of Himalayan glaciers [262].

To assess what the world will be like at higher temperatures is a very risky task, given the intricate pattern of positive and negative climate feedbacks that would be triggered [263, 264]. The conceivable ranges of temperature change do not give us a perception of what they would mean in practice. To this end it might help to know that doubling present levels of CO_2 and raising average temperatures by about 6 °C would have the same effect as magically moving the Earth's orbit by about 2 250 000 km closer to the Sun (assuming the Earth's albedo to be constant) [251]. On the other hand, about 18 000 years ago in the deepest period of the last glacial era, the average temperature was just 6 °C lower than now and ice sheets largely covered Northern Europe, the southern boundary passing through the present Germany and Poland [265].

7.3.2
Melting of Ice Sheets

The retreat of glaciers is widespread worldwide at mid to low latitudes [266, 267]. Ice in the Arctic Ocean has been shrinking since the late 1970s [268], as it experienced the five warmest decades in the last two millennia from 1950 to 2000 [269]. The perspective of an ice-free Arctic in summer time is looming [270]. The most recent satellite measurements suggest that also the two other largest bodies of ice, Antarctica and Greenland, continue to lose ice mass at an increased pace [271, 272].

If no action is taken to limit the greenhouse effect, the partial melting of the ice sheets will lead to an increase in the sea level: it has been estimated that Antarctic and Greenland melting will raise the sea level by 0.35 mm per year, a modest contribution to the present rate of sea level rise of 3.0 mm per year [273]. This is not surprising since the largest contribution is expected to come from thermal expansion of seawater to due global warming: the IPCC attributes to this latter factor the largest contribution to the estimated sea level rise of 18–59 cm during the present century [15]. Rapid and catastrophic collapse of inherently unstable large portions of ice sheets such as West Antarctica cannot be excluded, which could prompt a sea level rise of over 3 m [274]. Interestingly, human activities also counteracted this trend by trapping almost 11 000 km^3 of water in artificial reservoirs, which is estimated to have reduced the magnitude of the sea level rise by about 3 cm over recent decades [275].

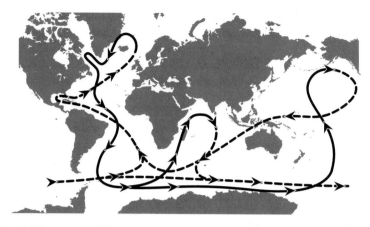

Cold water

Warm water

Figure 7.3 The global conveyor belt begins with the cold water near the North Pole and heads south between South America and Africa towards Antarctica. Here it gets recharged with more cold water and then splits in two directions, one heading to the Indian Ocean and the other to the Pacific Ocean. As the two sections near the Equator, they warm up and rise to the surface. Eventually, the two sections loop back to the South Atlantic Ocean and finally back to the North Atlantic Ocean, where the cycle starts again.

7.3.3
Interference with Ocean Currents

Oceans cover about 70% of the Earth's surface and have absorbed over 80% of the total human-induced heating of the Earth since 1960 [276]. In various ways [277] we are affecting the main player of the water cycle and the key regulator of global climate through oceanic currents, that are driven by winds and ocean density differences. The portion of ocean flow that is driven by density is called the *thermohaline circulation* (or global conveyor belt), that transports warm surface water towards the poles and cold deeper water back to the tropics (Figure 7.3). The Atlantic thermohaline circulation is listed among the tipping elements in the Earth's climate system, that is, one of those single factors with a critical threshold at which a small variation in human influence can have large, irreversible consequences for the whole planet [278, 279]. Indeed, paleoclimatic studies have recently evidenced that Greenland's ice cover is dramatically vulnerable also to natural forcing events [280] and, in the past, experienced abrupt climate changes within 1–3 years [281]. The growing concern about the stability of the thermohaline circulation has spurred more intensive research to unravel its complex and still largely unknown mechanisms [282].

7.3.4
Ocean Acidification

Another consequence of a CO_2-rich atmosphere is ocean acidification [283], because carbon dioxide dissolution in water generates carbonic acid that, according to the following equilibria, dissolves carbonate shells and skeletons in living marine organisms [41]:

$$CO_2 + H_2O \rightleftharpoons H_2CO_3$$

$$H_2CO_3 \rightleftharpoons H^+ + HCO_3^-$$

$$H^+ + CO_3^{2-} \rightleftharpoons HCO_3^-$$

About half of the CO_2 emitted since the industrial revolution has been absorbed by the oceans [284]. It has been estimated that the neutralization of the entire mass of CO_2 produced by combustion of fossil fuels will be largely accomplished (70–80%) by ocean sequestration, but will be completed in about 200 000 years [285].

At the current emitting rate, about 30% of anthropogenic CO_2 is absorbed annually by oceans (22 Mt per day) and 20% by the terrestrial biosphere [286], the rest being accumulated in the atmosphere. Ocean uptake of anthropogenic carbon dioxide has led to a decrease of seawater pH by about 0.1 unit from its average level of 8.2 ± 0.3 (the range is due to local and seasonal variations) [287]. This may seem to be a negligible variation, but pH is a logarithmic scale, hence 0.1 unit represents nearly a 30% increase in hydrogen ions in seawater. As the surface ocean will absorb increasing amounts of CO_2 in the following decades, its pH drop may become as high as 0.4, corresponding to a 150% increase in concentration of H^+ relative to preindustrial levels [288].

The most serious threat of ocean acidification is interference with the production of calcium carbonate shells and microstructures by a variety of marine organisms [288]. Other negative effects of increased CO_2 uptake by oceans include oxygen depletion of tropical seawater due to enhanced flourishing of marine algae followed by oxygen-consuming bacterial decomposition [289–291] and reduction of the availability of iron to marine phytoplankton [292].

7.3.5
Permafrost Melting

Permafrost is a layer of soil or rock, at some depth beneath the surface, in which the temperature has been continuously below 0 °C for at least 2 years. It covers 20% of the Earth's land surface primarily at the highest altitudes of the Northern Hemisphere but it is also found at lower latitudes, including Alps, Tibetan plateau, and Upper Mongolia [293–296]. Such a frozen soil is the foundation of forests, roads, houses, and other infrastructures that are at risk of physical collapse if temperature rises, a phenomenon already happening to an increasing extent [297]. Permafrost soils in boreal and Arctic regions store almost twice as much carbon as is currently present in the atmosphere, in the form of dead plants and animal remains accumulated and sequestered over tens of thousands of years. Hence accelerated melting leads to great concern about the exacerbation of the planet's warming due to the sudden release of carbon into the atmosphere [298]. The carbon loss following permafrost melting can occur either as CO_2, due to the action of soil bacteria on dried organic matter, or as CH_4, as a consequence of anaerobic decomposition in wetlands [299].

7.3.6
Effects on Weather and Ecosystems

The increase in the Earth's surface temperature is expected to have increasingly relevant effects on weather patterns across the planet, with an increase in extreme events such as severe thunderstorms and cyclones [300, 301], intense precipitations [302], frequent heatwaves [303, 304], droughts [305], and floods [15, 306]. Poor countries will likely face increased hunger and malnutrition due to reduced agricultural productivity, also in parallel with spreading of diseases [307]; this might even increase the risks of civil wars in Africa, as observed in the past during temporary climatic stress [308].

Global warming is expected to impact food security [309–311], water availability [312], fish production [40], and global forests [18]. Increased frequency of wildfires in several areas of the world is directly connected to higher temperatures and changes in precipitation patterns [313]. The extensive wildfires occurred in the Russian forests during the 2010 summer have likely the same origin. However, more subtle indirect effects can be also traced, such as the occurrence of more frequent natural wildfires in the Alaskan tundra which are attributed to a 10-fold increase in lightning observed in the last decade, [314].

The effects of climate change on global natural systems [315] and biodiversity as a whole are much more complicated to assess given the enormous number of variables to consider, primarily the extensive and diverse habitat fragmentation that most biota have experienced in the last few decades due to land use change by human activities.

Notably, the increased atmospheric carbon dioxide concentration should act as fertilizer for vegetation across the planet and, in principle, if all the other conditions were to remain constant (e.g., water availability, soil moisture), crop yields should increase [316]. In field experiments this has turned out to be true, but to a lesser extent than expected, suggesting that it is rather unlikely that global warming will provide greater food availability to our increasingly crowded planet [317].

7.4
Air Pollution and Global Warming

7.4.1
Globalizing Smog

Massive combustions are increasingly widespread over the planet. In affluent nations they are mainly fed by fossil fuels for transportation and electricity production, whereas in the developing world, excluding large metropolitan areas, they mainly consist of uncontrolled burning of biomass for cooking, heating, farming, and deforesting. Combustions generate a complex mixture of air pollutants (nitrogen and sulfur oxides, volatile organic compounds, fly ash, black carbon, etc.) that, historically, has been termed *smog*. Starting form the 1950s, it was realized that air pollutants, by the action of sunlight, undergo a variety of complex chemical reactions which generate secondary polluting agents like ozone and particulate matter (aerosols), originating the so-called *photochemical smog*. Large metropolitan areas in North America and Europe have been beleaguered by photochemical smog for decades but, finally, more and more stringent regulations on air pollution have contracted a phenomenon that, however, is now on the rise in the developing world, even if it has typically only a local influence.

In more recent times it has been recognized that pollution is a global problem, redistributed across the planet by atmospheric circulation [318]. At the end of 1990s it was first realized that, for many months over the year, a huge swath of brownish air covered part of South Asia, including the northern Indian Ocean, India, and Pakistan. It contains nanometer- to micrometer-sized soot particles, sulfates, fly ash, black carbon, ozone, and other gas pollutants and is originated by combustion of both biomass and fossil fuels [319]. Its brown color, also related to the presence of nitrogen oxides and soil particles that are suspended in air by land erosion and human activities, suggested calling it *brown cloud* [320]. Further studies have shown that persistent brown clouds are also present over many other world regions and that prevailing winds tend to transport brown clouds up to the Arctic regions, which are increasingly affected by yellow hazes and air pollution [321].

7.4.2
Aerosols and Black Carbon

Airborne aerosols have essentially a cooling effect on the planet because they reflect sunlight back to space. This created the paradox that combustions alter the Earth's radiative balance by producing CO_2 but, at the same time, they self-limit this effect by means of the air pollution they also generate [322]. This is true, but it is only one part of the story: some airborne aerosols such as sulfates and organic carbon only reflect sunlight back to space and do have a cooling effect [323], but black carbon aerosol strongly absorbs solar radiation and has a heating effect on the atmosphere [324]. Unlike the greenhouse effect of CO_2, which leads to a positive radiative forcing both of the atmosphere and at the surface, black carbon adds heat energy to the atmosphere, but reduces it at the surface.

The global mean effect of anthropogenic aerosols is cooling, but the relative contribution of the different types of aerosols determines the magnitude of this cooling [15, 325–327]. The dual role of aerosols as both cooling and heating agents is still far from being satisfactorily elucidated and also the detailed mechanism by which black carbon warms the atmosphere is still not completely clear. The overall picture is complicated by the fact that aerosols influence the formation of clouds

[328]. Therefore, changing aerosol production and distribution may also affect the Earth's radiative balance in which clouds are key players [329], as well as modifying local cloud cover and precipitation patterns [330–332].

The rapid warming of the Arctic region and the retreat of mountain glaciers might not be due only to atmospheric factors [333], but might also have been exacerbated by black carbon deposition on ice sheets which reduces their reflectance (albedo), leading to accelerated melting [334, 335]. Therefore, limiting black carbon, a not impossible task, could provide a ready-to-use handle to combat global warming [336].

7.4.3
Ozone, Ozone Depleting Substances and N_2O

Another important climate forcing effect is provided by ozone (O_3). The ground-level ozone concentration has approximately doubled since the industrial revolution and nowadays it constitutes one of the major areas of concern for air quality in industrialized countries [337, 338]. O_3 is the third largest contributor to positive radiative forcing after CO_2 and CH_4: according to IPCC its contribution is about 25% that of CO_2 [15]. Stratospheric ozone depletion has slightly contributed to cooling [15] and its recovery might accelerate surface warming over Antarctica [339]. Climate effects of the ozone hole in the southern polar region are extremely complex to evaluate [340, 341].

The Montreal Protocol, now signed by 191 countries, was aimed at the phasing out of ozone depleting substances, whose production has now decreased by 95% since then. Hydrofluorocarbons, that were developed to "safely" replace chlorofluorocarbons, exhibit a 1000 times more powerful greenhouse potential than CO_2 [342]. Another increasingly dangerous substance, both as a greenhouse and a climate forcing agent, is N_2O [343], produced by bacteria in soils and oceans (70%) and by burning fossil fuels [344].

7.4.4
A Complicated Picture

In the last few years it has become increasingly clear that the ongoing warming of our planet is an astoundingly intricate process which is not only driven by the increasing atmospheric concentration of CO_2 and other greenhouse gases. Accurate prediction of future climate changes will have to include other equally important forcing agents, in particular changes in ozone with altitude and longitude, changes in ozone depleting substances, and changes in aerosol concentration and chemical composition [345–348]. The above presented entangled issues highlight that scientific research is unambiguously showing a more and more complex pathway to put anthropogenic climate change under control. This intricate and unpredictable situation well exemplifies the no longer sustainable burden that our fossil fuel civilization is imposing over present and, even more, future generations [57].

7.5
Counterbalancing our Climate Influence

The ongoing modification of the Earth's climate due to human activities, particularly the combustion of massive amounts of fossil fuels, is by itself an uncontrolled planetary engineering experiment: we have already affected the natural climatic cycle of our planet. Our alteration of the global climate is poised to last for centuries [349] and this could severely interfere with the stability of the biosphere, potentially harming future generations. This scenario has led some scientists to consider deliberate counteractions capable of slowing or reversing the global warming trend, to be deployed on the global scale (Figure 7.4). These measures, which would have the advantage of prompt effectiveness, are termed geoengineering [350].

Geoengineering proposals fall into two main categories. The first one is reducing the amount of solar radiation reaching the Earth's surface; the second category involves removal of CO_2 from the atmosphere by means of specific technologies or natural processes (essentially, natural photosynthesis).

The idea of scattering sunlight and decreasing the amount of solar radiation reaching the Earth's surface have been somehow validated with a natural experiment that took place in the Philippines in June 1991, when the catastrophic eruption of Mount Pinatubo blew a plume of molten rock, ash, and gas as high as 40 km into the atmosphere. About 20 Mt of SO_2 were also released and partly transformed into sulfate aerosols that had a cooling effect over the entire planet of about 0.5 °C for the following 2 years. The maximum reduction in global temperature occurred in August 1992 with an average reduction of 0.73 °C. The decrease of the Earth's average temperature within a few years of major volcanic eruptions is well established over the last few centuries [351]. A further test might be provided by the eruption of Eyjafjallajökull in Iceland in April 2010.

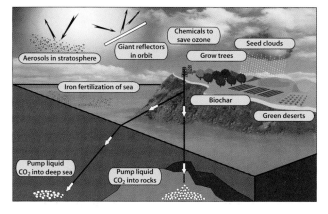

Figure 7.4 Schematic representation of the main geoengineering options.

The Pinatubo natural experiment showed that aerosol injection into the atmosphere not only has a cooling effect but also triggers remarkable changes in weather patterns. A few months after the event, a marked decrease in global land precipitations was observed [352]. This is not surprising because the hydrological cycle is very sensitive to temperature variations due to insolation even more than those due to increases of greenhouse gas concentration [353].

Despite many technical hurdles, several plans for intentional climate manipulation have been proposed as a relatively cheap geoengineering solution to global warming. They include dispersion of aerosols in the stratosphere [354], promotion of the formation of clouds [355], and placing a sunshade over 1.5 million km from the Earth at the so-called Lagrangian point, where the Sun and the Earth exert the same gravitational pull [356] (see Box).

Concerning proposals to foster CO_2 sequestration from the atmosphere by taking advantage of natural processes, the most promising strategy seemed to be enhancing marine productivity by fertilizing the world's oceans with iron in some selected iron-poor areas [362]. However, the latest experiments seem to be disappointing [363, 364] and skepticism about iron fertilization of oceans as a viable strategy for carbon sequestration is mounting [365].

Geoengineering: Ethical and Political Problems

At the present stage it is not clear yet if geoegineering plans will ever be implemented. At any rate, it would be utterly unwise to attempt these Herculean and costly efforts instead of reducing the injection of greenhouse gases into the atmosphere. It would be like trying relieving symptoms without curing the illness. Notably, the immense scientific and technical challenges posed by these proposals are probably dwarfed relative to ethical and political issues [357, 358]. First of all the efforts of releasing chemicals into the atmosphere should be protracted for several centuries, the time needed to offset anthropogenic climate change. But who can guarantee that our civilization will be able to sustain this effort over the long term considering the technical, economical and political points of view? If suddenly stopped, a shocking climate warming would occur right away, with unpredictable consequences. And what will happen if our artificial cooling is superimposed on a natural one as a consequence of, for instance, a massive volcano eruption? Additionally, in a fragile and globalized world, who would govern geoengineering actions that can severely affect climate [359] and, for this reason, might be potentially used as weapons [360, 361]? It is worth noting that an Environmental Modification Convention (ENMOD) has already existed since 1977, that prohibits the hostile use of techniques that modify the dynamics, composition, or structure of the Earth (including the atmosphere) or of outer space. Until now it has been ratified by only 75 states.

Increasing attention is being devoted to wood sequestration through pyrolytic techniques, that is, heating of wood in the absence of oxygen to make the so-called biochar [366, 367]. This material is not easily broken down by biotic or physical processes and can be used for soil amendment or otherwise safely stored. The International Biochar Initiative [368] estimates that biochar production has the potential to provide 1 Gt C per year in climate mitigation by 2040 using only waste biomass. Optimistic estimates forecast 5.5–9.5 Gt C per year of sequestered carbon by 2100 through this route [366]. So far, biochar pyrolytic sequestration has not been tested on a large scale.

7.6
Putting a Limit to CO$_2$

7.6.1
Regulatory Efforts to Curb Greenhouse Emissions

The first talks aiming at starting international climate negotiations were initiated in 1988. The first scientific report by IPCC was issued in 1990 and updated in 1995, 2001, and 2007. In 1997, over 150 countries signed the Kyoto Protocol, which bound 38 industrialized countries – including the US, EU, Japan, and Russia but not China and India – to reduce greenhouse gas emissions by an average of 5.2% below 1990 levels for the period of 2008–2012 [369]. To become law, at least 55 countries had to ratify the Protocol and 55% of inventoried emissions had to be covered. Though details were not finalized, the agreement included "flexibility" mechanisms that allowed industrialized nations to get credit for actions to reduce greenhouse gas emissions in other countries.

In the following years, the Kyoto Protocol underwent hot national and international debate and its survival was often endangered. The EU was strongly in favor of its implementation while the US, Canada, Australia, and Japan had a less resolute position. In 2001 President George W. Bush announced withdrawal of the US from the Kyoto Protocol. In December 2002 Canada ratified the Protocol which, finally, became international law on February 16, 2005 thanks to ratification by Russia that pushed the emissions of industrialized countries over the 55% mark. In July 2009, at the annual meeting in Italy, G8 countries agreed that 2°C of average global warming is a limit which should not be exceeded. To reach this goal, by 2050 global greenhouse gas emissions should be reduced by at least 50%, whereas emissions from developed countries should be reduced by 80% or more. In December 2009 in Copenhagen, despite great diplomatic efforts, only a generic commitment to limit global warming within 2°C was adopted.

In order to promote CO$_2$ emission reductions, the Kyoto Protocol has set the creation of carbon markets. These instruments enable a choice for polluters: either find solutions capable of cutting the emissions of their own sources or continue releasing CO$_2$, but pay others to lower their greenhouse gas output (typically in

Cap-and-trade market

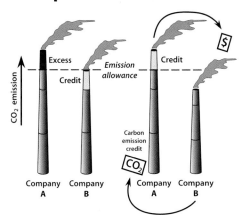

Offset exchange

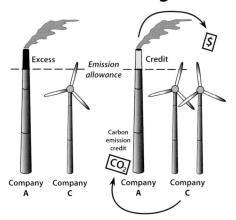

Figure 7.5 The two market-based mechanisms that should contribute to cutting emissions of greenhouse gases. In the cap-and-trade mechanism, company A is above its emission allowance and pays another "virtuous" emitting company, B, that is below. In the offset exchange scheme, A makes a deal with C, a company involved in renewable energy projects, typically in developing nations where costs for A are lower, but results are anyway effective since CO_2 is a global pollutant.

developing countries where costs are lower). Such carbon exchanges occur in two ways: the first is a cap-and-trade market whereby emissions are limited and polluters receive tradable emission credits or permits for each ton of allowed CO_2 emissions [370] (Figure 7.5). In this way the EU established the first emission trading scheme with a mandatory cap-and-trade system for European companies, creating the world's largest carbon market. The second strategy to trade carbon is through credits from projects that offset for emissions elsewhere. For instance, industrialized countries can gain credits by financing low-carbon-output projects in developing countries (e.g., reforestation or renewable energy plants).

Over the years criticism has mounted on market mechanisms as a means for the reduction of carbon emissions, for instance the "compensate elsewhere" approach sometimes resulted to be deceiving: offset credits have been literally fabricated by means of accounting tricks and CO_2 emissions grew without real compensation on the global scale [371]. Another problem is that in the Kyoto Protocol and in related climate legislations bioenergy production is uncritically considered carbon neutral. Accordingly, for instance, the clearing of a long-established forest to burn its wood or to grow energy crops is counted as a 100% reduction in energy emissions, although it causes a large release of carbon [372].

Despite these hurdles and uncertainties [373–375], volumes of the global carbon market are growing substantially: in 2009 they increased by 68% with overall transactions amounting to $136 billion [376]. It is not yet possible, however, to predict if this approach will be successful as it was in the case of SO_2 emissions, which were reduced thanks to a cap-and-trade mechanism [377]. The SO_2 market, however, was immensely smaller and less complex to manage and control than the carbon market.

Obviously, efforts to limit carbon emissions cannot be left to economic mechanisms only. Even greater importance must be given to natural CO_2 sinks, that must be better understood and preserved [245, 378, 379]; fortunately, some encouraging trends seem to appear [380]. Of course, personal behavior and lifestyle choices in the affluent world would provide the cheapest, fastest, and most rational route to reduce carbon emissions.

7.6.2
ppm or Teratons?

Scientific evidence that anthropogenic greenhouse gas emissions are affecting the Earth's climate is compelling [15, 267] and concern is growing [381, 382]. Now the big question is: what is the limit we should place to our emissions, in order to avoid catastrophic climate change [383]? At present, answering this question is simply impossible because we cannot predict with precision which temperature will be caused by a given level of CO_2 and, even if we were able to do so, we would be incapable of predicting in detail the real effects of such temperature on global ecosystems and human civilization. Among the countless factors of uncertainty, the thermal inertia of oceans plays an important part: the huge mass of ocean water is delaying the rate of climate warming. Then, once it has warmed, it will retard the Earth's cooling when emissions will be eventually stopped.

Early science and policy debates focused on doubling the atmospheric CO_2 concentration at twice the pre-industrial levels, that is, around 550 ppm [384]. The latest IPCC report argues that this would imply an average global temperature rise in the range 2.0–4.5 °C, with 3.0 °C as the most likely value [15]. Most scientists believe that such warming would be too much and set as acceptable a lower limit of 450 ppm [385], which would lead to a more reasonable increase of "only" about 2 °C globally. Other authoritative researchers think that the only acceptable level is (or, better, was) 350 ppm (we are now at 390 ppm) [386–388]. A scenario to stabilize the world's CO_2 emission at present levels until 2050 has been proposed. It suggests that, in principle, stabilization might be achieved with current or soon available technologies (mainly efficiency, renewables and carbon sequestration) with the chance of choosing among several pathways (wedges) [389]. Indeed such stabilization wedges sound more a warning of the enormous challenges posed by a swift energy transition rather than a real possibility.

Recently, several scientists have proposed to frame the discussion on CO_2 emission limits in terms of cumulative emissions, or "stocks," rather than emission rates or "flows," putting forward the concept of the carbon pie [390]. In this context, different approaches come essentially to the same conclusion: an insurmountable (as well as easy-to-remember and highly symbolic) limit is one trillion tons (1000 Gt) [391–395]. Beyond this level, the temperature increase would be higher than 2 °C, making the Earth warmer than it has been for millions of years. So far humanity has already emitted roughly 530 Gt of C: we have less than half of the pie left and we must decide shortly how to manage our remaining carbon budget.

The idea of a mass (Gt) rather than a concentration (ppm) target is less elusive and simpler to understand by decision makers and the general public, probably facilitating political debate and negotiations. Just to give an idea, two scenarios with cuts of 80% in emissions by 2050, in developed countries and globally, give an additional 325 and 216 Gt of carbon released in the atmosphere, respectively. This has to be compared with the 470 Gt left in the pie [396]. In this way it is not difficult to understand that CO_2 emissions have to be eliminated within a few decades in order to stabilize global mean temperatures [397], taking into account that human-induced damage already done will persist after emissions stop, because the excess CO_2 remains in the atmosphere for several centuries [349]. Therefore it is advisable to plan adaptation and recovery strategies for a warmer world, in order to limit human and economic suffering [398]. The economic cost of climate change and climate adaptation is a hotly debated matter and figures vary widely between a few tens to several hundred billion dollars per year [399, 400].

Once a limit, whatever it is, has been set, the next issue will be how to distribute reduction commitments. And this is a further conundrum, because the greenhouse gas production is not only unevenly allocated between affluent and poor nations, but also within each single country. In the UK, for instance, the top 10% of emitters are responsible for 43% of emissions and the bottom 10% for only 1% [401]; on the other hand, even in the poorest countries there are highly emitting individuals. Therefore the proposal of sharing CO_2 emission reductions among the top one billion emitters, irrespective of their geographic location and without compromising the efforts to alleviate extreme poverty, seems to be worth pursuing [402]. Other original ideas to limit carbon pollution are coming out, such as setting a mandatory link between fossil fuel extraction and carbon sequestration. In other words the sale and use of fossil carbon should be allowed only if it is certified that an adequate fraction of its carbon content will be permanently sequestered [403].

The road to emission reductions is still very steep but global awareness is rising and political action seems to follow: some of the top emitting nations are finally taking important decisions to reduce carbon pollution. The US will hopefully regulate CO_2 as an air pollutant under the Clean Air Act, China has approved resolutions to target energy saving and emission reductions, and India has passed a national trading scheme for carbon credits and energy efficiency [404–406].

7.7
Air Pollution and Human Health

7.7.1
A Complex Atmospheric Mix

Besides the complex pattern of damage inflicted to climate stability, the burning of fossil fuels produces atmospheric pollution that causes or exacerbates multiple diseases over short and long time periods [407], produces acid precipitations, and can reduce visibility.

It may be interesting to point out that the relentless burning of gigantic amounts of fossil fuels has no effect on the global concentration of oxygen. The level of atmospheric oxygen has remained constant at 20.95% v/v since the beginning of the twentieth century over the whole planet. Complete combustion of all reserves of fossil fuels would reduce the oxygen concentration by less than 0.3% [3].

7.7.2
NO_x

Combustion processes normally do not occur with pure oxygen but with air, and consequently burning of fossil fuels for transportation, electricity production, and other industrial processes generates nitrogen oxides. The major nitrogenous byproduct of combustion is NO, which is then oxidized to NO_2 after dilution in air and eventually to HNO_3 generating acid precipitations. NO and NO_2 (commonly labeled NO_x), are ubiquitous pollutants in all the major industrialized regions of the world. In recent years, a significant reduction of NO_x pollution over Europe (up to 7% per year) and parts of the eastern US has been detected, most likely as a consequence of enhanced diffusion of catalytic converters in cars and trucks [408, 409]. In contrast, a strong increase occurred in Asia, most particularly in China (up to 29% per year) and also in Iran and Russia.

Satellite measurements have made it possible to draw a detailed map of NO_x pollution and sources over the entire planet, also showing that other relevant sources of NO_x are biomass burning, soil emissions and lightning [408, 410]. Scientific evidence links short-term NO_2 exposures with adverse respiratory effects and shows a connection between breathing elevated short-term NO_2 concentrations and hospital admissions for respiratory issues, especially asthma [411].

7.7.3
Ozone

In the presence of high temperatures and intense solar radiation, NO_x react with volatile organic compounds (VOCs), emitted in the atmosphere by both anthropogenic and natural sources, to produce ozone. These photochemical processes are extremely complex and involve a variety of radical species [412, 413]. Ozone is removed from the atmosphere through dry deposition to surfaces (including the human respiratory tracts), uptake by vegetation, and chemical reactions in the atmosphere with a timescale of hours to weeks. Very importantly, ozone formation is related to precursor emissions in a non-linear fashion.

Ground-level ozone is one of the most harmful air pollutants across affluent countries and highly industrialized areas of the world [338]. Breathing ozone can trigger a variety of health problems, including chest pain, coughing, and throat irritation; it can worsen bronchitis, emphysema, and asthma and repeated exposure may permanently scar lung tissue. Particularly at risk are elderly people and children [414]. The World Health Organization has estimated 21 000 premature deaths annually in 25 EU countries on and after a few days with high ozone levels [415]. Ozone can also have detrimental effects on plants and ecosystems. Unfortunately, given the very complex processes governing ozone formation, dispersion, and removal, it is difficult to assess ozone concentration trends over the long term, even in intensively monitored regions such as Europe [338, 416].

7.7.4
Particulate Matter

Particulate matter (PM) is a complex mixture of extremely small particles and liquid droplets, which can be of both primary and secondary origin; it is also often termed atmospheric aerosol [417]. Primary PM comes from sources such as vehicle exhaust, road dust, smokestacks, forest fires, windblown soil, volcanic emissions, and sea spray. The most abundant species contained in secondary PM include sulfate, nitrate, ammonium, and the so-called secondary organic aerosols, that consist of hundreds or thousands of individual organic species, only a small fraction of which (10–20%) is routinely identified or measured within PM [418]. As organic aerosols from different sources age in the atmosphere, their chemical and physical properties tend to become remarkably similar. Hence their relevant role in atmospheric pollution and climate forcing can be discerned without knowing every one of its components [419].

PM can be classified into coarse, fine, and ultrafine particles. Coarse particles (PM_{10}) have a diameter between 10 and 2.5 µm and are mostly derived from soil and sea salts. Fine particles (0.1 to 2.5 µm in diameter, $PM_{2.5}$) and ultrafines (<0.1 µm) are predominantly of secondary origin or derived from combustion of fossil fuels. Combustion particles have a core of elemental carbon that is coated with a layer of chemicals, including organic hydrocarbons, metals, nitrates, and sulfates. Particle size, surface area, and chemical composition determine the health risk posed by PM [417]; $PM_{2.5}$ and ultrafines are the most harmful because can penetrate deep into the lung and pass directly into the circulatory system [420]. Worldwide epidemiological studies show a consistent increase in cardiac and respiratory morbidity and mortality from short-term exposure to particulate matter with peaks of hospital admissions [407, 421–424]. A substantial increase in deaths related to PM and ozone pollution was observed across Europe during the 2003 heatwave [425, 426]. Very interestingly, in August 2003, right after the big North American blackout, a substantial decrease of PM and O_3 pollution was monitored, highlighting the burden imposed on public health by electricity production through combustion of fossil fuels [427]. Air quality standards have been constantly made more stringent in the US and Europe over the last 20 years to take into account the increasing evidence of the strong connection between fine PM ($PM_{2.5}$) and adverse health effects.

It is worth emphasizing that PM air pollution is not only a problem of affluent and industrialized nations. More than three billion people in poor countries continue to depend on solid fuels such as wood, dung, agricultural residues, and coal for their energy needs for cooking and heating, often on open fires or on improperly vented stoves [428]. For social reasons, children and women are mainly affected. Typical adverse effects include acute lower respiratory tract infections, that are the single most important cause of death in children under 5 years old worldwide (at least 2 million deaths annually) in this age group [429].

7.7.5
Carbon Monoxide (CO)

Carbon monoxide, that is produced in a variety of combustion processes, is odorless, colorless, and toxic. It does not represent a danger in open air, but participates in photochemical processes leading to the formation of secondary PM. Despite its reactive character, it can remain in the atmosphere for up to 1–2 months, thus becoming a major transboundary air pollutant with a key role in global atmospheric chemistry, as determined since early satellite measurements [318]. CO can represent a severe danger in indoor space in the presence of worn or poorly adjusted and maintained combustion devices such as boilers and furnaces or, in poor dwellings, stoves or open fires [429]. It is fatal at very high concentrations, due to the formation of carboxyhemoglobin in the blood, which inhibits oxygen intake.

7.7.6
Sulfur Dioxide (SO$_2$) and Acidic Precipitations

SO$_2$ is a colorless gas with a characteristic pungent and irritating smell at relatively high concentrations. During combustion processes, SO$_2$ is produced by the oxidation of sulfur contained in coals and oils, which is instead removed by natural gas to allow distribution in pipelines (Chapter 5). Nowadays the wide diffusion of low-content sulfur fuels for road transportation has made negligible the production of SO$_2$ from this sector. Hence SO$_2$ is mainly generated by fossil fuel combustion at power plants and other industrial facilities, with a minor contribution from burning of high sulfur containing fuels by locomotives, large ships, and non-road equipment. Short-term exposures to SO$_2$, for up to 24h, can cause several adverse respiratory effects. Connection between short-term exposure to SO$_2$ and increased visits to emergency departments has been evidenced.

High concentrations of SO$_2$ or NO$_2$ in highly polluted areas can decrease, even by several units, the pH of atmospheric water from its regular value of about 5.2 (slightly acidic due to the presence of CO$_2$). The reaction chain of SO$_2$ entails ready oxidation to SO$_3$ in the atmosphere by the ubiquitous hydroxyl radical, followed by transformation into sulfuric acid (H$_2$SO$_4$) upon dissolution in water; similarly, NO$_2$ is transformed into nitric acid (HNO$_3$). Thus SO$_2$ and NO$_2$ are particularly harmful pollutants because they are not only precursors of secondary sulfate and nitrate PM, but also key ingredients for acid precipitations (rain, snow, fog) and acid dry deposition (Figure 7.6). These processes have been shown to have adverse impacts on forests and freshwaters, killing plants, insects, and aquatic life forms

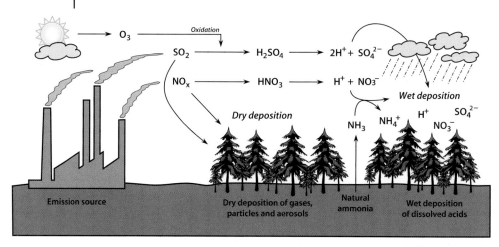

Figure 7.6 Mechanisms of acid formation and deposition upon production of SO_2 and NO_x. Wet deposition can also form when ammonia gas (NH_3) from natural sources is converted into ammonium (NH_4^+).

[430, 431]. They are extremely dangerous for soils because they can leach away essential nutrients and elements such as magnesium, calcium, and potassium releasing, at the same time, toxic elements like aluminum or heavy metals [432]. They also cause deterioration of buildings and accelerated corrosion of metals.

Desulfurization and denitrification devices (along with electrostatic precipitators to remove fly ash) are very expensive and increase the capital and maintenance costs of power facilities, but are a first step toward the internalization of the hidden costs of electricity production. In affluent countries, these technical advancements have substantially reduced SO_x and NO_x pollution [410, 433, 434], but these phenomena are on the rise in developing countries, particularly in China [435].

7.7.7
Heavy Metals

Mankind has produced local pollution by heavy metals for millennia, due to mining and related activities [436]. These elements became global pollutants only after the industrial revolution, mainly due to the combustion of fossil fuels, particularly coal. This phenomenon is unambiguously deducible from Arctic and Himalayan ice records, which show the rise of atmospheric concentrations of several toxic heavy metals such as mercury, lead, cadmium, thallium, arsenic, antimony, tin, and bismuth since the end of the eighteenth century [437–439]. Among these, mercury is currently the main environmental concern [440]. The over 1100 coal-fired power plants in the US release about 50 tons of mercury into the atmosphere per year, a minuscule amount in absolute terms but enough to be harmful for human health: several states in the US are going to set mercury emission limits on coal-fired power facilities [441].

Mercury and its compounds are highly toxic to humans, ecosystems, and wildlife. High doses can be fatal to humans, but even relatively low doses can have serious adverse neurodevelopmental impacts; possible harmful effects on the cardiovascular, immune, and reproductive systems can also occur [440]. Mercury is persistent and can change in the environment into methylmercury cation, the most toxic form. Methylmercury readily passes both the placental barrier and the blood–brain barrier, inhibiting potential mental development even before birth. Methylmercury collects and concentrates especially in the aquatic food chain, making populations with a high intake of fish and seafood particularly vulnerable. Mercury is also contained in small amounts in compact fluorescent lamps (Section 13.2) [442].

The wider adoption of catalytic converters in road vehicles has greatly improved air quality in urban areas but, as a side effect, has prompted an unprecedented spreading of platinum group heavy metals (Pt, Rh, Pd, Os) in the environment with potential negative effects, not fully predictable at present [443–445]. This, once again, shows the intricate complexity of problems that our fossil fuel civilization has created.

7.8
Land and Water Degradation

Environmental pressures arising from the exploration, extraction, processing, distribution, and combustion of fossil fuels are manifold [446, 447] and several of them have been presented in the chapters discussing each single fossil source. New and unexpected consequences sometimes occur. For instance, the gigantic mud volcano eruption in East Java, which started in 2006 and is still ongoing, might have been triggered by drilling activities for gas exploration [448, 449]. Probably less unpredictable was the observation that oil and gas extraction is largely responsible for the sinking of river deltas worldwide. These ecologically fragile areas have experienced temporary submergence of at least 260 000 km² during the last decade and might be more affected by sea level rise induced by global warming [450, 451].

7.8.1
Oil Spills

Oil pollution from coastal refineries, offshore installations, and sea transport put a significant environmental burden on the marine ecosystem. The consistency of spilled oil brings about surface contamination and asphyxiates marine biota, and its chemical components can cause acute toxic effects and long-term impacts. Most human-made oil pollution comes from land-based activities, but attention is more commonly focused on oil spills from off-shore extraction and seagoing oil tankers because their massive consequences on beaches and wildlife are utterly evident and thus extensively covered by the media.

The Exxon Valdez oil spill in Prince William Sound, Alaska, on March 24, 1989 is considered one of the most devastating man-made environmental disasters ever to occur at sea [452, 453]. This is not due to the amount of oil released, about 40 000 tons, which was much less relative to major accidents, but to the fact that the disaster occurred in the Arctic region, which is extremely fragile from the ecological point of view and so remote that response efforts were slow and extremely difficult. The crew of the Exxon Valdez noticed icebergs in their planned route but, instead of trying to weave through the them, they decided to go in another direction. On the new route, the oil tanker hit the Bligh Reef and sank. The oil released covered 28 000 km^2 of ocean. Thousands of animals died immediately; the best estimates include about 250 000 seabirds, 22 whales, 2800 sea otters, and 300 harbor seals and the destruction of billions of fish eggs [454]. In 1990, in response to the Exxon Valdez disaster, the US Congress passed the Oil Pollution Act, which was an incentive to introduce legislation requiring tankers to be double hulled. In 1991 Exxon agreed with the federal and Alaskan governments to pay $900 million for restoring Prince William Sound. As with each spill oil accident, the Exxon Valdez case has given rise to long and complicated legal actions and, 20 years after the disaster, there was still disagreement between researchers from ExxonMobil and those of government agencies about the ecological consequences of the disaster [454]. Despite the fact that today the area has turquoise waters, sizeable amounts of subsurface oil are still present and will likely persist for decades [455, 456]. This shows that the widely available array of cleaning techniques for the sea after massive oil pollution events (use of microorganisms, dispersants, detergents, sorbents, chemical agents, controlled burning, dredging, skimming) cannot be a panacea. Since the melting of the Arctic ocean is attracting more shipping and energy exploration, the possibility of oil spills in this pristine environment is poised to increase in the next future, calling for enhanced research efforts to prevent and recover from these disastrous events [457].

Public attention to oil pollution is essentially focussed on major events caused by massive spills over sea or land, such as the BP Macondo well in the Gulf Mexico in 2010. However, the daily pollution caused by illegal oil discharges associated with tank washings or other illicit operational activities during sea transportation is not less harmful [458].

7.8.2
Coal Combustion Residues (CCRs)

The substantial progress achieved in recent decades by coal-fired power plants, in terms of reduced atmospheric pollution, comes at a price: the tougher air pollution controls become, the higher the solid waste piles obtained from flue gases grow [459]. In the EU, despite relatively low electricity production from coal, the wastes represent about 4% by weight of the total generation of waste and residues from all economic activities [460].

The most utilized devices to remove SO_2 from the flue gases of coal power plants are wet scrubbers, that are reaction towers containing lime or limestone in slurry form, to be mixed with the stack gases containing SO_2. The calcium carbonate of the limestone produces pH-neutral calcium sulfate, that is physically removed, thus turning gaseous sulfur pollution into an inert solid material. When the purity of calcium sulfate is sufficiently high, it is sold to chemical companies as gypsum but, very often, it is simply placed in landfills. Besides desulfurization residues, CCRs consist of fly ash, bottom ash, and coal slag and contain a broad range of toxic heavy metals. If not properly managed, CCRs may pose a severe risk to human health and the environment. More than 40% of these waste materials are somehow recycled, essentially in the construction industry. For instance, fly ash is bound in concrete and cement, SO_2 waste is used to make gypsum wallboard, and a substantial fraction of ash is placed on land as a structural fill and road base; finally some CCRs are spread on farmland to "amend" soil or are used to fill in depleted mines. The remaining part of this material (75 million tons) has to be disposed of in landfills [459].

CCR reservoirs constitute an increasingly hard to manage environmental legacy. In December 2008, a huge volume of coal-ash sludge broke through a dam at a coal-fired power plant in Kingston, TN, USA. The reservoir covered $340\,000\,m^2$ and was one of the plant's three holding ponds for the ash waste of the combustion process. About half of the reservoir content was released, namely over 4 million m^3 of materials, which flowed through farmland and into the Emory and Tennessee Rivers. No casualties were counted but the ash entirely covered an area of more than $1\,km^2$ in a 1.8 m deep mass of sludge containing a mix of toxic elements (Figure 7.7). These types of accidents seem to occur at an increasing rate in recent times, also due to the mounting stock of waste residues [459]. Each year, 131 million tons of CCRs are generated by the 460 coal-fired power plants in the US.

Figure 7.7 The aftermath of the coal-ash spill that took place in Kingston, TN, in December 2008.

If loaded on to railcars, these residues would fill a train stretching from New York City to Los Angeles 3.5 times over.

7.9
So, What?

Despite recurrent alarms about incumbent supply shortages, it seems that at present environmental legacy, rather than consistency of exploitable reserves, is the most dramatic problem posed by the relentless increase of fossil fuel global demand. In affluent countries, advanced environmental legislation has generally reduced air and water pollution, as well as better preserved local ecosystems. However, as we have seen, the footprint of fossil fuel extraction and transportation is global and, for example, energy resources used in an environmentally sound fashion in Europe may carry an unseen dirty legacy in other continents. Additionally, better environmental performance of rich countries is due to partial shifting of the dirtiest and most energy intensive industrial activities in underdeveloped countries, which are characterized by weaker environmental legislation or enforcement. Hence the problem can be thoroughly analyzed only on a global scale.

Life on Earth is made possible thanks to the incessant cycling of water and materials that are needed to assemble living bodies as well as to provide suitable environmental support for their sustainment. Three key life-building elements (carbon, nitrogen, sulfur) have been processed incessantly for billions of years through the Earth's biogeochemical cycles that are powered by solar radiation [3, 461, 462]. Throughout millennia, mankind had no discernible effects on these natural processes but, during the last century, anthropogenic interference in global biospheric cycles became a matter of scientific concern and eventually of public policy debate. These interferences have been brought directly (e.g., CO_2, SO_2 and NO_x emissions) or indirectly (e.g., nitrogen enrichment of soils [463]) by the burning of gigantic amounts of fossil fuels which, albeit in a context of global injustice and uneven wealth distribution, made possible a 20-fold increase of the world economic product during the twentieth century.

After over 100 years, our fossil fueled civilization is endangering the habitability of the biosphere, and this calls for a problematic but not impossible energy transition. Scientists have the moral duty to increase the awareness of citizens all over the world about the gloomy future of human civilization and to speak the plain truth: time is running short for taking action and avoiding collapse.

Part Three
Nuclear Energy

Energy for a Sustainable World: From the Oil Age to a Sun-Powered Future. Nicola Armaroli and
Vincenzo Balzani
© 2011 WILEY-VCH Verlag GmbH & Co. KGaA, Weinheim
ISBN: 978-3-527-32540-5

8
Nuclear Energy

"We nuclear engineers of the first nuclear era have had success,
but the generation that follows us must resolve the profound
technical and social questions that are convulsing nuclear energy."

Alvin M. Weinberg

8.1
Principles of Nuclear Fission and Fusion

8.1.1
Radioactivity, Mass and Energy

As discussed in Section 2.1, energy comes in a variety of forms that can be transformed one into another, but these transformations obey a rule without exceptions:
the total energy is conserved. The law of conservation of energy was put to a severe
test in 1896 by the French scientist Henri Becquerel. He discovered that uranium
salts emitted radiation that, unlike phosphorescence, did not depend on an external
source of energy but seemed to arise spontaneously from uranium itself. Two years
later Pierre and Marie Curie isolated two new elements, named polonium and
radium, that exhibited an intense, spontaneous emission of radiation, a process
that for the first time was called "radioactivity." Later it was shown that an electric
or magnetic field could split radioactive emission into three types of rays, that were
given the alphabetic names alpha, beta, and gamma, still in use today. It was also
shown that α rays are helium nuclei, β rays are electrons, and γ rays are high-energy
electromagnetic radiation, like X-rays. Where does the energy arise when an atom
spontaneously ejects high-energy α, β, and γ radiation? The answer is that some
elements have "excess" energy in their nuclei and are unstable over time. The
radioactive decay is therefore the process of spontaneous nuclear disintegration.

In 1905 Albert Einstein, in developing his "special theory of relativity," came to
the conclusion that all varieties of energy, E, are associated with changes of mass,
m, by the simple universal law

$$E = mc^2 \tag{8.1}$$

Energy for a Sustainable World: From the Oil Age to a Sun-Powered Future. Nicola Armaroli and
Vincenzo Balzani
© 2011 WILEY-VCH Verlag GmbH & Co. KGaA, Weinheim
ISBN: 978-3-527-32540-5

where c is the speed of light in vacuum. This equation tells that potential energy is stored as mass and that, because of the speed of light ($c = 3.0 \times 10^8\,\mathrm{m\,s^{-1}}$), very small changes in mass correspond to huge changes in the energy of the system, and vice versa. In chemical reactions, the change in mass cannot be appreciated because the force involved is weak and the energy changes are thus very small. In nuclear reactions, however, the energy changes are huge and therefore they are accompanied by measurable changes in mass.

Einstein's equation is easy to use. It shows, for example, that the disappearance of 1 kg of mass in a nuclear reaction would liberate $3.0 \times 10^8 \times 3.0 \times 10^8$ or $9 \times 10^{16}\,\mathrm{J}$, equivalent to 25 billion kWh of energy. Such an enormous amount of energy would be sufficient to cover the average primary energy consumption of about 580 000 European citizens (~120 kWh per person per day) for 1 year [7]. To produce the same amount of energy by burning coal (calorific value ~8.0 kWh kg^{-1}) one would need 3.1×10^9 kg of coal.

8.1.2
Structure of Matter

Matter is an assembly of atoms, very small particles whose volume is of the order of 10^{-24} cm^3. The complex structure of atoms (Figure 8.1) can be described as consisting of a very tiny central part, the nucleus, in which is concentrated nearly all the mass and which occupies about one millionth of a billionth (10^{-15}) of the volume of the atom [464]. The nucleus contains protons and neutrons, particles

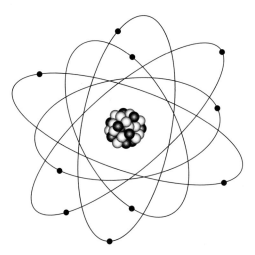

Figure 8.1 Schematic picture of the atomic structure. The nucleus, constituted by proton and neutrons, is surrounded by electron shells. Note that the picture is not to scale since in reality the radius of an atom is about 100 000 times the radius of its nucleus.

having about the same mass (~1.66×10^{-24} g). The proton carries a positive electric charge, whereas the neutron is electrically neutral. Protons and neutrons are usually referred to as nucleons. The nucleus is surrounded by electron shells, whose diameter is that of the atom (~10^{-8} cm). The electron has a mass about 2000 times smaller than that of the proton and neutron, and carries an electric charge, of the same magnitude as but opposite in type to that of the proton. Atoms are electrically neutral since they contain an equal number of protons and electrons.

An atom is classified according to the number of protons and neutrons in its nucleus. The number of protons (atomic number, Z) determines the chemical element. For example, all atoms of hydrogen have one proton ($Z = 1$) and all atoms of uranium have 92 protons ($Z = 92$). The number of neutrons (N) in the nucleus, and therefore the atomic mass ($Z + N$), may vary. Atoms of an element that have certain but different numbers of neutrons in their respective nuclei are called isotopes. For example, the hydrogen nucleus exists in three different forms: the single proton alone, the proton with one neutron, and the proton with two neutrons. The heaviest isotopes of hydrogen, with atomic mass 2 and 3, are represented by ^{2}H and ^{3}H and are often called deuterium and tritium, respectively; this is an exception, however, since generally isotopes have no specific names. The uranium atom that, in addition to the 92 protons, contains 146 neutrons has atomic mass 238, whereas the uranium atom that contains only 143 neutrons has atomic mass 235. The two different isotopes are represented by ^{238}U (or uranium-238) and ^{235}U (or uranium-235). The electrons, which are negatively charged, are bound to their nucleus, which is positively charged, by electrical energy. Any type of chemical reaction, for example, combustion of carbon with oxygen, has only to do with the electrons of the atoms involved and changes of their energies. The nuclei are completely indifferent to the occurrence of chemical reactions.

According to classical physics, two like-charged bodies repel each other inversely by the square of their distance apart. Therefore, one would expect that protons repel each other in the very tight structure of nuclei. For example, the helium nucleus, which contains two protons and two neutrons, should split into two deuterium nuclei:

$$^4\text{He}^{2+} \rightarrow {}^2\text{H}^+ + {}^2\text{H}^+ \tag{8.2}$$

Such a nuclear reaction, however, does not happen, and it can be shown that the reverse reaction releases energy. This clearly proves that there must be a very short-range nuclear force, strong enough to overcome the Coulomb repulsion in nuclei. Nuclear reactions, which involve participation of such a strong nuclear force, are associated with enormous changes in the energy of the system.

Experiments show that a nucleus is lighter than the total for the protons and neutrons that constitute it [465]. The amount by which the mass of the atom falls below the mass of the same number of hydrogen atoms and neutrons constituents (Z hydrogen atoms and N neutrons) is the nuclear binding energy (BE), which is

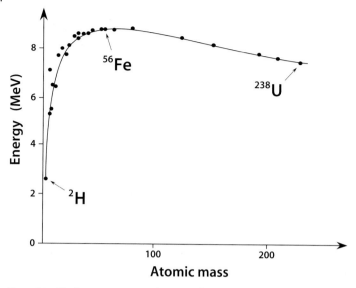

Figure 8.2 Binding energy per nucleon as a function of atomic mass number.

usually measured in million electronvolts (MeV). For deuterium ^{2}H, BE = 2.22 MeV, and for ^{4}He, BE = 28.29 MeV. For comparison purposes, it is convenient to make reference to the average binding energy per nucleon, BE/(Z + N), which, for example, is 28.29/4 = 7.07 MeV for ^{4}He. If the average binding energies per nucleon are plotted as a function of the mass number over the full range of nuclei, the graph in Figure 8.2 is obtained. Clearly, both the very heavy and very light nuclei are unstable compared with the nuclei of the elements with intermediate atomic mass around iron, which have a BE of about 8 MeV per nucleon. It can be calculated [465] that if 56 ^{1}H atoms could be induced to form a single ^{56}Fe atom, 56 × 8 = 469 MeV would be liberated according to Einstein's equation, which is about half the mass of a single proton. On the other hand, if ^{238}U were to be broken up into four nuclei of mass similar to that of iron, the energy liberated would be about 311 MeV, equivalent of about one-third of the mass of a single proton.

From the above discussion, it is clear that enormous amounts of energy can be liberated by *fission* of heavy nuclei or by *fusion* of light nuclei.

8.1.3
Nuclear Fission

Nuclear fission is a nuclear reaction in which the nucleus of an atom splits into smaller parts, releasing large amounts of energy as electromagnetic radiation and kinetic energy of the fragments. Some isotopes of heavy elements, such as ^{235}U, undergo radioactive decay, which is a form of spontaneous fission. The spontaneous fission follows an exponential decay rate and can thus be characterized by its half-life, the time needed to reduce the number of atoms to 50% of their original

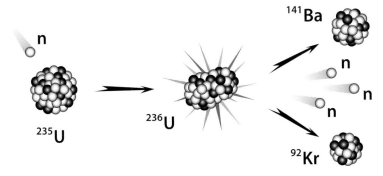

Figure 8.3 The fission process as it occurs in ^{235}U upon absorption of a slow neutron. Formation of ^{141}Ba and ^{92}Kr is accompanied by the emission of three neutrons, which can cause the fission of other ^{235}U atoms, generating a chain reaction.

number. The recognition of nuclear fission occurred in 1938 when the formation of barium was detected after bombarding uranium with neutrons [466]. Further studies showed that only the ^{235}U isotope, which constitutes 0.7% of natural uranium, undergoes nuclear fission upon collision with a neutron, whereas the much more abundant ^{238}U (99.3%) is not fissile. The fission process of ^{235}U takes place as shown in the scheme in Figure 8.3. The total mass of the fragments produced is about one-thousandth less than the mass of the original ^{235}U and an equivalent amount of energy is produced.

8.1.4
Controlled and Uncontrolled Chain Fission Reactions

In 1939 it was shown that ^{235}U fission by slow neutrons causes the formation of two or three neutrons per event (Figure 8.3). This result opened the possibility of obtaining chain reactions that would allow the passage from the infinitesimally small individual nuclear reaction to a reaction involving millions of billions of billions of nuclei, generating enormous amounts of energy. It was also clear that such a chain reaction could lead to two very different results. If the system is engineered to allow each fission to cause precisely one subsequent fission (and so on), the process continues at a constant rate and energy is continuously generated. This is what happens in nuclear power reactors. If more than one of the generated neutrons is allowed to cause further fissions, the rate of fission grows exponentially and the system becomes an atomic bomb explosive in less than a microsecond.

On December 2, 1942 in Chicago, Enrico Fermi succeeded in controlling the propagation fission of ^{235}U in an atomic "pile," opening the way to the construction of nuclear reactors for power production. In the following years, a large team of scientists under the scientific direction of J. Robert Oppenheimer, within the frame of the Manhattan Project [465], found the conditions to create an exponential chain reaction and prepared the material needed to construct the first atomic bombs. Since the discovery of nuclear energy, mankind has increased his power and spaceship Earth has become more fragile.

Nuclear Fusion

As we have seen above (Figure 8.2), combination of two light nuclei into a larger nucleus, that is less massive than the total mass of the light nuclei being combined, must produce an amount of energy equal to the difference in the binding energy of the nucleons [465]. For example, the fusion of two deuterium nuclei (binding energy 1.11 MeV per nucleon) to give an He nucleus (binding energy 7.07 MeV per nucleon) would releases 23.84 MeV of energy as gamma radiation:

$$^{2}H^{+} + {}^{2}H^{+} \rightarrow {}^{4}He^{2+} + 23.84 \, \text{MeV} \tag{8.3}$$

Fusion reactions taking place in the Sun at temperatures above 10×10^{6} K are the source of solar energy (Chapter 9). Fusion reactions, however, are very difficult to obtain on the Earth because, as discussed above, the nuclei, having positive electric charge, mutually repel each other more and more strongly as they approach, while nuclear forces operate only over small distances. For the very short-range nuclear forces to take over and overcome the electric repulsion, the two nuclei have to be knocked against each other with much energy, that in a laboratory may be obtained by special instruments that accelerate individual nuclei. Fusion reactions take place in the Sun because the temperature is so high that the average kinetic energy of the atoms approaches the energy of a nucleus in a high-energy accelerator's beam and matter is confined by intense gravitational forces.

The most accessible and practical fusion reaction is that between deuterium and tritium:

$$^{2}H^{+} + {}^{3}H^{+} \rightarrow {}^{4}He^{2+} (3.5 \, \text{MeV}) + \text{neutron} \, (14.1 \, \text{MeV}) + 17.6 \, \text{MeV} \tag{8.4}$$

Other possible fusion reactions that could take place in a laboratory are those between two deuterium nuclei and a deuterium with ^{3}He. Nuclear fusion has since long been used to create thermonuclear (or hydrogen) bombs. Scientists, however, have not yet succeeded in exploiting the huge energy of fusion reactions in a controlled fashion for useful purposes (Section 8.3).

8.2
Power from Nuclear Fission

8.2.1
Past and Present

In a nuclear power plant, the heat of a nuclear reaction produces high pressure steam which is used to turn the turbine of an electric generator. On June 26, 1954, the USSR's Obninsk Plant became the world's first nuclear power plant connected

Nuclear reactors & net operating capacity in the world
(from 1954 to 2010)

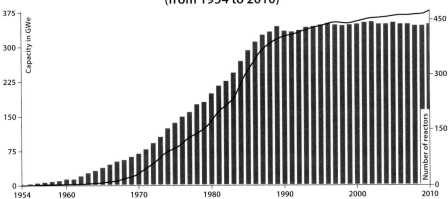

Figure 8.4 Historical deployment of nuclear energy. Bars, number of reactors; line, operating capacity.

to an electricity grid. It provided around 5 MW of electric power, enough to feed 2000 homes [467]. The first nuclear power station in the western world, Calder Hall in England, was opened on October 17, 1956 with an initial capacity of 50 MW. In 1957, the European Atomic Energy Community (EURATOM) and the International Atomic Energy Agency (IAEA) were founded.

Installed nuclear capacity initially increased relatively quickly, rising from less than 1 GW in 1960 to 100 GW in the late 1970s (Figure 8.4). From 1970 to 1975 growth averaged 30% per year, but just as the nuclear industry was achieving dominance as a major source of electricity generation, it stopped growing and more than two-thirds of all nuclear plants ordered after January 1970 were eventually canceled [468]. A much slower increase then led to a capacity of 330 GW in the 1990s and then to 373 GW in 2010 [469]. In the last few years, while the installed capacity increased slightly through technical improvements at existing plants ("uprating"), nuclear electricity generation decreased (2%) from 2658 TWh in 2006 to 2608 TWh in 2007, with a particularly large decrease (6%) in the EU. A further decrease of 7 TWh occurred from 2007 to 2008, and of 43 TWh from 2008 to 2009 [470]. In 2009, nuclear power plants generated about 13.5% of the world's commercial electricity (down from 15% in 2006, 16% in 2005 and 18% in 1993). Since roughly 16% of the world's end use comes from electric energy, nuclear energy now contributes to less than 2.5% of the world's end-use demand. Note that the three times higher figure concerning the nuclear energy contribution to world energy consumption (Chapter 2) refers to the heat production by nuclear power plant, without taking into account that the thermal-to-electric energy conversion efficiency is only 33%.

The stopping of the development of nuclear energy in the last two decades is due to various reasons, including rising economic costs (partly related to extended construction times), oscillations of fossil fuel prices, energy conservation policies, environmentalist opposition, waste disposal and dismantling problems, and the accidents at Three Mile Island (1979, US) and Chernobyl (1986, Ukraine, former USSR).

8.2.2
Nuclear Fuel

Elemental isotopes that undergo fission when struck by a thermal, slow moving neutron are called "fissile" [465]. Fissile isotopes can sustain a chain reaction and are therefore used as nuclear fuel. Since ^{235}U is the only fissile isotope existing in Nature (in any appreciable amount), all commercial nuclear plants presently use uranium as fuel. Natural uranium, however, contains only 0.7% of ^{235}U, mixed with 99.3% ^{238}U.

Uranium is a toxic, relatively common metal which is present in the Earth's crust with an average concentration of $2.7\,mg\,kg^{-1}$ (parts per million by mass, ppm), and also in seawater ($3.2 \times 10^{-3}\,mg\,l^{-1}$). Nuclear fuel is obtained from ores in which uranium is contained, mainly as the oxide U_3O_8, in a sufficiently high concentration. Such "rich" ores are available in Canada and in a few other countries. The worldwide production of uranium in 2009 amounted to $50\,772\,t$ [471]. Three countries, Canada, Kazakhstan, and Australia, together account for about 65% of world uranium production (Figure 8.5).

Most uranium mining is very volume intensive, and thus tends to be undertaken as open-pit mining. In some locations, a process called "in situ leach mining" is used. Although uranium itself is not very radioactive, uranium minerals are always associated with more radioactive elements such as radium and radon [471]. Therefore, the ore which is mined has to be handled with care for safety reasons.

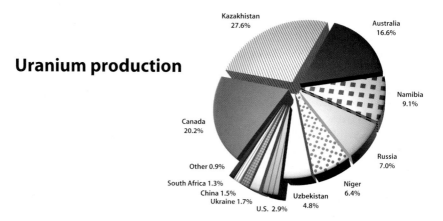

Figure 8.5 World production of uranium in 2009 [471].

Depending on the nature of the ores, uranium is extracted by using sulfuric acid, alkali, and oxidants like hydrogen peroxide, besides huge amounts of water. Runoff from the mine stockpiles and waste liquors from the milling operation must be collected in secure retention ponds for isolation and recovery of any heavy metals or other contaminants. At the conclusion of mining, equipment is usually buried with the tailings.

After extraction, uranium is chemically converted to dry, purified uranium concentrate called "yellowcake" (U_3O_8). Then, it has to be converted to UF_6 gas for enrichment. This process, which leads to the 3–5% ^{235}U concentration needed for nuclear reactors, can be carried out by two different, energy consuming techniques, diffusion and centrifugation, both of which utilize the slight mass difference between ^{235}U and ^{238}U. The depleted uranium (DU), which is available in huge quantities as a byproduct of the nuclear power industry, is extensively used in military applications because of its high density (1.7 times that of lead). DU weapons are especially dangerous because when the bullet strikes a hard surface up to 70% of the uranium burns and vaporizes into a fine mist of toxic particles which can be spread for miles downwind, and are small enough to be inhaled into the lungs.

8.2.3
Uranium Supply

As happens for oil and any other energy resource, it is difficult to estimate the amount of uranium that it is worth mining. In particular, two limits have to be taken into account: an economic limit, imposed by the competition with other resources, and an absolute limit, imposed by energy balance analysis of the whole nuclear process (EROI, Chapter 2). U_3O_8 spot prices have increased dramatically since 2003, when the commodity cost under $22 kg^{-1}. Its peak spot price in 2007 was over $275 kg^{-1} and has since then fallen to much lower values. Currently (August 2010), the price of uranium on spot market is around $50 kg^{-1}.

In the case of nuclear power, energy balance studies are extremely difficult to do and have seldom been performed. Since uranium fission generates much energy, the tendency is to assume that EROI is positive for all ore grades. Some studies, however, indicate that below 0.01–0.02% (100–200 ppm) in ore, mining, milling, and the clean-up of the mine site become the main components of the total energy consumption and the energy required by such a nuclear fuel cycle comes close to the energy gained by burning the uranium in the reactor [472, 473]. There are also other kinds of physical limits. For example, the mining process consumes or uses huge masses of water [58], and extracting 1 kg of uranium out of a 0.01% ore needs the processing of 10 t of material. According to IAEA [469], WNA [471], and NEA [474] in 2009 known recoverable resources, intended as reasonably assured resources (RARs) plus inferred resources (IRs), recoverable at less than $130 per kg of uranium, amounted to 5.4 Mt of uranium, distributed as shown in Figure 8.6. Undiscovered resources (URs), that is, "uranium deposits

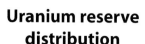

Uranium reserve distribution

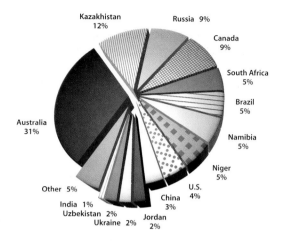

Figure 8.6 Distribution of uranium reasonably assured resources (RAR) plus inferred resources (IR) recoverable at less than $130 per kilogram of uranium, year 2009 [471].

that can be expected to be found based on the geological characteristics of already discovered resources," are assumed to be 10.5 Mt. These data are cited also by a recent MIT study pointing out, however, that its 2003 recommendation for a significant global uranium resource evaluation program has not been followed [475]. The above-mentioned agencies affirm that the identified resource base will remain sufficient for 100 years at current consumption.

The IAEA/WNA/NEA classification [469, 471, 474] of uranium resources into various categories (RAR, IR, UR) and cost classes gives the impression of a high data quality and reliability. According to Energy Watch Group (EWG) [472] and Oil Drum (OD) [476], in fact, large national and private interests are involved so that the objectivity and the accuracy of the presented data are not assured. Several contradictions have also been evidenced by an accurate analysis, and the low scientific level of the data is shown, for example, by the fact that only roughly known numbers are given with an unbelievable precision of 0.1% or better. EWG and OD estimates of uranium availability are much lower than those of IAEA/WNA/NEA: the discovered available RARs would be somewhere between 1.9 and 3.3 Mt, depending on the cost class, and the IRs, with lower data quality, are between 0.8 and 1.4 Mt. At current, reactor uranium demand (65.5 kt per year), the 3.3 million tons of discovered available RARs with cost below $130 kg^{-1} would last for about 50 years. Inclusion of IRs up to $130 kg^{-1} would extend the availability to 70 years. Even in this case, peak production could be achieved as early as 2035, with all the related implications. In conclusion, uranium supply can be expected to be a bottleneck for nuclear energy in the next few years because the new uranium mines will not be capable of compensating for the diminishing of secondary uranium resources (coming from dismantling of nuclear arsenals), and in the medium term because of the limited amount of recoverable uranium [472, 476]. Therefore, countries interested in the construction of new nuclear power plants within the next 10–20 years, in particular European countries, before investing huge amounts of money should consider that it could be difficult to find enough uranium fuel for the 40–60 years of operation of new plants.

It should be pointed out that the use of uranium creates huge safety, security, ecological, economic, and political problems, not even mentioned in the WNA and related reports. It can be noted from Figure 8.6 that 64% of the resources are concentrated in six countries and that among the 14 countries shown in the figure, none belongs to the EU. This means that for the European countries the uranium import dependence is now almost 100% and, as such, much larger than their relative dependence on oil and gas imports. At present, the US production amounts to about 1500 t per year, with a 15% decrease in recent years [469]. In 2009, the US produced 14% of the uranium consumed and imported about 23% of its uranium from Russia, which also supplies about 25% of the uranium needed to European countries. Clearly, a possible shortage of uranium, for any reasons, would be a problem for the US and for the EU countries, particularly for France, which relies almost exclusively on uranium electricity. The French company AREVA owns two uranium mines in Niger, that accounts for 30% of French consumption and 32% of Niger's exports, but less than 5% of Niger's GDP. In the past few years uranium mining in Niger has been associated with the Tuareg rebellion and several political problems, including a controversy between AREVA and Niger's government. It is easy to understand that uranium, as oil, drives the internal and foreign policy of several countries, expands corruption, blurs the line between trade relations and neo-colonialism, and endangers peace.

Currently, the production of uranium falls short of demand by about 25 kt per year (37%). Since the end of the Cold War, the gap between production and demand has closed with uranium drawn from stockpiles of military origin which were accumulated before 1980. Continuing with 45 kt per year production and 67 kt per year demand will presumably exhaust the military stocks by 2015 [476].

8.2.4
Nuclear Reactor Technologies

Various types of nuclear reactors are in operation [5, 58, 465]. Most of them contain neutron moderator materials that slow neutrons until they are thermalized. These reactors are called thermal reactors. Thermalized neutrons have a far higher probability of fissioning the fissile nuclei ^{235}U and a relatively lower probability of neutron capture by ^{238}U compared to the faster neutrons that originally result from fission [465]. This allows the use of low-enriched uranium or even natural uranium fuel. The moderator (which sometimes also plays the role of a coolant) is usually water under high pressure to increase the boiling point. As a reactor operates, it generates increasing amounts of ^{239}Pu from the reaction of ^{238}U with neutrons. ^{239}Pu is fissionable, just as ^{235}U. Over time an operating reactor produces an increasing proportion of its total energy from the fissioning of ^{239}Pu and less from fissioning ^{235}U. In the first few days and weeks a reactor runs, the plutonium it produces can be chemically extracted and used as is to produce atomic bombs.

The fuel rods, that must be replaced after 2–3 years of operation, are extremely radioactive and contain unburned uranium and plutonium as well as many other radioactive isotopes.

Most of the current reactors, built before 1990, are considered Generation II reactors. Generation III reactors, sometimes called "evolutionary" reactors [58], are those incorporating developments in design which include improved fuel technology, superior thermal efficiency, and passive safety systems. The term Generation III+ is often used for the design of reactors which offer further improvements in safety and economics. The Advanced Candu Reactor (ACR), Advanced Pressurized Water Reactor (APWR), and European Pressurized Reactor (EPR) are Generation III+ systems [58]. Generation IV refers to an international forum involving 10 countries, started in 2000 to lay out a path for the development of nuclear plants for electricity generation and hydrogen production after 2030 [58]. Generation IV reactors exist only on paper as a set of theoretical nuclear reactor designs currently being researched [477].

Nuclear fuel cycles are divided into two major categories: once-through, in which the fuel is removed from the reactor after use and destined to permanent disposal (Figure 8.7); and closed cycle, in which the used fuel is recycled to extract more energy (breeder reactors) or to destroy undesirable isotopes (burner reactors) (Figure 8.8). Recycling used fuel requires several repeated steps, including chemical or electrochemical processing. The vast majority of commercial reactors adopt a once-through fuel cycle. Closed fuel technologies are used in a few countries, including France and Japan. Although fuel recycling may look appealing on the paper, in practice it is part of the problem, not the solution [478]. Closed fuel technologies are more expensive and raise unsolved proliferation issues [479, 480]. For these reasons, they have never been adopted in the US, and recent authoritative reports affirm that there is no need for closed fuel technologies to enable expansion of nuclear power, at least until 2050 [58, 475]. Current generation technology for recycling, such as MOX (mixed oxide) and PUREX (plutonium and uranium extraction), as well as other proposed separation techniques (e.g. those called UREX) can reduce the volume of long-lived high-level radioactive waste, but

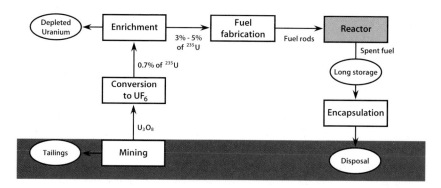

Figure 8.7 Once-through (open) nuclear fuel cycle.

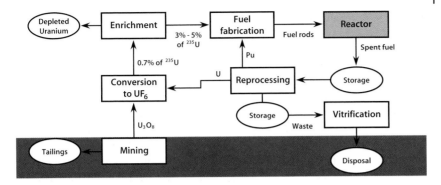

Figure 8.8 Closed nuclear fuel cycle.

repositories capable of sequestering high-level waste for thousands of years would still be needed and a larger quantity of lower-level waste would be produced.

India, which has about 32% of world's thorium reserves, is particularly interested in once-through and closed thorium fuel cycles, which involve conversion of thorium-232 into the fissile uranium-233 isotope [481].

8.2.5
Cost and Time Issues

The construction of a nuclear plant is a complex enterprise, starting from preparing an accurate financial plan, finding a suitable site, obtaining a bunch of licenses, recruiting expert engineers, beginning the construction, receiving the material on time, and proceeding under safety and security controls. Furthermore, nuclear power is extremely capital intensive, and the cost of a nuclear plant is difficult to evaluate, especially because of the long timescale required for construction. It has been pointed out that the history of nuclear power tells us three lessons [468]: (i) the estimated costs are invariably lower than actual costs incurred during plant construction; (ii) "generic" estimates, such as those regularly provided by industry trade associations, are always lower than estimates provided by companies that are involved in real projects such as vendors or utilities; (iii) comparisons of reactor costs across the world are not always normalized properly, sometimes resulting in inaccurate trend comparisons. Even for large utilities, a nuclear plant can represent an important fraction of the company's net worth, potentially putting the entire company at risk should the project be delayed substantially or costs escalate significantly. The unsolved problem of nuclear waste depositories (Section 8.2.8) and the long time and high cost of the complex decommissioning procedures (Section 8.2.9) introduce further uncertainties. Of course, most of these problems are hidden in those nations where the nuclear plants are constructed directly or indirectly by the government.

The 2009 report of the American Academies [58] recognizes the difficulty of estimating the cost of new nuclear reactors. It is reported to vary by as much as a factor of two, with "overnight" costs between $3.0 and 6.0 billion per gigawatt. In fact, the costs are much higher and continue to increase. In February 2008, Florida Power & Light calculated an overall cost from $5.7 to $8.1 billion per gigawatt, but a more recent estimate shows that it will spend from $18 to $24 billion, depending on the technology it selects for its two reactors [482]. The 2005 estimated cost of $5.8 billion by Ontario Power Authority to build a new 2 GW reactor in Ontario was revised to $15 billion by Moody's in 2008. But in July 2009 the only compliant bid received, from Atomic Energy of Canada for two 1.2 GW Advanced Candu Reactors, was around $26 billion, that is, $10.8 billion per gigawatt, more than three times higher than the assumed cost [483].

The Finnish Olkiluoto-3 reactor, still under construction by the French company AREVA, is a particularly important case because it seemed to contradict the conventional wisdom that liberalization and nuclear power orders were incompatible [470]. The Olkiluoto-3 order, the first nuclear order in Western Europe and North America since 1993 when it was placed in December 2003, was a huge boost for the nuclear industry. AREVA anticipated that, once completed, the plant would provide a demonstration and reference for other prospective buyers of the EPR technology. The turnkey contract €3 billion signed by the Finnish utility TVO with AREVA and Siemens had a target construction of 48 months. The cornerstone ceremony took place in September 2005. In July 2006, the work already had a 1 year delay, which increased to 1.5 years in December 2006, 3 years in October 2008, and 3.5 years in January 2009 [484]. In March 2009, AREVA admitted a cost over-run of €1.7 billion. In the meantime, Siemens had announced withdrawal from AREVA and the contract is now the subject of a dispute between AREVA and the customer, TVO. In September 2009 the cost had risen to €5.3 billion. In November 2009, the nuclear safety authorities of Finland, France, and the UK (where AREVA would like to build nuclear plants) in a joint statement asked AREVA to make changes to the reactor's control and instrumentation systems [485]. Clearly, turnkey contracts in the field of nuclear plants are very risky. AREVA was almost forced to make such an advantageous offer to TVO because there were fears that if an order for an EPR was not placed soon, AREVA would start to lose key staff and the design would become obsolete [470]. In July 2009, the same type of reactor had been offered to the Ontario Government at about twice the cost of the Finnish reactor [483].

Moody's has stated that it is unable to make a finite determination of the range of the all-in costs associated with new nuclear plants and it has also stated that it will likely take "a more negative rating position for most issuers actively seeking to build new nuclear plants" [486]. According to Synapse Energy Economics [487], the real cost for a 1000 MW unit can be as much as $15 billion or more. Such a high cost would likely make electricity from nuclear energy non-competitive with electricity obtained from other sources [488]. Under these conditions, banks are worried about lending money to build nuclear plants, US utilities are not encouraged enough by federal incentives [58, 470], and taxpayers fear the risk of being involved in an unsuccessful nuclear energy renaissance [487].

Besides cost and uncertain construction times, several other hurdles are present along the road of nuclear renaissance [58].

8.2.6
Proliferation

Today, at least nine nations (US, Russia, UK, France, China, India, Pakistan, North Korea, and Israel) possess nuclear weapons, three more nations (Iran, Syria, and Myanmar) might have nuclear weapon programs, and some of the countries that possess nuclear weapons (India, Pakistan, and Israel) have not signed the Treaty on the Non-Proliferation of Nuclear Weapons (NPT) [489]. Indeed, it is difficult to explain to a nation why it should not build nuclear weapons if more or less nearby countries possess them.

In principle, proliferation and terrorist use of nuclear material can be impeded by denying knowledge or by denying access to fissile material. Both actions, however, are practically impossible. The denial of knowledge is less and less effective with the passage of time, particularly in the Internet age. Restricting access to fissile material is more and more difficult as the amount of this material increases in a larger number of countries. Therefore, there is only a radical, admittedly utopian, solution to this problem: the total elimination of nuclear weapons, as outlined by President Obama in a speech in Prague, April 2009. The agreement of April 2010 between the US and Russia to reduce further the number of nuclear weapons is a step in the right direction. To realize this dream, which could be extended to finding reasonable solutions to other problems in our society, we would need a larger number of political leaders capable of looking far ahead in space and time: at the welfare of their own country, but also of the entire spaceship Earth; at the needs of people living now, but also of future generations.

Nuclear fuel cycle technologies introduce the risk that weapons-usable material could be produced. This is possible because nuclear plants produce plutonium, and the same technologies used to enrich uranium (typically 3–5% ^{235}U) can be applied to achieve higher enrichments (>20% ^{235}U) for producing weapons-usable material [490]. In addition, conventional fuel recycling technologies (i.e. the PUREX process) produce plutonium, separated from uranium and other transuranics, that can be used to produce powerful and reliable nuclear weapons. No reprocessing technology is completely proliferation resistant, and none of the technologies currently under development would be deployable in non-nuclear weapons states without causing significant proliferation concerns [58, 480].

International plans have been suggested to mitigate proliferation risks associated with the fuel cycle, even while increasing the use of nuclear power worldwide. An example is the "fuel bank" proposal, based on a multilateral center for the production and distribution of civilian nuclear power plant. Kazakhstan is considering housing such a site, but some states have expressed concerns that in the frame of such a bank they would lose all access to their own enrichment and reprocessing technology. Furthermore, there are hurdles related to the positions of Egypt against Israel and of Russia against India, since Israel and India are among the few countries that did not sign the NPT [491].

Extension of civil nuclear power, unfortunately, favors nuclear proliferation in several ways and, more generally, adds significantly to the vulnerability of society. The more fissile material is around, the more we should avoid the spread of terrorism. Any attempt should also be made so as to reduce the level of conflict in the domestic and international scenes rather than exacerbating groups or small nations and taking revenge. Even recent history shows that the often quoted sentence "One does not export democracy in an armored vehicle" is indeed true.

8.2.7
Safety and Security

In the field of nuclear energy production, safety can be defined as measures that would protect nuclear facilities against failure, damage, human error, or other accidents that would disperse radioactivity into the environment; security concerns measures that would protect facilities against sabotage, attacks, or theft. Major fears concern [465]: (i) explosion of the reactor like a bomb; (ii) melt-down of the reactor because of loss of coolant and failure of the emergency core cooling system; (iii) massive release of radioactive material due to an accident; (iv) continuous release of radioactive material during normal operation; (v) radioactive waste disposal; (vi) problems related to transportation of radioactive material; (vii) ecological damage caused by waste cooling heat; (viii) theft of plutonium or other enriched uranium by terrorists and rogue nations; (ix) terrorist attack, particularly vulnerability to aircraft attack; and (x) proliferation risks [492]. The policies for safety and security are promoted by the IAEA [469] whose role is often undervalued [493].

Much of the complexity of more recent nuclear plants is due to redundancy of systems, extensive backups and specific strategies to reduce the risk that a single failure of a critical system could cause a core meltdown or a catastrophic failure of reactor containment. There is no doubt that excellent work has been done in this field, but any complex system, no matter how well it is designed and engineered, cannot be deemed failure proof. This is the reason why insurance agencies do not wish to put their own survival in balance and refuse the full indemnity insurance of nuclear activities. From 1950 to 2006, 23 major civilian nuclear accidents have been listed, the most important one being the Chernobyl catastrophe in 1986 [465], when the meltdown of a reactor released about 50 t of radioactive material into the atmosphere. This accident caused 56 direct deaths and contaminated an area inhabited by 5 million people. In 2005, a panel of scientists estimated that 4000 people or even more will die from cancer caused by radiation on the approximately 600 000 most highly exposed people [494, 495]. The report pointed out that the accident caused several other problems to the population, including frequent mental disease among the 350 000 residents who were forcibly relocated.

8.2.8
Waste Management

One of the biggest (and perhaps the single biggest) problem of the nuclear power industry and of the entire society is the storage of nuclear waste. The radioactivity of spent fuel decreases over time as the various elements decay; that of some fission products decreases in a few hundred years. Plutonium isotopes and minor actinides are responsible for the bulk of radioactivity in the medium term (300–20000 years). Other radioisotopes (e.g. neptunium-237 and technetium-99) have a half-life greater than 1 million years. Reprocessing of nuclear fuel reduces the mass of used fuel generated, but increases the radioactivity per unit mass [58]. Additional waste is also produced during operations (such as contaminated gloves, tools, and water purification filters) and when the plant itself is decommissioned. The low-level waste is often buried near the surface of the Earth. The radioactivity of high-level waste, however, is so great that it would be irresponsible to store it in accessible sites or in those exposed to flowing groundwater that can return to the Earth's surface. Today the amount of high-level waste (Greater than Class-C waste) worldwide amounts to about 250000 t and is increasing by about 12000 t every year, which is the equivalent to about 100 double-decker buses.

The ultimate disposal of high-level long-lived radioactive material is a very difficult task. If the crazy proposals of burying the waste under the ocean floor or shooting it into space are excluded, one remains with the apparently reasonable option of disposing of these materials in an underground repository. Even this solution, however, is difficult to carry out because, according to studies based on the effect of estimated radiation doses, the time frame in question ranges from 10000 to 1000000 years, which practically means "forever". More than 30 years ago Hannes Alfvén, winner of the Nobel Prize for Physics, warned [496]: "The problem is how to keep radioactive waste in storage until it decays after hundreds of thousands of years. The geologic deposit must be absolutely reliable as the quantities of poison are tremendous. It is very difficult to satisfy these requirements for the simple reason that we have had no practical experience with such a long term project. Moreover permanently guarded storage requires a society with unprecedented stability." Concerns associated with managing high-level waste are, indeed, intrinsically intergenerational.

The effort of the US to build the first of such repositories at Yucca Mountain (Nevada) failed after more than 25 years' work [497, 498] and expenditure of at least $13.5 billion out of the overall $96 billion estimated [499]. Therefore, used fuel is, and will be for many years, subjected to interim storage in dry casks at plant sites, until permanent disposal facilities become available [58, 480]. A broad program to create an understanding of the range of waste management options has been solicited [475]. Several other nations have long since had projects to construct permanent geologic depositories, but progress, if any, is very slow everywhere [500]. For example, in Japan the final selection of a repository location is expected between 2023 and 2027, and in China a final site should be selected by 2020 and actual disposal should begin by about 2050 [501]. The failure of the Yucca Mountain project will further slow all the current plans.

Practically, the problem of disposition of used nuclear fuel remains unsolved [58], contrary to the recurring declarations of the nuclear lobby. A corollary of this situation is, of course, that nobody knows how much the real cost of nuclear waste storage will be. The reason why this issue is hardly mentioned by the nuclear renaissance supporters is that such a cost is simply passed on to the shoulders of the next generations. Production of high-level long-lived radioactive waste in the absence of safe ultimate depositories violates the "polluter pays" principle. Creating problems without knowing how, when, and whether they can be solved does not contribute to the progress of mankind.

8.2.9
Decommissioning and Dismantling

Even if there is a trend to extend the life of nuclear plants to 60 years, the fact remains that, eventually, any plant has to be closed and itself becomes a form of nuclear waste. Some of the metal components of the reactor core and the used fuel must continue to be stored on-site until they can be removed to yet non-existing permanent disposal facilities [58]. Decommissioning and dismantling a nuclear power plant are a complicated issue that requires much time and costs much money because of the complexity and risks involved in working with contaminated materials.

In the official estimations of the cost of nuclear plants, decommissioning is always considered a very minor fraction of the overall expenses [502]. Although information on times and costs of decommissioning is most often hidden [503], the few available data show that decommissioning is indeed a very costly operation.

The IAEA has defined three options for decommissioning nuclear plants [469]: immediate dismantling, safe enclosure, and entombment. Regardless of the selected option, there are several hurdles and many inconvenient features. Entombment, which simply means encasing the radioactive plant in a permanent sarcophagus as in the case of the Chernobyl reactor, is a frantic attempt to conceal the intrinsically dangerous nature of nuclear power plants. Immediate dismantling sounds a more appealing option, but, in fact, "immediate" means waiting several years to allow cooling and decay, at least of the short-lived radionuclides. Since highly contaminated sites can only be dismantled using remotely operated tools, this option is very expensive. Immediate dismantling, has been used for only a few plants and has always required huge amounts of money [503]. In France, dismantling of the small Brennilis power plant, following a "transparent procedure," began in 1985 and ended in 2005 with a cost of €482 million, 20 times higher than the initial estimate [504]. In the case of the Yankee Rowe (US) power plant, which had been completed in 1960 at a cost of $39 million and closed in 1992, the initially estimated cost for dismantling was $368 million, but after 8 years of operation, the cost was $508 million. Costly, yes, but looking at the pictures of the site before and after decommissioning (Figure 8.9) the reactor site appears to have been converted into a garden [505]. Another, less advertised picture (Figure 8.10), however, must be added to the frame [506]: that of the nearby site, hidden "in a slight valley between natural ridgelines out of the sight of nearby rivers," where high-level radioactive spent fuel and sections of the

Figure 8.9 The Yankee Rowe nuclear station (Massachusetts) prior to (inset) and after decommissioning. The arrow shows the location of the deposit of the high-level radioactive material depicted in Figure 8.10.

Figure 8.10 The 43 dry storage casks of the Yankee Rowe decommissioned nuclear facility. They contain 1019 spent fuel assemblies and sections of the reactor vessel (high-level radioactive Greater than Class-C waste), waiting to be transported to a still non-existing permanent repository.

reactor vessel have been placed "until the Department of Energy meets its legal obligation to remove it." Therefore, it is unfair to say that that "decommissioning of the former Yankee Rowe plant was completed in 2007" [505]. This is the general situation for the "immediately dismantled" power plants.

The third and most followed option, namely safe enclosure, means postponing the final removal of the facility until 100 or more years into the future, waiting for a significant decay of radioactivity and hoping for technological developments in the field of remotely operated tools. This places an economic, environmental, and social burden on future generations to deal with waste from which they have received no direct benefit. Today the sector, public or private, has to deal with waste management and decommissioning expenses that far outweigh estimates of the past. In 2005, the UK Nuclear Decommissioning Authority estimated a cost of £55.8 billion for decommissioning the existing UK nuclear sites, but in 2006, the cost had risen to at least £70 billion [507]. Since the project should cover a time frame of a century, it could eventually be wrong, thereby leaving a certain financial liability for a future generation. For the Three Mile Island power plant, as of December 31, 2008, the radiological decommissioning cost estimate was $831.5 million, to be compared with a decommissioning trust fund of $484.5 million [508].

In conclusion, decommissioning is a complex, costly, and very long operation which will cause ecological problems and have severe economic and social impacts on future generations.

8.2.10
Other Limiting Factors

In several countries, governments and/or public opinion oppose the introduction or expansion of nuclear power. In some places, for example in Japan, there is a hostile environment due to volcanic and earthquake risks [509]. Other general hurdles to overcome for nuclear expansion are as follows.

- **Water shortage:** Nuclear power plants require more cooling water per kWh of electricity produced than fossil fuel plants due to their lower thermal efficiency. In 2008, a severe drought in the southeast US threatened the cooling water supplies of more than 24 of the 104 nuclear power reactors [510]. In 2003, a severe drought caused the loss of up to 15% of nuclear power generation capacity for 5 weeks in France. Climate change could make things worse, not only because of fresh water shortage, but also because reactors on coastlines could become inundated.

- **Territory occupation:** Uranium mines occupy extensive areas in several countries with severe ecological and social implications [511, 512], Figure 8.11.

- **Manufacturing:** Japan Steel Works (JSW) Limited is the only industry that can forge components from ingots up to 450t as needed for the EPR and other Generation III reactor pressure vessels [470]. Other manufacturers could

Figure 8.11 Ranger uranium open pit uranium mine in Northern Australia.

upgrade their capacity, but will not go ahead with investments of hundreds of millions of dollars if they do not have firm orders for several years ahead. The recent cancellations of advanced projects (e.g. in Canada) as well as repeated delays in many other projects (e.g. in Finland) do not help to establish the indispensable confidence level for the necessary capital-intensive investments.

- **Workforce and competence:** The lack of a trained workforce and massive lack of competence are internationally recognized problems [470].

8.2.11
Perspectives

The excitement that accompanied the early development of the nuclear energy age is well reflected in the famous prediction of the chairman of the US Atomic Energy Commission, Lewis L. Strauss, reported in the *New York Times* on September 17, 1954: "Our children will enjoy in their homes electrical energy too cheap to meter." As mentioned above, in 2009, after more than 50 years of heavily supported research and development (54.8% of the Government funds for energy R&D within IEA countries between 1974 and 2007 [470]), nuclear energy provides only 14% of the total amount of electric energy produced.

In the past decade, there has been renewed interest in nuclear power plants. The main factors driving this trend are (i) a perceived need to increase generation from low-carbon sources to combat climate change and (ii) the hope of achieving energy independence from unstable oil and gas producers. Other favorable factors are the increasing number of nuclear power plants in North America and Western

Europe which are close to their retirement age, and the development of a new generation of nuclear power plants (Generation III+, Section 8.2.4), claimed to be more reliable and less expensive [470]. But while nuclear supporters talk about a nuclear renaissance phase [469], nuclear opponents believe that nuclear energy is already phasing out [476].

Nuclear energy is often assumed to be a carbon-free electricity source [513] and thus it is considered the most important part of any solution aimed at fighting climate change. Once again, however, things are not so simple, for two reasons: (i) non negligible amounts of greenhouse gases are emitted in the many steps involved in the lifecycle of nuclear plants, and (ii) in order to have a substantial impact on the reduction of greenhouse gases production, new nuclear capacity should replace a substantial amount of the electricity presently generated by coal and gas plants. The lifecycle of a nuclear power plant involves many energy consuming activities: construction; uranium mining, milling, refining, and enrichment; reactor maintenance and operation; waste packaging, separation, temporary storage, and final disposal; decommissioning and dismantling. Estimates of lifecycle CO_2 emissions of nuclear plants are often very inaccurate or, on purpose, incomplete [514]. According to a recent report, indirect emissions of CO_2 are between 8 and more than $110 \, g \, kWh^{-1}$ [515]. Such different values result, at least in part, from the quite different specific assumptions that can be used regarding numerous relevant issues such as quality of uranium ore, type of mining, and method of enrichment. Other studies give mean values for emissions of CO_2 of 66 [514] and $40 \, g \, kWh^{-1}$ [58]. These data show that electricity from nuclear energy is in no way "emission free," even if it is much better (from purely a carbon-equivalent emissions standpoint) than fossil fuels (coal, 900; natural gas, $360 \, g \, kWh^{-1}$). On an average, however, it is worse than renewables (wind, ≤ 10; photovoltaic, 20–60; geothermal, $15–55 \, g \, kWh^{-1}$) [514, 516]. It should also be noted that whereas emissions from conventional power plants tend to decrease with time, emissions from nuclear plants tend to increase because of the decreasing grade of the uranium minerals.

The climate crisis is real and accelerating (Chapter 7). Therefore, addressing climate is an urgent need, not a long-term project. Nuclear energy can play only a very minor role in decreasing greenhouse gas emission in the near future because deployment of nuclear energy, if any, will be very slow. For example, in the US the very unlikely construction of 5–9 new plants by 2020 would avoid production of 40–150 million tons of CO_2 equivalent per year, or 0.04–6% of 2007 US emissions [58]. A recent MIT report points out that one must look at 2050 to hope that a substantial capacity of new nuclear reactors will be available worldwide [475]. But fighting the climate crisis cannot be based on hope and cannot wait so long. As far as energy independence is concerned, some people (and also some governments) believe that deployment of nuclear energy in a country could lead to independence from oil import. They forget that nuclear power plants currently produce only electric energy, which represents between 15 and 25% of the energy end-use of a nation. They also forget that electric energy and oil play very different roles in the present economic systems. Most of the oil is used for transportation and cannot be replaced by electric energy, at least in the short term. In fact, the amount of oil consumed by France (58 nuclear reactors) is higher than that of Italy (no nuclear plants) [7], although the two countries are comparable in terms of inhabitants and wealth.

8.2.12
Nuclear Industry Renaissance?

The already mentioned report of the US Academies [58] points out that there are many severe hurdles to overcome before constructing even a few plants, including high construction costs and storage and disposal of radioactive waste. The low level of hope in a renaissance of nuclear energy in the US clearly emerges from the following "if" sentences of the report: "If these hurdles are overcome, if the first new plants are constructed on budget and on schedule, and if the generated electricity is competitive in the marketplace, the committee judges that it is likely that many more plants could follow the first plants." [58]. Otherwise, the report concludes, the option of expanded nuclear deployment in the US will be closed. At present, applications for 26 reactors have been submitted, but, in spite of substantial federal subsidies (loan guarantees, insurance against delay and production of tax credits [470]), no new plans are under construction. Recently the US Government offered $8.33 billion in loan guarantees to back construction of two 1100 MW nuclear reactor power plants in Georgia. However, the Congressional Budget Office has warned that the default risk is high and more than half of the companies receiving loans could default because of nuclear power's high construction costs [517, 518]. This government's decision is probably related to the need to promote in the international market the AP1000 Westinghouse design, which, however, has not yet received approval from the Nuclear Regulatory Commission [517].

In recent years, several international assessments of the possible future of nuclear power in the world have been adjusted to optimistic prospects for 2030 [474, 519–521]. According to the World Nuclear Industry Status Report 2009 [470], however, a detailed examination of the situation shows that nuclear power will hardly increase in the next two decades. Without any significant new build for years, the average age of the world's operating nuclear power plants has been increasing steadily, standing at 26 years in 2010 [469]. Nuclear utilities envision reactor lifetimes of 40 or even 60 years, but experience shows that the average age of the 123 units that have already been shut down is 22 years [470]. In any case, one can calculate how many plants would be shut down year by year by 2056. This calculation allows the evaluation of the number of plants that would have to come online over the next several decades simply to maintain the current operating capacity (Figure 8.12) [470]. In addition to the 52 units under construction, 42 reactors (generating 15 900 MW) would have to be planned, completed, and started up by 2015, one every month and a half, and an additional 192 units (170 000 MW) over the subsequent decade, or one every 19 days. The achievement of the 2015 target is simply impossible given existing constraints on the fabrication of key reactor components. As a result, the number of reactors operating will decline over

Projection of nuclear reactors numbers and capacity
(Years 2009 - 2056)

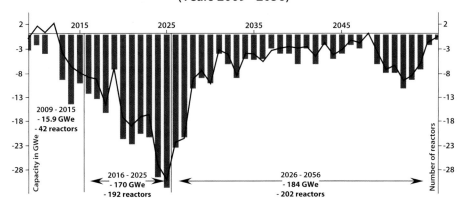

Figure 8.12 Projection of the world nuclear reactor number/capacity (added–shutdown), between 2009 and 2056. General assumptions of 40 years mean lifetime plus German phase-out [470].

the years to come–even if the installed capacity level could be maintained–unless lifetime extension beyond 40 years becomes standard. Were life extensions to become standard, many other questions regarding safety, maintenance costs and other issues would need to be more carefully addressed.

In conclusion, a nuclear renaissance is very unlikely because many factors militate against fast building of new nuclear power stations. In any case, trying to solve the energy and climate problems by choosing the nuclear option would be not only an impossible mission, but also a quite dangerous obstacle to the evolution of our civilization. Nuclear energy is characterized by so many uncertainties, unsolved problems, and risky features to endanger the most important values of our life: health, environmental sustainability, social stability, and international relationships. Last but not least, it should be pointed out that export of nuclear technology to developing countries, an activity in which Russia and France are actively engaged, leads to a new form of colonization.

8.3
Civilian Use of Nuclear Fusion?

8.3.1
A Difficult Problem

As discussed in Section 8.1.5, the process of fusing atoms together is radically different from that of breaking heavy atoms apart. To obtain fusion, two nuclei have to be knocked against each other with enough energy to overcome electrostatic repulsion until they get so close as to allow the very short-range nuclear force to prevail. This can be easily done in a weapon, where the energy for igniting fusion is supplied by radiation implosion of a primary fission device. But even after about 60 years of efforts, the possibility of exploiting nuclear fusion for civil applications is very far in the future and might also remain a dream. The two most important research approaches to controlled nuclear fusion are [464]: (i) magnetic confinement, where a system called a plasma, in which electrons are not bound to the nuclei, is created and contained at very high temperature by a strong magnetic field for suitable extended periods of time; (ii) inertial confinement, where fusion is realized in a small, concentrated volume of plasma heated and compressed extremely rapidly with high-energy lasers.

8.3.2
Magnetic Confinement Approach

Magnetic confinement is presently the most followed option in the attempt to obtain electricity from nuclear fusion. The idea is to use magnetic compression to get the plasma density and temperature high enough to produce energy by fusion faster than it is lost by radiation from the hot plasma. For the most favorable case of the deuterium–tritium fusion reaction (Equation 8.4), the temperature necessary to maintain the fusion process is of the order of 1×10^8 K.

Experimental studies on magnetic confinement have been conducted in several countries since the 1950s. A breakthrough came in 1968 when Russian scientists obtained for the first time 1×10^7 K in a device dubbed Tokamak, which is basically a donut-shaped cylinder where the trapped plasma circulates without touching the walls of the chamber [465]. In 2007, the EU, India, Japan, China, Russia, South Korea, and the US founded a consortium called ITER (International Thermonuclear Experimental Reactor) to demonstrate the feasibility of fusion power [522]. The starting of the deuterium–tritium fusion operation was originally scheduled for early 2025, but, as usually happens for very complex mega-projects, particularly in the field of nuclear energy, temporal delays and increasing costs have begun to float up [523–525]. If ITER will be successful, a demonstration power plant (DEMO) should be constructed by 2033, and then tested before 2038. In case of success of DEMO, an expanded and updated power plant should be constructed to prepare for the first commercial fusion reactor after 2050 [58]. Investing €10 billion on the first phases of an uncertain project that has a term of 30–50 years is not a good way to approach the urgent problems of climate crisis and phasing out from fossil fuels.

While the ITER project apparently works on paper and will certainly be useful for advancing the knowledge of physics, it has been pointed out that even if practical means of generating a sustained power producing fusion reaction were found, many, perhaps insurmountable, hurdles would remain before arriving at commercial use. For example, in the ITER experiment the generated heat will simply be dissipated, whereas in a real reactor the wall should also contain heat exchangers to remove the heat, which should then be used to produce steam and finally electricity. At present, interest from the utilities companies in nuclear fusion "is nonexistent and will probably remain so for the foreseeable future," since "the huge lump capital investment is beyond the risk level of any utility system" [526].

8.3.3
Inertial Confinement Approach

As discussed in Section 8.1.5, in a hydrogen bomb the fusion fuel is compressed and heated by the explosion of a separate fission bomb. The inertial confinement approach (ICF) to create a fusion reactor is based on the same concept, except that instead of using a fission bomb, the reaction is initiated by heating and compressing a fuel target, typically in the form of a pellet of deuterium and tritium, by using high-energy beams of laser light. The aim of the ICF is to produce a condition known as "ignition," where this heating process causes a chain reaction that burns a significant portion of the fuel. Typical fuel pellets are about the size of a pinhead and contain around 10 mg of fuel.

The primary problems in making a practical ICF device would be building a laser of the required energy and making its beams uniform enough to collapse a fuel target evenly. The most advanced site for ICF is the National Ignition Facility (NIF) at the Livermore National Laboratory in California [527], where fusion experiments should start soon. A recent development of ICF is the concept of "fast ignition," which decouples the heating and compression phases of the implosion.

Other approaches to ICF include the pellet implosion by X-rays produced from "Z-pinch" machines. Instead of using an external magnet to generate the induction field, tens of millions of amperes of current driven by 6×10^6 V of pulse power are used to vaporize a set of very fine tungsten wires running around the fuel pellet. The plasma formed in this way fills the target with X-rays, which implode the fuel pellet [528].

8.3.4
Wishful Thinking

Before closing with nuclear fusion, a couple of issues which will likely remain in the category of wishful thinking should be briefly recalled:

- **Cold fusion:** In 1989 a tabletop experiment involving electrolysis of heavy water on a palladium electrode was reported to generate anomalous heat production that was explained in terms of nuclear processes [529]. Generation of small amounts of nuclear reaction byproducts, including neutrons and tritium, was also reported. These results suggested the occurrence of a *cold fusion* and raised hopes of a cheap and abundant source of energy. Enthusiasm turned to skepticism as it was discovered that nuclear reaction byproducts had not been detected in the original experiment and early replication experiments suggested possible sources of error in heat production measurement. In the following years, a flurry of attempts to repeat the experiment around the world failed. Mainstream science does not believe in cold fusion [530] also because of a lack of an accepted explanation using conventional physics. However, since cold fusion would be such an extraordinary process from both a theoretical and a practical viewpoint, sessions on this field are regularly organized, under the umbrella name "low-energy nuclear reactions," at American Chemical Society meetings.

- **Lunar helium:** Most fusion processes produce highly energetic neutrons which bombard the reactor components, generating heat that is used to produce steam and finally electricity. The efficiency of electric power generation in this way is limited by the Carnot cycle. Furthermore, neutron bombardment induces radioactivity in the reactor's components.

A fusion reaction that does not show such drawbacks is that between deuterium and helium-3, which yields helium-4 and a proton:

$$^2H^+ + {}^3H^{2+} \rightarrow {}^4He^{2+}(3.67\ \text{MeV}) + p^+(14.68\ \text{MeV}) + 18.35\ \text{MeV} \qquad (8.5)$$

A uniform magnetic field applied perpendicularly to the beam produced could separate electrons from the high-energy protons and He nuclei. This charge separation could give rise to a high voltage electric current without going through steam generation, thereby bypassing the Carnot cycle limitations. The development of this concept, however, will find several hurdles [251]. The most important one is that ^3He is rare on Earth. Since the surface of the Moon is somewhat rich in ^3He, the idea would be to go to the Moon and carry back what appears to be the only valuable resource available there [531]. This project appears to belong to science fiction, but history shows that sometimes today's science fiction becomes tomorrow's science.

Atomic Bombs on Hiroshima and Nagasaki

The atomic bombs dropped on Hiroshima and Nagasaki in August 1945 caused the deaths of around 200 000 people (mostly civilians) from injuries sustained from the explosion and acute radiation sickness, and even more deaths from long-term effects of ionizing radiation. The decision to use atomic bombs in the war against Japan and to drop them on cities was taken, not without controversy, in a few convulsed weeks, as shown by classified documents now publicly available [532]. On June 11, 1945, when the bomb was ready and the war against Japan was still in progress, a panel of seven scientists chaired by James Frank wisely pointed out that "the use of nuclear bombs for an early, unannounced attack against Japan is inadvisable" and suggested first to demonstrate to the world the power of the bomb in an appropriately selected uninhabited area. But on June 16, a panel composed of Oppenheimer, Fermi, Compton and Lawrence wrote that there was "no acceptable alternative to direct military use" of the atomic bomb. On June 27, Undersecretary of the Navy Ralph A. Bard wrote that use of the bomb without warning was "contrary to the position of the United States as a great humanitarian nation". On July 3, Leo Szilard sent a petition to President Truman calling atomic bombs "a means for the ruthless annihilation of cities" and asking the President "to rule that the United States shall not, in the present phase of the war, resort to the use of atomic bombs." On July 4, Szilard sent his petition to colleagues at Oak Ridge and Los Alamos, with a covering letter in which he pointed out the need for scientists to take a moral stand on the use of the bomb. On the same day, General Leslie Groves, the military leader in charge of the Manhattan Project, sought ways to charge Szilard with violating the Espionage Act. In mid-July, 67 scientists at the Oak Ridge Laboratories sent a petition to the President recommending that "before this weapon be used without restriction in the present conflict, its powers should be adequately described and demonstrated, and the Japanese nation should be given the opportunity to consider the consequences of further refusal to surrender." On July 16 the first atomic bomb was tested at the Trinity Site in Alamogordo, New Mexico, under the eyes of several scientists, including Fermi, who estimated the explosion of the bomb "to correspond to the blast that would be produced by ten thousand tons of TNT." Even before the bomb was tested, a second bomb had secretly been dispatched to the Pacific for an attack on Japan. On July 17, 1945, Szilard and 69 co-signers at the Manhattan Project in Chicago petitioned the President, asking that "you exercise your power as Commander-in-Chief, to rule that the United States shall not resort to the use of atomic bombs in this war unless the terms which will be imposed upon Japan have been made public in detail and Japan knowing these terms has refused to surrender." On July 24 at the Potsdam Conference in defeated Germany, President Truman told Stalin that the US "had a new weapon of unusual destructive force." On July 25, Truman ordered the Air Force to "deliver its first special bomb as soon as weather will permit visual bombing after 3 August 1945". In his diary Truman wrote that he had ordered "the target will be a purely military one," but the official bombing order issued to General Spaatz on July 25 made no mention of targeting military objectives. The cities of Hiroshima, Kokura, Niigata, and Nagasaki were the targets, although General Groves had advocated the choice of Kyoto because of its tremendous cultural importance.

The bombs were deliberately used on areas inhabited by civilians to provoke the early surrender of Japan through the use of impressive power. The controversy on whether or not bombing was necessary to end the war and on the ethical problems related to bombing cities in such a cruel way has not yet ceased [533]. Supporters of the bombings assert that the Japanese surrender prevented massive casualties on both sides in the planned invasion of Japan. Others who oppose the bombings argue that it was militarily unnecessary and inherently immoral [534]. Even today American people have different opinions on this issue.

Besides the two atomic bombs dropped on Hiroshima and Nagasaki at the end of World War II, 2083 other nuclear weapons have been tested in the atmosphere or underground: US, 1054; Soviet Union/Russia, 715; France, 210; UK, 45; China, 45; India, 6; Pakistan, 6; North Korea, 2. There are still about 20 contaminated islands, as well as one that has entirely disappeared. Eleven bombs have been lost in accidents, and some have never been found. It has been estimated that the 528 nuclear tests in the atmosphere have spread radioactive material that has probably caused about 300000 deaths among the population of the world [465].

Unlike fission weapons, there are no inherent limits on the energy released by thermonuclear weapons based on atomic fusion (sometimes called hydrogen bombs, Section 8.1.5). The first hydrogen bomb was tested on 1952 by the US. The most powerful thermonuclear bomb ever exploded was tested in the atmosphere by the Soviet Union in 1961. Its power reached nearly 50 million tons of classical TNT explosive, that is, the equivalent of 3800 Hiroshima bombs, covering all imaginable, and even unimaginable, needs of military establishments.

Part Four
Renewable Energies

Energy for a Sustainable World: From the Oil Age to a Sun-Powered Future. Nicola Armaroli and Vincenzo Balzani
© 2011 WILEY-VCH Verlag GmbH & Co. KGaA, Weinheim
ISBN: 978-3-527-32540-5

9
Solar Energy Basics

"Scientists are called too see what everyone else has seen
and think what no one else has thought before."

Albert Szent-Györgyi

9.1
The Origin of Sunshine

The Sun is one of the roughly 100 billion radiant bodies of the Milky Way galaxy;
it constitutes 99.86% of the mass of the entire solar system. It is an almost perfectly
spherical star with a diameter of about 1 392 000 km (about 109 Earths). Approximately 75% of the Sun's mass consists of hydrogen, most of the rest is helium,
and less than 2% is accounted for by other elements such as oxygen, carbon, iron,
sulfur, and neon. Due to its temperature (13 600 000 K in the core and 5800 K at
the surface, on average) it is in a plasmatic state and has a density of 1.41 g cm^{-3}.
The Sun is aged about 4.5 billion years and is almost half way to a transformation
that will make it a so-called red giant, namely an agonizing luminous star with
vastly expanded diameter that will swallow the closest planets orbiting around it.
Life on Earth will disappear well before this terminal transformation, that is, in
about 1 billion years, when the Sun's brightness will increase because of a shortage
of hydrogen supply which will lead to core contraction and heating of outer layers.
The extra solar energy input will evaporate the Earth's oceans, devastating the
entire planet and likely letting survive only minor forms of life. Taking into
account that the history of complex civilizations has lasted so far only 5000 years,
this horrific perspective sounds like a minor concern. A scheme representing the
structure of the Sun is depicted in Figure 9.1.

In the late 1930s it was realized that the Sun's energy is generated within the
core by nuclear reactions, the most important one involving four protons which
combine to form a nucleus of helium. This process is termed proton–proton chain
and begins with two protons colliding to form a nucleus of deuterium. Upon
fusion, one of the protons turns into a neutron and releases energy in the form
of a positron and a neutrino, then the positron annihilates with an electron and
generates two γ-rays. The deuterium nucleus subsequently combines with another

Energy for a Sustainable World: From the Oil Age to a Sun-Powered Future. Nicola Armaroli and
Vincenzo Balzani
© 2011 WILEY-VCH Verlag GmbH & Co. KGaA, Weinheim
ISBN: 978-3-527-32540-5

proton, releasing a γ-ray and forming a nucleus of ³He. Finally, the ³He nucleus fuses with another ³He to form ⁴He and two free protons, which allows the starting of another proton–proton chain. The process occurs at temperatures in the range 10–14 MK.

The proton–proton chain and other minor nuclear reactions in the Sun's core consume about 4.3 Mt of matter per second and, according to Einstein's equation, deliver about 3.9×10^{26} J s⁻¹ of nuclear energy, that is, about one million times our current global primary energy consumption *per year* (as a further comparison, a 1000 MW coal or nuclear power plant converts 0.130 kg of matter into energy in 1 year). Solar fusion reactions emit high-energy γ-radiation which travels about 1 μm before being absorbed by adjacent matter within the so-called radiative zone (Figure 9.1). This absorption heats the neighboring atoms, which re-emit other photons, which, in turn, travel a short distance before being absorbed again. The process is repeated countless times before a photon is finally emitted to outer space at the Sun's surface. The last 20% of the journey takes place in the so-called convective zone, where the energy is transported primarily by convection rather than radiation. It is estimated that it takes between 10 000 and 170 000 years for the initial γ-ray photon to be worked out as lower energy radiation and emitted through the Sun's surface (photosphere, Figure 9.1). By contrast, a photon takes only about 8 min to travel from the photosphere to the Earth. Once it gets to our planet, its

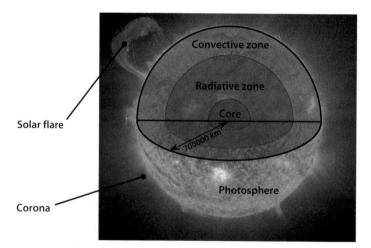

Figure 9.1 Main components of the Sun. The Sun's reactor is the *core*, where thermonuclear fusion occurs. The released electromagnetic energy is then transferred (and reduced in energy) via radiation and convection to its visible external gaseous surface, termed *photosphere*. On the photosphere there are localized cool areas called *sunspots* and, from this layer, *solar flares* composed of gas, electrons, and radiation may also erupt. The overwhelming brilliance of the photosphere hampers the observation of the faint, thin solar atmosphere which can be seen from Earth only during total eclipses. The solar atmosphere consists of two main regions: (i) the *chromosphere*, a 5000 km layer of hydrogen and helium gas, and (ii) the *corona*, a plasmatic atmosphere extending millions of kilometers into outer space. For reasons not yet fully understood, the corona is hundreds of times hotter than the photosphere.

Earth's energy budget

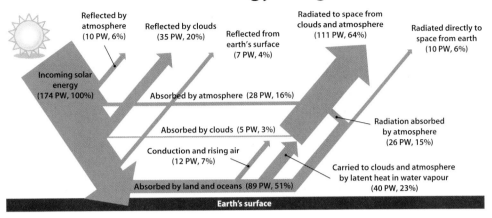

Figure 9.2 The Earth's radiation budget.

unique atmosphere and peculiar sea and land surface work out the incoming radiation according to the scheme described in Figure 9.2. This complex pattern of reflection, absorption, scattering, and re-absorption effects is one of the countless prerequisites that makes possible life on Earth because, among others, it prevents the highest energy radiation from reaching the surface and it warms the planet to a comfortable 15 °C average temperature. As discussed in Chapter 7, the burning of massive amounts of fossil fuels and other related human activities are altering the radiative balance depicted in Figure 9.2.

9.2
Solar Radiation and Attenuation

Assuming isotropic radiation, the power emitted through every square meter of the Sun's photosphere is about 64 MW, a value which undergoes virtually no attenuation as radiation travels through space [51]. Only a tiny amount of this power, approximately 1368 W m^{-2} (~174 PW globally), reaches the outer boundary of the atmosphere that protects the minuscule spaceship Earth. This value represents the so-called *solar constant*, which is defined as the total radiant energy received vertically from the Sun, per unit area per unit of time, just outside the Earth's atmosphere and assuming the planet is at its average distance from the Sun. The solar constant fluctuates by about 6.9% during the year (from 1412 W m^{-2} in early January to 1321 W m^{-2} in early July) due to the Earth's varying distance from the Sun. Short-term variations (days to weeks) can be associated with the appearance of sunspots across the solar disk, whose frequency is subject to an 11-year solar cycle [535] in which the Sun's irradiance varies by about 0.1%. Notably, these minor oscillations of solar output have virtually no effect on the Earth's ongoing climate change [260, 261].

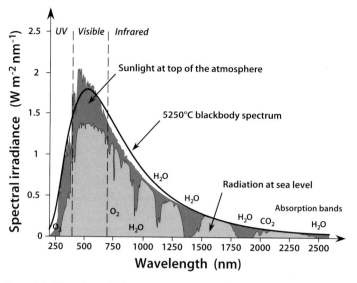

Figure 9.3 The solar radiation spectrum at the external boundary of the atmosphere (AM0, grey) compared to that at sea level (AM1.5, light grey). The former profile matches that of a black body at 5250 °C. Fairly evident is the crucial role of stratospheric ozone in preventing higher energy UV radiation from reaching the biosphere. AM1.5 corresponds to the ASTM G173-03 direct standard (for details, see text).

The wavelengths observed in the Sun's electromagnetic spectrum span over 10 orders of magnitude: from the shortest γ- and X-rays (<0.1 nm) to radio waves longer than 1 m. However, the overwhelming majority of radiative energy reaching our planet is placed in the ultraviolet (UV, ~8%), visible (Vis, ~40%) and infrared (IR, ~52%) regions. The quantity and quality of solar radiation reaching the Earth's surface are rather different from the extraterrestrial flux (Figure 9.3). This is due to several effects, including the reflectivity (*albedo*) of clouds, the absorption and scattering by atmospheric gases and particles, the spherical shape, inclination, rotation and elliptic orbit of the planet. Accordingly, the solar radiation at the Earth's surface displays very complex spatial (i.e. latitudinal), temporal (over days and seasons), and climatic (depending on average extent of cloud cover) patterns. These effects combined reduce the average power density of solar radiation to about one-eighth of the solar constant, that is, $170 \, \mathrm{W \, m^{-2}}$ for oceans and $180 \, \mathrm{W \, m^{-2}}$ for continents. To give an idea of the intricacy of these effects, it is interesting to note that, for instance, the equatorial city of Jakarta in Indonesia receives peak midday summer fluxes of solar radiation similar to subarctic cities like Edmonton in Canada or Irkutsk in Siberia.

Given the relevant role of the atmosphere in attenuating the Earth's incoming electromagnetic flux, a convenient way to define solar radiation is the AM (air-mass) scale, where air-mass generally indicates the optical pathlength of extraterrestrial light through the Earth's atmosphere. AM0 (air-mass zero) indicates solar radiation just outside the Earth's atmosphere and is used to estimate the performance of photovoltaic (PV) cells on orbiting spacecraft. AM1 is the term used to quantify solar radiation that reaches sea level at the zenith (i.e. directly overhead) in a clear sky. A more practical indicator of solar radiation is AM1.5, which corresponds to a solar zenith angle of 48.2°, representative of mid-latitudes. Practically, the number 1.5 means that the length of the path of light through the atmosphere is 1.5 times that of the shortest path when the Sun is directly overhead. The AM1.5 spectrum, which has been standardized and refined many times over time, is almost universally used to characterize terrestrial solar panels. It has to be noted that slightly different solar radiation spectra are currently used: AM1.5G, where G stands for "global" and includes both direct and diffuse radiation, and AM1.5D, which includes direct radiation only. The AM1.5G solar spectrum entails a power flux of about $970\,W\,m^{-2}$, which is usually normalized to give $1000\,W\,m^{-2}$, a quantity sometimes termed "one Sun."

9.3
Abundant, Fairly Distributed, Vital

Integration of the average solar flux of $170\,W\,m^{-2}$ over the entire Earth's surface ($\sim510\,Tm^2$) yields about 90 PW of relentless incoming solar power. Comparison with the rate of primary energy demand of the early twenty-first century, which is about 15 TW, shows that the Sun provides about 6000 times more energy than the global current energy consumption. Therefore, solar energy is not only inexhaustible on any conceivable timescale for human civilization (as pointed out above) but is also widely abundant even in a scenario of doubling or tripling current energy demand.

Of course it has to be considered that a portion of solar energy is indispensable to sustain key natural processes in the biosphere, but these consume a very small part of the Sun's vital input. For instance, only a small fraction of 1% of the solar flux is captured by land and ocean vegetation to power photosynthesis [536]. Curiously, this tiny fraction of a vast number is still vast: the sunlight consumed every day by plants worldwide is comparable to the energy stored in all the world's nuclear weapons put together. This energy is utilized to turn hundreds of millions of tons of carbon dioxide, every single day, into food and living tissue (Chapter 11).

The thermal effect of the large IR portion of the solar radiation is an essential player in the biosphere. Heating IR radiation warms living organisms and their shelters, sustaining metabolic processes. It also triggers atmospheric circulation and winds, which are estimated to utilize about 1% of the solar flux [3]; additionally, solar IR radiation powers the water cycle. Water covers almost 70% of the planet and thus absorbs about 65 PW of solar radiation hitting the Earth's surface. Seawater also absorbs about two-thirds of the IR radiation re-emitted downwards by the atmosphere, thus reaching a huge overall annual heating rate of about 175 PW; about one-fourth of this power sustains the global water cycle [3]. Water, because of its remarkable properties, is the key thermoregulator of our planet by means of, for instance, ocean currents.

After having discounted the energy used by natural phenomena occurring on the planet, we still have a plentiful share of solar energy, some thousands of times current consumption, that can be converted and exploited by human civilization. Definitely, the latest problem with solar energy is abundance.

The distribution of solar radiation on the Earths' surface is now available with great precision [537]. The highest insolation is at low latitudes and over the highest mountain chains (e.g. Himalayas and Andes). Notably, the most populated areas of the world, including Europe and North America, which are at mid-high latitudes, enjoy relatively uniform sunshine. After all, the difference between sunny Rome and cloudy London is a factor of 1.6, fairly similar to the difference between freezing Fairbanks (Alaska) and torrid Albuquerque (New Mexico) [538]. These data have to be compared with the distribution of conventional energy resources (oil, gas, coal, uranium ores), which are extremely concentrated and virtually absent in large areas of the planet, creating increasing tensions among producing and consuming countries. Accordingly, it is hard to dismiss the suggestion that the exploitation of plentiful and free solar energy may constitute a formidable opportunity to initiate an unprecedented and beneficial economic, political, and social change across the planet.

Solar radiation energizes most of the chemical and physical phenomena that shape our planet Earth, mainly via thermal and photochemical processes. There are essentially two exceptions to solar radiation control: the gravitational forces moving huge masses of water in tidal phenomena (Section 12.3) and the reworking of the Earth's ocean floor and continents by geotectonic forces, accompanied by earthquakes and volcanic eruptions [3]. Notably, fossil fuels are a form of buried sunshine [88], hence also powering modern civilization is ultimately carried out by solar energy. Indeed, terming the Sun as an "alternative" energy source is superlatively ridiculous: our existence is totally dependent, each and every instant, on solar energy.

It is worth emphasizing that sunlight, while providing its fundamental energetic services, gives blue to the sky and green to the leaves, generates rain and snow, and enables the recycling of carbon dioxide, the life's exhaust. Sunlight not only makes life, but also the beauty that frames it.

9.4
Sun's Limits: Dilution and Intermittency

As has been emphasized in the previous paragraph, solar energy is overwhelmingly abundant. There is no doubt that, in principle, it may provide the present and future energy supply of the billions of people who populate spaceship Earth. But this consideration alone is too simplistic to define the Sun's potential and draw conclusions about its capability to satisfy, one day, the energy demand of our civilization.

We have pointed out that solar radiation hits the Earth's surface at the rate of about $170\,\mathrm{W\,m^{-2}}$. This value represents the so-called *power density* and this parameter, more generally, is an extremely useful tool to describe energy flows in terms of supply or demand and make straightforward and useful comparisons.

Power densities of final energy uses in modern societies range between 20 and $100\,W\,m^{-2}$ for houses and low-energy intensity manufacturing buildings. Supermarkets and office buildings use 200–$400\,W\,m^{-2}$, energy intensive industrial activities such as steel mills and refineries require 300–$900\,W\,m^{-2}$, whereas high-rise buildings may need up to $3000\,W\,m^{-2}$ [52]. These figures clearly show that it is certainly possible to power a house with the amount of electromagnetic energy intercepted by its roof, but this will never be possible for an oil refinery or a skyscraper. The mismatch between the fixed solar energy flow and the kind of society that we have shaped over the last century, which (hopefully ameliorated) we would like to transmit to future generations is, in essence, the biggest drawback we have to face along the path to a solar energy world.

In order to supply the power densities of 10–$10^3\,W\,m^{-2}$ which are demanded by affluent societies, the current fossil-fueled energy system extracts and produces energy flows (e.g. in power stations) at rates of 10^3–$10^4\,W\,m^{-2}$, that is, up to three orders of magnitude higher; then it dilutes it to the final consumer, for instance through the gas or electric grids (Chapters 5 and 13). If we want to switch to a solar-based society we have to do exactly the opposite, that is, to concentrate dilute energy flows from the harnessing site to the final consumer. In fact, solar renewable energies have power densities ranging between 50–$60\,W\,m^{-2}$ (photovoltaic panels) and 0.1–$1.2\,W\,m^{-2}$ (biomass) [52]. The main strategies to "squeeze" solar energy resources and match required power densities are (i) sunlight concentration and (ii) transformation of the electromagnetic energy into heat or other storable forms of mechanical, potential, electric, or chemical energy [539]. All of these approaches will be described in the following chapters.

Another problem with solar energy is intermittency and intensity fluctuations due to diurnal cycles and atmospheric conditions. This problem has also to be overcome with accumulation strategies but, notably, it is physically less of a concern than the constraint of power density. After all, in fact, we have got some compelling evidence that half of our planet is constantly illuminated by the Sun. Probably tomorrow's energy system, hopefully more advanced than the present one, will take advantage of this granitic certainty.

9.5
The Conversion of Solar Energy: Heat, Fuels, Electricity

In everyday life we use essentially three forms of energy: heat (for cooking, washing, space heating), fuels (typically for transportation), and electricity (well, the list would be too long). Nowadays we primarily get these "usable" energies by burning fossil fuels in suitably designed machines that human ingenuity has manufactured and optimized over the last 150 years: internal combustion engines, turbines, furnaces, and so on. The solar energy pouring over our heads has the same versatility as exhaustible fossil fuels that are stored underground: with

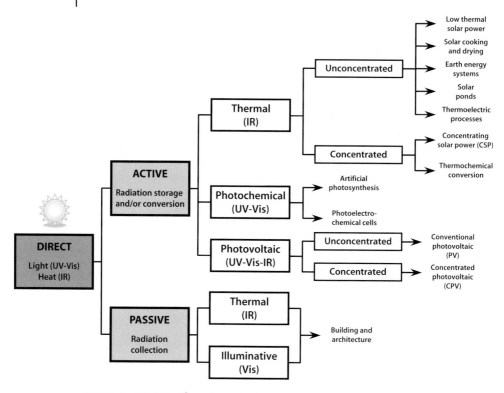

Figure 9.4 Direct transformations of solar energy.

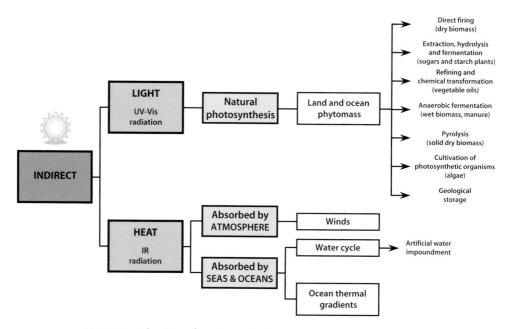

Figure 9.5 Indirect transformations of solar energy.

properly designed technologies we can convert the solar electromagnetic flux into heat, electricity, and fuels (Chapters 10 and 11). The key difference is that solar energy is continuously generated and delivered in real time, whereas fossil fuels were formed over geologic eons and are being consumed at a rate at least 100 000 times the rate they were produced: a trend that is inexorably impossible to perpetuate.

The electromagnetic radiation generated by the Sun can be converted in many ways, directly or indirectly, which are schematized in Figures 9.4 and 9.5. In the following two chapters we will describe the main technologies, already available or under development, that enable the exploitation of the immense potential of solar energy. Their pros and cons will be addressed, along with data on current deployment and projections for future expansion.

10
Solar Heat and Electricity

"I have no doubt that we will be successful
in harnessing the Sun's energy.
If sunbeams were weapons of war,
we would have had solar energy centuries ago."

George Porter

10.1
Passive Solar Harnessing in Buildings

The simplest means of direct solar energy conversion is *passive harnessing*, when heat and light are not converted to other forms of energy by means of mechanical systems, but are just collected. An example of this concept is a greenhouse, in which light and heat are captured to maintain year-round growing conditions for plants.

Passive solar features in building rely on direct heating or convection to collect or move thermal energy carried by the Sun. These design concepts were used for thousands of years by several civilizations all over the world (Egyptians, Romans, and Native Americans, among others) allowing living conditions even in the harshest climates; a valuable cultural heritage completely ignored during the ephemeral decades of fossil fuel addiction, which has given the illusion of an inexhaustible energy supply. Passive solar architectural design is now back on the rise because it provides an effective approach to energy efficiency [540]. It is achievable via thorough site selection, use of suitable construction materials, orientation of buildings towards the Sun, inclusion of architectural features that absorb light where it is needed and exclude it were it is not useful, and placing vents and ducts allowing warm air circulation through buildings.

Passive solar heating techniques belong to three general categories: direct gain, indirect gain, and isolated gain. Direct gain is solar radiation that directly penetrates and is stored in the living space. Indirect gain takes advantage of thermal storage materials which are capable of absorbing, storing, and later releasing significant amounts of heat, then distributed indoors via conduction, radiation, or convection. Isolated gain systems collect solar radiation in an area

Energy for a Sustainable World: From the Oil Age to a Sun-Powered Future. Nicola Armaroli and Vincenzo Balzani
© 2011 WILEY-VCH Verlag GmbH & Co. KGaA, Weinheim
ISBN: 978-3-527-32540-5

that can be selectively closed off from or opened to, when needed, the rest of the house.

Indirect gain is probably the most popular approach, a classical example being the Trombé wall, named after the French inventor F. Trombé. It consists of a south-facing glass window and a blackened wall made of a high-capacity material (e.g. concrete, water in oil drums), with an air space between them. The blackened wall is heated by sunlight passing through the glass; at night it releases the absorbed heat, causing warm air to rise in the empty space and circulate all across the building.

Notably, passive solar design can be used not only for heating, but also for cooling. Appropriate use of outdoor air can often refresh a home without the need for mechanical cooling, especially when effective shading, insulation, window selection, and other means already reduce the cooling load. In many climates, opening windows at night to flush the house with cooler outdoor air and then closing windows and shades by day can greatly reduce the need for artificial cooling. Additionally, the removal of naturally rising warmer air through upper-level fans or openings (stack effect) fosters lower-level openings and admits cooler replacement air. Passive solar design techniques have a higher first cost but are normally less expensive when the lower annual energy expenditures are factored in over the life of the building. They should be designed into the original plan, because retrofitting existing buildings is less effective and quite expensive.

10.2
Thermal Conversion: Unconcentrated Solar Flux

The management of the heat carried by non-concentrated IR solar radiation is usually accomplished by relatively simple devices that rely on a conducting fluid such as water or air [541]. Harvested thermal energy can be exploited not only for heating but for also other purposes such as refrigeration or desalinization [542].

10.2.1
Solar Thermal Panels

10.2.1.1 Collectors

A classic flat solar thermal collector (Figure 10.1) is made of a thin absorbing sheet of polymer or conductive metal (aluminum, copper, steel) to which a radiation-absorbing coating is applied, typically black. Copper pipes, through which a thermal fluid is circulated, wind back and forth through the absorbing collector sheet and, to maximize heat absorption, are often painted black and bonded to the absorbing material. The system is placed in a case with an insulated backing and a cover (glazing) made of glass or polycarbonate to limit heat dispersion. The working fluid removes the heat from the absorber and transport it to an insulated water tank. The fluid can be the water itself to be utilized (open loop system) or an anti-freeze

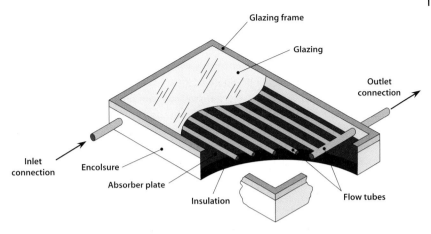

Figure 10.1 The structure of a classic flat solar thermal collector.

liquid (propylene glycol) which, by flowing through a heat exchanger, warms the water (closed loop system). The latter option must be used in colder locations where temperatures may often go below 0 °C and cause breaking of the pipes. Sometimes the circulating fluid is air, to be directly utilized for space heating.

Another type of solar collector is based on evacuated tubes, in order to eliminate heat loss through convection and radiation. They perform better than flat plate collectors in cold climates because they do not rely on the outside temperature. However, when snow accumulates above them, melting is rather slow due to their insulating external cage.

10.2.1.2 Water Management

Solar collectors can be connected to their storage tanks by passive harnessing or a circulating pump. Direct passive systems rely on the thermosyphon effect, that is, expansion and upward movement of the warmer portion of a fluid mass. In passive harnessing systems the fluid is stored in a tank placed at a higher elevation than the collector. Hot water circulation is thus activated spontaneously. Such systems are popular in hot regions such as southern Europe and Israel where flat roofs are the norm and the tank is simply placed on a frame in the open air. The main advantage of these thermosyphon systems is that they do not require electricity to operate. In colder climates, a thermostatic pump moves an antifreeze fluid through the solar collector and then into a heat-exchanger unit where the fluid warms cool water stored into a tank.

Solar and conventional heating can operate in concert, with solar heating providing some temperature gain even during periods of weak insolation. The best designed solar heating systems can also be connected directly to dishwashers and washing machines, where hot water is usually produced by using electricity. Accordingly, solar thermal panels can substantially contribute to reducing residential electricity consumption.

10.2.1.3 Sun Exposure

Whatever the specific design, it is essential to get the maximum possible exposure of the collector to sunlight, by choosing the best inclination for the specific latitude. As a general rule of thumb, tilting of the panel at an angle equal to the local latitude affords an almost perpendicular orientation at midday in March and September and therefore a good compromise solution year-round. Classic flat plate collectors have a life expectancy of over 25 years with minimum maintenance. Evacuated tubes are less robust and have a shorter life expectancy of around 15 years. Energy payback times are estimated to be much shorter than the device life expectancy. For instance, a life-cycle analysis of a passive heating water system in southern Italy (Sicily) is estimated to have, in the most pessimistic scenario, a payback time of 4 years [543].

10.2.2
Current Deployment and Trends of Solar Thermal Panels

Existing solar hot water and heating capacity increased by 20% in 2009 reaching 180 GW$_{th}$, doubling the capacity since 2005 [544]. China installed more than 80% of 2009 global added capacity (22 GW$_{th}$) and remains the world leader (Figure 10.2), with more than 70% of existing global low-temperature solar thermal energy systems.

China started an aggressive policy on solar water heaters in the 1970s but water heating is nowadays still largely accomplished by electricity or by burning coal, with a heavy impact on air quality. Hot water accounted for only 10% of the water consumption by Chinese residents in 2004 (90% in Western countries) and a survey carried out in 2006 revealed that 86% of Chinese housewives had rheuma-

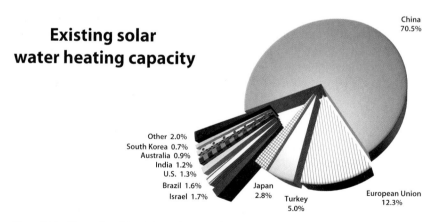

Figure 10.2 Share of world existing solar water/heating capacity, top 10 countries/regions in 2008.

toid arthritis, largely attributable to the use of cold water for laundry and dishwashing [545]. In such a situation, in a country with tens of millions of citizens living in remote areas with limited access to electricity or conventional fuels, solar thermal technology has a wide potential for further expansion. China has a collector area of $150 \, km^2$ (2010) and plans to double it by 2020. It can be roughly estimated that the latter target would avoid 60 Mt of CO_2 emissions per year [545], about 0.1% of China's current annual emissions or the entire CO_2 production of nations such as Portugal or Hungary.

The highest density of solar collectors in the world is located in Cyprus and Israel. In 1980, in the aftermath of the energy crisis of the 1970s, the Israeli legislative assembly passed a law requiring the installation of solar water heaters in all new homes, except high buildings with insufficient roof area. As a result, Israel became a world leader in the use of solar thermal energy and today 85% of households are equipped with solar thermal systems, which provide about 3% of the national primary energy supply.

Europe is the second largest world market for solar thermal energy. Wide expansion potential is available not only in the southern Mediterranean basin, where at present only Greece has reached a good level of deployment, but also at higher latitudes, as shown by Austria and Germany in recent years [546]: in 2008, 200 000 systems (1.5 GW$_{th}$) were installed in Germany, setting a record growth [544]. The EU is committed to reach the so-called "20–20–20 target": a 20% cut in emissions of greenhouse gases by 2020, compared with 1990 levels; a 20% increase in the share of renewables in the energy mix; and a 20% cut in energy consumption. To achieve this ambitious goal, and taking into account that almost 50% of the final energy demand in the EU is heat, a wide expansion of solar thermal energy technologies throughout the continent is indispensable.

Assuming there will be a 9% reduction (relative to the year 2006) of the overall final energy demand in 2020 due to energy efficiency measures, the contribution of solar thermal to the EU's 20% renewable energy target could be between about 2.5 and 6.5%, if an annual growth rate of 15 and 26%, respectively, is consolidated [547]. The most optimistic scenario of a 26% yearly growth rate would lead to the deployment of $388 \, km^2$ of solar thermal collectors by 2020 (272 GW$_{th}$), an area slightly larger than Malta, with $0.8 \, m^2$ per inhabitant, to be compared to the current EU share of $0.04 \, m^2$ per inhabitant. This achievement would create nearly half a million new jobs across the continent and avoid the emission of about 70 Mt of CO_2 per year [547].

In North America, swimming pool heating is the dominant application of solar thermal energy, with an installed capacity of 13 GW$_{th}$ of unglazed plastic collectors [548]. In June 1979, President Jimmy Carter proposed a "new solar strategy" for the US, in order to "move our Nation toward true energy security and abundant, readily available energy supplies." In an effort to set an example for the country, Carter installed solar panels on the roof of the White House, which were used to heat water for the staff mess and other areas of the presidential headquarters. Behind that famous symbolic event the federal administration proposed a package of research and development funding, tax incentives, and loans. Afterwards, between 1981 and 1986, the Reagan administration cut by 90% funding for renewable energy research as well as public renewable energy tax credits. President Reagan took the solar panels down from the White House in 1986. Nowadays the US accounts for a miserable 1.3% of the world installed solar flat panel and evacuated tubes collectors, corresponding to a capacity of 1.9 GW_{th}, that is, about ten and three times less than in the EU and Japan, respectively [548]. Perhaps the time has finally come to assess which was the wiser and more far-sighted energy policy between the Carter and Reagan administrations.

Solar thermal hot water capacity is on the rise in developing countries such as Brazil, India, Morocco, and Tunisia [544]; Turkey, with 75 GW_{th} of installed capacity, ranks second in the world after China [548]. As far as market penetration is concerned, measured as kW_{th} capacity in operation per 1000 inhabitants, the leading countries are Cyprus (527), Israel (371), Austria (285), Greece (253), Barbados (201), Jordan (102), Turkey (98), Germany (88), Australia (67), and China (66) [548].

Solar thermal energy exploitation is not limited to small residential applications, since there are about 150 large-scale plants (>500 m^2 or 350 kW_{th}) in operation in Europe with a total installed capacity of 0.16 GW_{th}. The biggest plants for solar assisted district heating are located in Denmark with 13 MW_{th} (18 300 m^2) and Sweden with 7 MW_{th} (10 000 m^2). The biggest reported solar thermal system for providing industrial process heat was installed in 2007 in China, where a 9 MW_{th} (13 000 m^2) plant generates heat for a textile company [548].

10.2.3
Earth Energy Systems (EES)

As shown in Figure 9.3, about half of the solar radiation is absorbed by land and oceans; therefore, this represents the largest energy input received by our planet. The exploitation of this abundant resource is made possible by means of the so-called Earth Energy Systems (EESs), which are essentially a type of heat pump, that is, a device circulating a fluid which exchanges heat from a lower temperature body to a higher temperature reservoir or vice versa, thus providing space heating or cooling. In general, there are two main types of heat pumps, namely compression and absorption. Compression systems always operate on mechanical energy through electricity, while absorption heat pumps may also run on heat as an energy source, through either electricity or fuels.

Soil is a bad heat conductor and, a few meters below ground, its temperature is kept between about 5 and 15 °C year-round (depending on latitude), as well known by the owners of good wine cellars. These values are higher (cold season) and lower (hot season) than the surface temperature; accordingly, soil is used by EES heat pumps as a heat reservoir in winter and as a heat dump in summer. The overall result is year-round space conditioning at greater efficiencies than those of conventional systems.

Typically, in domestic air-conditioning systems, the heat removed from the house is discharged into the outdoor air, which is of course at a higher temperature than the air inside the building. Such an uphill (i.e. higher temperature) heat transfer is more energy demanding than dumping heat into the cold surrounding ground at lower temperature, although the insulating character of the ground requires the dispersive coils and piping to be much more extended than those that transfer heat into the air. These can be spread horizontally a couple of meters deep across an area of several tens of square meters or, when space is limited, can be placed vertically down to 60 m deep [549]. A ground-heat exchange system can provide a substantial gain in energy efficiency compared to conventional systems with reductions in energy consumption of 30–70% in the heating mode and 20–95% in the cooling mode, but its installation cost is 2–3 times higher. Good performance is possible even with extreme temperatures in the winter or summer because the components of the system are not exposed to the outdoor conditions.

EESs are known under various names; the most common is probably geothermal heat pumps, but this name can be misinterpreted because, at shallow depth, geothermally generated heat contributes to the overall ground heat to only a minor extent, the overwhelming contribution being of solar origin.

The earliest EES applications date back to 1912, when the first patent for a system using a ground loop was recorded in Switzerland. However, it was not until the 1970s that EESs gained significant market acceptance. By the mid-1980s, advances in heat pump efficiencies and operating ranges, combined with better materials for ground loops, allowed ground-coupled heat pumps to enter the market. Currently, tens of thousands of EES heat pump units are sold each year, mainly in North America and Europe, in a wide variety of configurations and sizes that use the ground, ground water or surface water as a heat source and sink. The diffusion of this technology is currently on the rise also in China [550].

The largest EES system currently operated in the world draws water from Lake Ontario through tubes extending 5 km into the lake, reaching to a depth of 83 m. It is part of an integrated cooling system that covers Toronto's financial district and has a cooling power of 207 MW. It has enough capacity to cool 3 200 000 m^2 of office space and is particularly simple, since it uses the water itself as heat-exchange fluid. Notably, the cold water drawn from Lake Ontario is not returned directly to the lake after being run through the heat-exchange system, but is utilized to meet the city's domestic water needs, avoiding heat pollution of the lake.

10.2.4
Solar Thermoelectrics

The thermoelectric effect is the direct conversion of temperature differences into electric voltage. The first experimental observations attributable to this effect were made almost two centuries ago, but only in the last 20 years have thermoelectric materials with a potential for practical applications been made available [551, 552].

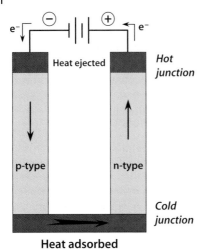

Figure 10.3 Scheme of functioning of a thermoelectric device.

A simple thermoelectric cell (Figure 10.3) consists of two semiconductors (p-type and n-type) joined together in two places (the junctions). One of the junctions is then heated (or cooled) with respect to the other. As the temperatures at the junctions begin to differ, a current starts to flow through the circuit. The strength of the current depends on the kinds of materials used as conductors and on the temperature difference between the two junctions. In order to increase heat-to-electricity conversion, materials with minimized thermal conductivity and maximized electrical conductivity are needed [551]. Another crucial feature is the intrinsic thermoelectric power, also termed thermopower or Seebeck coefficient, which is a measure of the magnitude of an induced thermoelectric voltage in response to a temperature difference across a given material, expressed in volts per kelvin. At present, the most promising thermoelectric materials are nanostructured composites of metal and semiconducting elements such as lead, antimony, bismuth, silver, tellurium, and selenium [553–555]. However, their high thermopower has been proved only at the nanoscale and typically at very high temperatures (hundreds of kelvin); a lot of research is still needed to upscale such tiny devices into the bulk materials needed for large-scale applications [556].

The successful development of cheap thermoelectric materials with conversion efficiencies exceeding 30% [556] would be a major breakthrough in energy transition for at least two reasons. First, they could directly convert the Sun's energy to electricity in a distributed fashion and offer an alternative to photovoltaics [554]. Second, taking into account that over 55% of the primary energy supply is currently wasted as heat without providing any useful service [557], they could offer a formidable tool for energy efficiency. Solar thermoelectric generators have been used so far to supply electric power in orbiting spacecraft especially in near-Sun missions where the performance and reliability of photovoltaic systems cannot be guaranteed due to high temperatures [558].

Notably, thermoelectrics can also operate in reverse, using electricity to cool things down or heat them up, as discovered by the French physicist Jean Peltier in 1834. Nowadays a number of electronic devices that need mild cooling to −30 °C take advantage of the so-called Peltier effect; among them worth mentioning are photodetectors in spectroscopic equipment.

Solar Cookers and Solar Dryers

Solar Cookers

Firewood is still the primary energy source in large areas of the developing world. It is primarily used for domestic cooking and heating and, in the poorest regions, it may account for as much as 90% of the total personal energy supply [559]. In arid regions such as Sahel in Africa, the consumption of firewood exceeds the growth rate of the forests, leading to deforestation and soil degradation. In the urban agglomerates of this region, the distance from harvesting areas can be of several kilometers, forcing people, particularly women and children, to hours of marching in remote and unsafe areas to collect energy resources free of charge [428]; in fact, for many people, buying firewood on the market is simply unaffordable. Additionally, indoor biomass burning in improperly vented cooking stoves and fireplaces has dramatic health consequences, with hundred of thousands of deaths annually due to pulmonary diseases [429], as pointed out in Section 7.7.

Though it might seem naive for affluent Westerners, food can be conveniently cooked using solar radiation as fuel, by taking advantage of devices called solar cookers or ovens. China started distributing subsidized cookers in rural areas in 1981 and, coming to more recent years, both governmental and non-governmental organizations (NGOs) in North America and Europe have conducted solar cooking promotions in the developing world [560]. There are several types of collectors which can bake, braise, stew, and fry food based on flat plate or parabolic collectors [559].

At least 85% of the countries in Africa, the Americas, and Oceania are highly appropriate for solar cooking promotion efforts. A list of 20 countries with the highest potential for solar cooking has recently been compiled, the top five being India, China, Pakistan, Ethiopia, and Nigeria; criteria for this ranking include annual average sunlight, cooking fuel scarcity, and population size [561]. Of the estimated 500 million people who have abundant sunshine and suffer from fuel scarcity, 85% of them live in just 10 countries.

Solar cooking holds the promise of alleviating environmental, economic, and health problems associated with cooking by means of biomass (wood, shrubs, dung, charcoal) in the developing world but, so far, efforts to promote the practice have fallen far short of their potential [560]. According to the experience gained in South Africa, four issues have been the strongest obstacles to wider deployment: lack of financial mechanisms (e.g. microcredit) to enable purchasing by very poor households; difficulty in spreading information and confidence, as most people simply do not believe that a solar cooker can actually cook; poor commercial distribution; and difficulty in delivery and reliable assistance in remote areas [562].

Solar Dryers for Food Conservation

Many fruit and vegetables contain a large quantity of moisture and are therefore highly susceptible to rapid quality degradation and then spoilage. In the affluent world, if not delivered right away, these products are promptly stored in thermally controlled storage facilities, but this is simply unaffordable in the poorest areas of the world where fruit and vegetables constitute the dietary base for hundreds of millions of people. For these populations it is vital to convert these perishable products into more stabilized and safe foodstuffs that can be kept in an easily controlled environment for an extended period of time, also alleviating problems in case of prolonged supply shortages.

Sun drying under the open sky for preserving food and crops, practiced since ancient times, has many disadvantages because products can be spoiled by atmospheric agents, eaten by animals, or deteriorate due to the action of insects or fungi. Additionally, the process is labor intensive, time consuming, and requires a large area for spreading the product. Solar thermal systems are now available to dry food in a practical and economical fashion. Such systems can broadly be categorized into direct, indirect, and specialized solar dryers, designed with a specific product in mind, and may include hybrid systems where other forms of energy may also be used [563].

The deployment of solar drying facilities requires careful understanding of the local situation into which the technology is to be transferred, including lifestyles of the people and their ability for technology uptake. It is often advisable to set up large centralized facilities run by trained individuals rather than provide solar dryers to single households [564].

10.3
Thermal Conversion: Concentrated Solar Flux

10.3.1
Concentrating Solar Power (CSP)

Low-temperature heat is very useful, but has an intrinsic limit of power: to scale up solar thermal energy at an industrial level and produce high-temperature fluids for electricity generation or other uses, the unsurpassable Sun's dilution limit has to be overcome (Chapter 9). This can only be accomplished by concentrating sunlight with the help of mirrors or lenses, so as to obtain temperatures of at least 300 °C.

Concentrating solar power (CSP) has a long tradition which, in the Western world, dates back to at least the third century BC when the Greek mathematician

Archimedes allegedly torched a fleet of Roman ships that had attacked the Greek colony of Syracuse, Sicily, by reflecting the sun's rays with a mirror device made of glass or bronze. This is most likely a myth [565] but this famous tale shows how long the exploitation of concentrated sunshine has intrigued and fascinated humans. For sure, in antiquity, rudimentary solar concentrating mirrors were utilized on a smaller scale by the Greeks, Romans, and Chinese to light torches for religious purposes. In 1866, the French inventor Auguste Mouchout used a parabolic trough to produce steam for the first solar steam engine, but the first patent for a concentrating solar collector was obtained by Alessandro Battaglia in Genoa, Italy, in 1886. In the same city, the engineer and mathematician Giovanni Francia (1911–1980), a pioneer of solar energy technologies, designed and built the first solar concentrating plant which began operation in Sant'Ilario, near Genoa, in 1965 with a power output of almost 1 MW.

CSP can be operated economically on a large scale only within the huge sunbelt which extends between the 35th northern and 35th southern latitudes, approximately the whole north–south extension of the African continent. It is a fast-growing solar technology which exploits several light harvesting concepts. Their common feature is the gathering of sunlight over a wide area, to be concentrated on a much smaller absorbing surface. The ratio of collecting to absorbing surface is termed the concentration ratio (CR); current systems are characterized by a CR between about 10 and 10 000, the higher value being obtainable only with proto-typical double light concentration systems called "solar tower reflectors" [566, 567]. As far as electricity production is concerned, CSP uses exactly the same technology as conventional fossil fuel or nuclear power plants (see Chapter 13): they simply replace the steam-producing heat source with the Sun. Overnight or on cloudy days, CSP facilities either may switch to fossil fuel combustion or use the heat which had been accumulated in suitable heat storage systems during periods of intense sunshine [568].

The main environmental concern of CSP systems producing steam to generate electricity is the need for cooling water (except for dish-Stirling systems, Section 10.3.4), which is scarcely available in the arid areas where these facilities are typically located; some water is also needed for periodic cleaning of reflective surfaces. In addition, there may be some pollution of water resources through thermal discharges or accidental release of plant chemicals such as heating fluids (synthetic oils, salts), although the latter is a remote possibility and can be avoided by good operating practice [569].

The CSP industry is growing rapidly with over 1 GW under construction, another 14 GW announced globally through 2014 (Figure 10.4) [570], and a potential global deployment of 25 GW by 2020 [571]. Spain and the US are by far the leading countries, but the market is expected to expand in North Africa, the Middle East, India, and China [572]. CSP is envisaged as a key technology to provide the huge amounts of energy needed to carry out seawater desalination, a looming perspective in many areas of the world under increasing freshwater stress [573, 574].

CSP entails four main different concepts for generating electricity and one for generating fuels, that are illustrated below.

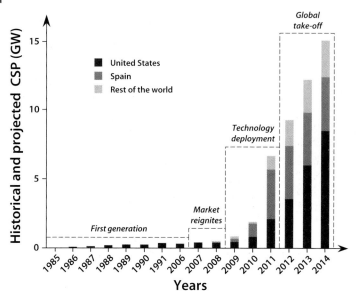

Figure 10.4 Historical and projected growth of CSP through 2014. Rest of the world includes Dubai, Australia, Libya, Qatar, Chile, India, Morocco, Algeria, Tunisia, South Africa, Italy, China, Egypt, France, Portugal, Iran, South Korea, and Israel.

10.3.2
Parabolic Trough Collectors

This approach involves an extended trough-like parabolic mirror which focuses sunlight along a pipe, suspended above the mirror, through which a heat-transfer fluid is circulated (water, synthetic oils, or molten salts) (Figure 10.5). During the day, the system is moved along one axis only, tracing the Sun while it travels across the sky. Initially 50–100 m long solar collectors were constructed, but now lighter and stiffer designs allow 150 m long collectors to be made with an opening width of almost 6 m and a total surface area exceeding 800 m² for each single unit [575]. The mirror surface is made of special thick glass which is almost 95% reflective (regular mirrors are only 75% reflective) while the absorber pipe is made of steel and covered with an optically selective surface coating to allow optimal absorption of the solar spectrum with minimal re-radiation of lower wavelength light. To reduce heat losses further the steel pipe is encased in an evacuated glass tube. The greatest source of mirror breakage is wind, but operators can turn the mirrors to protect them in case of intense wind storms. An automated washing mechanism is used to clean the parabolic reflective panels periodically, in order to keep sunlight harvesting at optimal levels. The concentrating ratio of parabolic trough systems is in the range 30–100.

The largest solar installation of any kind in the world uses this technology and is called the Solar Energy Generating Systems (SEGS). It is a cluster of nine solar

Figure 10.5 An array of parabolic trough solar concentrators.

power plants located in California's Mojave Desert, with a peak output of 354 MW thanks to about 2 km² of collectors (approximately 250 football fields) over an area of 6.5 km² [575]. This surface is certainly very large compared to a conventional power station; however, when considered in a life-cycle perspective, renewable technologies often turn out to be less land demanding than fossil fuel or nuclear technologies [576]. This is not surprising because, for instance, solar technologies do not disturb land by extracting and transporting fuel to the power plants and eliminate the necessity for reclaiming land for mining or for the disposal of waste products, as happens, for instance, with coal or nuclear installations (see Chapters 7 and 8). The average gross solar output for the nine plants at SEGS is around 75 MW, corresponding to a capacity factor of 21%. The turbines are allowed to burn natural gas up to 25% of their thermal input but, in a working experience of over two decades, only 10% of electricity production has been obtained from the fossil resource. The heating fluid is synthetic oil that is brought to over 400 °C.

Currently there are three other large operating CSP power stations in the world: in Boulder City (Nevada, USA, 64 MW), near Ciudad Real (Castilla-La Mancha, Spain, 50 MW), and in Guadix (Andalucia, Spain, 100 MW). The last facility is equipped with a heat storage system in which a molten mixture of 60% sodium nitrate and 40% potassium nitrate absorbs part of the heat produced in the solar field during the day. A turbine produces electricity using this heat during dark hours or when the sky is overcast; this almost doubles the number of operational hours at the solar thermal power plant during the year [577].

Notably, a new type of collector has been developed in which flat mirror strips (linear Fresnel reflectors) are used instead of curved mirrors [568, 578]. They are less efficient compared to conventional troughs but, due to the lower cost of flat mirrors, they may abate the initial investment cost.

10.3.3
Solar Towers

Stronger solar concentration compared to trough concentrators is achieved by an array of two-axis Sun-tracking mirrors (heliostats) arranged to focus sunlight onto the same central point on the top of a "power tower" where an absorbing fluid is placed. Concentrating ratios in the range 300–1500 can be obtained, reaching temperatures exceeding 1500 °C [579]. After 20 years of studies on prototypes, however, it has proven too complicated to control superheated steam fluxes under conditions of fluctuating solar power, thus the recently erected commercial solar tower systems use steam at moderated temperature and pressure [575].

After the failure of two large demonstration projects in southern California during the 1980s and 1990s (Solar One and Solar Two) [580, 581], the first solar tower power plant in the world to generate electricity in a commercial way (PS10) started operation in 2007 in southwestern Spain, near Seville. The solar tower is 115 m high and uses water as operating fluid to create steam at 260 °C and 40 atm, that generates up to 10 MW of electricity, with a conversion efficiency of solar energy of about 16%. The surrounding field contains 624 heliostats, each with a 120 m^2 reflecting surface. Right next to PS10, an even larger facility (PS20) went on-line in early 2009 which, by means of 1255 heliostats, produces 20 MW of electricity, Figure 10.6. Another plant is under construction in the same Spanish region that will use molten salt as the working fluid, as initially tested in the abandoned Californian facility mentioned above. This will permit a daily thermal storage of up to 15 h of turbine load, compared to just 30 min for PS10 and PS20 that are operated with water [571]. During the year, facilities operated with molten salt can operate about 70% of the time, compared to 30% for conventional plants.

Figure 10.6 The PS10 and PS20 CSP solar tower facilities in Andalucia, southern Spain.

Figure 10.7 A concentrating solar dish collector with a Stirling engine at the focal point.

10.3.4
Dish Collectors with Stirling Engines

A third, less mature, approach to exploit concentrated solar radiation entails a radial concentrator dish structure that supports an array of curved glass mirror facets, designed to track the Sun automatically and focus on a single point where a Stirling engine is placed [568, 582], Figure 10.7. Concentrating ratios in the range 1000–5000, that is, even higher than those of solar towers, can be obtained.

Stirling engines are highly efficient (up to 30% heat conversion into useful work) and do not require water for cooling, but they are expensive. Development projects for several hundred megawatts of CSP electricity by Stirling dish collectors are at an advanced stage in Arizona [583]. This technology is also an interesting option for stand-alone off-grid power generation in remote areas, with the added value of working with other fuels if needed (unlike photovoltaic systems) [575].

10.3.5
Solar Updraft Towers (Chimneys)

A solar updraft tower (or solar chimney) is a large facility that might convert solar radiation into electricity by combining the greenhouse effect with the chimney effect [584, 585]. A few prototypes have been tested, mainly in Spain and Australia, but this concept is not yet ready for commercial applications. Solar radiation heats air under a low circular transparent or translucent roof open at the periphery. Towards its center, the roof curves upwards to join a wide and tall chimney, creating a funnel: the heated air underneath the collector roof is conveyed to the

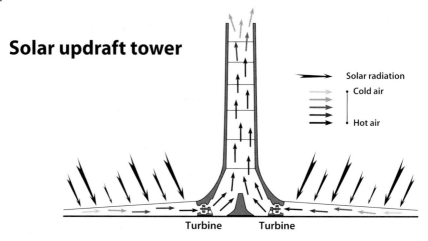

Figure 10.8 Working principle of a solar updraft tower (or solar chimney).

chimney, whereby it is expelled after having powered multiple wind turbines at its bottom Figure 10.8. Suction from the tower then draws in more hot air from the collector, while cold air comes in from the outer perimeter. The ground under the collector roof is the heat storage medium and hot air can thus be delivered for an extended time after sunset [586].

The efficiency of solar chimney power plants is poor (<2%) and depends mainly on the height of the tower; furthermore, large plants require huge areas that in principle might be utilized as a greenhouse for agricultural purposes [584].

10.3.6
Cost Considerations and Carbon Footprint of Solar Thermal Electricity

The cost of electricity generation with concentrated solar energy has substantially decreased since the 1980s. SEGS, the large commercial facility located in southern California (Section 10.3.2), initially produced at about $0.30 kWh^{-1}, but now the cost is $0.12–0.15 kWh^{-1} [575]. Of course, profitability strongly depends on local insolation: in this regard the Mojave Desert is an ideal place (even better than southern Spain) and, moreover, is relatively close to one of the largest metropolitan areas in the world.

Assuming the same irradiation conditions and comparing with the best sites for wind power generation, solar thermal electricity on the tens of megawatts power scale is roughly twice as expensive as wind energy and almost comparable in price to photovoltaics [575], which underwent a substantial cost reduction after 2008 [587]. Obviously, in order to be competitive in the electricity markets, solar thermal electricity has to be subsidized with tax incentives or feed-in tariffs (FITs). The latter guarantee long-term contracts (typically 15–25 years) to purchase electricity generated from renewable energy sources, at a price higher than that produced with conventional sources [588]. Nowadays feed-in tariff policies are enacted in tens of countries around the world, after having being introduced in Germany in the 1990s [589]. This regime has turned out to be very effective for promoting renewable electricity worldwide and has been crucial to sustaining the boom of solar thermal electricity in Spain [590].

It has to be emphasized that the amount of public money which is being mobilized to promote these and other renewable technologies, although often criticized by uniformed people, is minimal compared to the huge flow of money that has been poured into the fossil fuel and nuclear energy sectors in past decades in terms of tax breaks, tax incentives, and many other instruments. In the EU about US$30 billion was spent in 2005 on fossil fuel subsidies and, if environmental and health externalities were included (see Chapter 7), the total subsidy is estimated to be in the range $65–100 billion [591]. The US government provided $72 billion in subsidies to the fossil fuel industry between 2002 and 2008 [592], whereas worldwide, the fossil fuel subsidies currently amount to about $330 billion per year [593], with producing countries particularly munificent with their citizens. In Germany, despite its generous feed-in tariff to foster renewable energy, the coal-mining industry is still subsidized annually with about €2 billion [594]. In the US, nuclear energy continues to receive the largest proportion of tax breaks (76%), followed by coal at 12%, with renewables trailing at 3%. At the meeting held in Pittsburgh in 2009, the G20 leaders took a generic commitment to phase out fossil fuel subsidies [592]. Historically, from 1948 until 2000, 59% of US energy R&D expenditure was on nuclear energy, 23% on fossil fuels, 11% on renewables and 7% on energy efficiency [591].

Finally, it is important to point out that solar thermal technologies are also particularly effective in terms of CO_2 emissions because life-cycle analysis (LCA) suggests that CSP-generated electricity produces about $\sim 10\,g\,CO_2\,kWh^{-1}$ [516], a value comparable to wind energy and substantially lower than hydroelectric ($\sim 20\,g$) and photovoltaics ($20–60\,g$) [516].

10.3.7
Solar Thermochemical Conversion

All of the solar technologies presented so far are concerned with the production of heat or electricity but, as pointed out elsewhere in this book, we cannot conceive a realistic energy transition without implementing innovative approaches to the production of storable energy in the form of fuels.

Biomass can be an alternative feedstock to fossil resources for making syngas and then producing chemicals and fuels [595]. This can be accomplished via thermochemical conversion, which entails a variety of processes such as gasification, liquefaction, hydrogenation, and pyrolysis (Section 11.3). Pyrolysis has received special attention because it can convert directly biomass into chemicals and solid, liquid, and gaseous fuels, via thermal decomposition in the absence of oxygen [596].

Biomass thermochemical processes are particularly attractive for regions with vastly available agricultural by-products, but so far they have been powered by burning fossil fuels or the biomass itself, limiting attractiveness in terms of

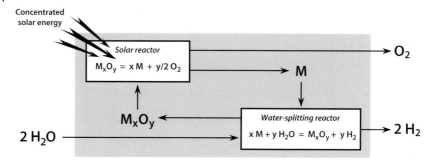

Figure 10.9 Scheme of a water splitting thermochemical reactor using metal oxides.

sustainability. The sector, however, is now on the verge of a breakthrough since the first commercial plant fueled by concentrated solar power has been announced recently in the US [597]. The 60 MW facility will use an array of 2799 mirrors to concentrate sunlight on a 20 m tall tower where solar heat will drive a chemical reactor producing syngas from biomass, to be then converted into 19 million liters of gasoline annually.

The next goal for solar thermochemical conversion is to obtain syngas by replacing the biomass feedstock with H_2O and CO_2 [598]. This, unfortunately, is extremely complicated because the splitting of H_2O and CO_2 requires specialized catalysts that must be transferred between two separate stages to carry out different reactions (see below). Thermochemical water and carbon dioxide splitting using concentrated solar light are being pursued with a variety of approaches [556, 566] and one of the most promising is illustrated in Figure 10.9. This thermochemical cycle is composed of two steps. The first is an endothermic "solar" step, in which a metal oxide M_xO_y is thermally decomposed into the metal M and oxygen, taking advantage of concentrated solar radiation that heats the reactor at around 1500 °C. In the second exothermic dark step, the metal M reacts with water to produce hydrogen, while the resulting metal oxide is recycled back to the first step. The net reaction is $2H_2O \rightarrow 2H_2 + O_2$. A key feature of this process is that H_2 and O_2 are produced in different compartments, eliminating the need for gas separation that would be technically complex and energy demanding.

These cycles have been examined thermodynamically and tested in solar reactors for Fe_3O_4/FeO and ZnO/Zn redox pairs. Other pairs, such as TiO_2/TiO_x, Mn_3O_4/MnO, and Co_3O_4/CoO, were also considered, but the yield of the hydrogen producing step was too low to be of any practical interest [599]. A very similar approach is feasible by using CO_2 instead of H_2O, thus generating CO, the other component of the syngas mixture to be utilized for making liquid fuels and chemicals [598].

The peculiarities of concentrated solar radiation, in particular very high thermal flux and temporal fluctuation of the solar input, make these processes much more complicated to run than in conventional reactors. A chemical process is operated most efficiently at constant temperature without interruption; therefore, solar thermochemical reactors should work with a thermal storage system. However, keeping temperatures above 1200 °C for hours is technically very challenging. Another issue is the behavior and stability of the metal oxides at high temperatures: currently iron, zinc, and cerium materials are used but none of them has turned out to be a silver bullet [598].

The best reported efficiency in converting sunlight into fuel via solar thermochemical conversion is 3%, with a theoretical maximum Carnot efficiency of 75% [598]. A reasonable goal is to achieve 20% efficiency by 2020 at a cost of less than $2.5 per liter of liquid fuel produced. This would likely start to be competitive, at that time, with fossil combustibles.

10.4
The Birth and Rise of Photovoltaics

The very beginning of photovoltaics (PV) dates back to 1839 when the French physicist Edmond Becquerel observed that a voltage and a current were produced when a silver chloride electrode, immersed in an electrolytic solution and connected to a metal counter electrode, was illuminated with white light [600]. In 1876 William Adams and Richard Day published the first scientific paper based on the photovoltaic effect [601] and the first true solar cell was constructed around 1883 by Charles Fritts, who used junctions formed by coating selenium with a thin layer of gold [602].

In the first half of the twentieth century some progress was made in the US and the Soviet Union, but it was not until the 1950s that practical solar cells were manufactured [603]. The year of birth of modern photovoltaic technology is 1954, when Daryl Chapin, Calvin Fuller, and Gerald Pearson at Bell Telephone Laboratories in the US developed the silicon PV cell, the first device capable of converting enough of the Sun's energy into power, so as to run some electrical equipment [604]. They produced a silicon solar cell with 4% efficiency, then achieved 11% in 1957 and reached 14% by 1960 [605].

The boom of space exploration in the 1950s and 1960s was extremely beneficial to PV technology, as it is the ideal tool to provide reliable and long-lasting energy supply in the space environment. On March 17, 1958 the first space solar array was carried aloft on the US satellite Vanguard I. It consisted of six single crystal silicon PV panels, about 10% efficient, mounted on the outer surface of the satellite, which produced about 1 W of power for over 6 years. Since then, space solar arrays and power systems have grown in size and complexity, and are now the primary source of power for all Earth orbiting satellites [606].

In the 1970s, in the aftermath of the first oil crisis, large PV companies were established mainly in the US and Japan, prompting the first golden age of solar energy. In that period of time the first projects to deliver solar PV electricity to remote, off-grid areas were implemented. This was conceived either to benefit underdeveloped countries or to power sophisticated instrumentation for environmental monitoring around the globe. In 1977, the world production of PV modules was around 500 kW, it rose to 9.3 MW in 1982 and was more than twice as much the subsequent year. In the 1980s, although plummeting oil prices ephemerally decreased the attractiveness of solar technologies, important achievements were accomplished: the first amorphous silicon solar module (1984), the first systems with efficiency exceeding 20% (1985), and the first commercial thin film PV module (1986).

In the 1990s big PV companies initiated some cooperative projects and German firms vigorously entered the business; several 4–5 MW PV plants went online in Germany between 2002 and 2004. In 2002 the number of solar PV manufacturers around the world was about 80, then it mushroomed to about 800 at the end of the decade, with the large majority of new companies based in China and Taiwan. In 2005–2007 a silicon shortage occurred that caused a temporary price increase [607], which was followed by a market oversupply that, in 2009, halved the price of solar modules relative to 2008 and the industry revenues plunged by 40% [587]. In the next few years a restructuring of the global PV industry is expected, with the smaller manufacturers probably disappearing or being acquired by larger firms [587].

10.5
Inorganic Photovoltaics: Key Principles

The creation of a voltage (or a corresponding electric current) in a material, upon exposure to electromagnetic radiation, is termed the photovoltaic effect; although it is related to the photoelectric effect, the two processes are different and should be distinguished. In the photoelectric effect electrons are *ejected* from a material's surface upon exposure to radiation of sufficient energy. In the PV effect, instead, electrons are *transferred* from one material to another.

Inorganic PV cells are made of semiconducting materials containing silicon or other combinations of elements such as gallium, arsenic, and tellurium, sometimes thoroughly combined with metals (e.g. copper or cadmium). Semiconducting materials do very little on their own as they are electrically neutral and, even under the action of light, they will never be able to create an electrical potential and move their valence electrons in a fixed direction. The key prerequisite to generate electron flows in semiconducting photovoltaic materials is the creation of an electric potential inside them, which will force all electrons to move in the same direction. This is obtained by means of two strategies, one technically complex, the other quite simple: doping and physical contact.

Silicon, the most widely utilized material in PV cells, has four electrons in its outer shell, which are used to establish covalent chemical bonds with four other silicon atoms in the crystalline lattice, hence there are no "free" electrons. Upon absorption of heat or light of suitable wavelength, some electrons can be pumped from the valence to the conduction band of the crystal, which thus becomes conductive. Doping allows one to exploit in a useful fashion such an externally stimulated conductivity, by providing two modified forms of semiconductor: n-type and p-type, where n stands for negative and p for positive. To produce n-type silicon a tiny number of atoms (around one in a million) are replaced by an atom with five electrons in the outer shell, typically phosphorus. Statistically, this atom will be surrounded by tetravalent silicon atoms only and will thus have one electron spare, which will require much less energy to be released and roam through the crystal. An analogous concept drives the production of p-doped silicon: trivalent boron will be used and this will promote the insurgence of electron "holes" in the silicon lattice. It is worth emphasizing, since it often causes confusion, that n- and p-type materials, as such, are neutral.

To make a solar cell, a slice of n-type material (e.g. phosphorus doped) must be placed in physical contact with a slice of p-type material (e.g. boron doped), so that the "free" electrons in the n-side will rush to fill holes placed in the p-side: at the interface, the initially neutral doping atoms will become positively or negatively ionized. Practically, an electron migration over the p–n junction occurs, which creates a net negative charge on the p-doped material (the side initially tagged "positive" becomes negative!), counterbalanced by an identical positive charge on the n-doped material (Figure 10.10). This spontaneous process creates an electric field that, at a given point, will prevent more electrons from crossing the junction and filling all the holes in the p-side. In essence, a device is obtained that stops electrons flowing in one direction, but lets them rush in the other direction (a diode). For silicon diodes, the built-in potential is approximately 0.6 V.

When incoming solar light frees an electron of a silicon atom in the vicinity of the junction, the negative particle will be attracted on the positive side of the field, flowing through the junction towards the n-type material. By connecting together the two sides of the cell, *away from the junction*, electrons will be forced to travel outside the cell, generating an electric current from the n-side to the p-side, from which useful work can be extracted.

The conversion efficiency of a practical solar cell is primarily related to the width of the specific bandgap. Photons carrying energy below the bandgap of the cell material are totally wasted: they may either pass through the cell or be converted into heat within. Photons above the bandgap fruitfully utilize only a portion of their energy to release electrons, the remainder also being lost as heat (Shockley and Queisser limit [608]). This physical limit lowers the theoretical maximum efficiency of crystalline silicon solar cells to about 30%.

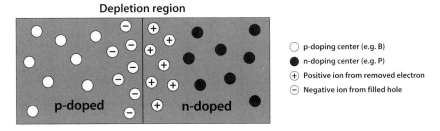

Figure 10.10 The core a photovoltaic cell: the built-in electric field at a p–n junction in the so-called "depletion region." The density of doping centers in real devices is much smaller than schematized here; there is a doping atom every about one million silicon atoms.

The performance benchmark of solar cells is the power conversion efficiency η, which is defined as

$$\eta = V_{oc} \times J_{sc} \times FF/P_{in}$$

where V_{oc} is the open circuit voltage, that is, the maximum possible voltage across the cell under sunlight when no current is flowing through the external circuit, J_{sc} is the short-circuit current, that is, the maximum possible current flowing freely through the external circuit without load or resistance, FF is the material fill factor, that is, the ratio of the actual power of the PV cell to its theoretical power if both current and voltage were at their maxima, and P_{in} is the incident solar power integrated over all the available wavelengths, generally fixed at $1000 \, \text{Wm}^{-2}$ when solar simulators are used. Small bandgap materials absorb a broader part of the solar spectrum but at the expense of lowering the V_{oc}; on the other hand, large bandgap semiconductors sacrifice the low-energy wavelengths in the red part of the spectrum, thereby lowering J_{sc}. All in all, for a single junction PV cell, there is an optimal bandgap of ~1.4 eV that yields a maximum η of 33%, as set forth by Shockley and Queisser [608].

One of the most promising strategies to overcome the above limits and exploit the solar energy flow with higher efficiency is the use of the so-called multi-junction (or tandem) cells, in which two or more cell junctions are combined, each of which has a bandgap optimized for a particular part of the solar spectrum [556].

There exist various solar cells made of different types of materials, but all are similarly designed to generate the movement of electrons and produce useful electricity. Cells connected in series have a higher voltage, while those connected in parallel produce more electric current. In order to make the appropriate voltages and outputs available for different applications, single solar cells are interconnected to form larger units, called *modules*. These can rarely provide enough power for a home or a business, so the modules are linked together to form an *array*; in smaller systems, an array can consist of a single module. Solar arrays are usually embedded in transparent ethyl vinyl acetate, fitted with an aluminum or stainless-steel frame and covered with transparent glass on the front side.

PV arrays produce direct current (DC) that, to be plugged into the grid or to be directly utilized by appliances, has to be converted into alternating current (AC) by means of a device termed an inverter. Solar arrays are ranked by the peak electrical power they produce, measured in watts and denoted W_p (watt peak). The peak output of a solar module is obtained under standardized conditions of $1000 \, \text{W m}^{-2}$ of sunlight (AM1.5) at 20 °C.

10.6
Silicon Solar Cells

Although the efficiency of practical solar cells has been improving steadily since their first appearance in 1954, over 80% of today's commercial solar cells are still based on the same material and basic concepts developed at the Bell Laboratories in the early 1950s: light-induced charge separation at the p–n junction between two slices (wafers) of doped silicon in either single-crystal silicon (sc-Si) or poly-crystalline silicon (poly-Si) form. Below we provide an overview of the fundamentals of this technology.

10.6.1
Manufacturing of Poly- and Single-crystalline Silicon Cells

The starting material to fabricate silicon solar cells is quartz (SiO_2), a fairly abundant mineral in the Earth's crust. However, the fabrication of solar modules containing Si wafers of sufficient purity for solar energy conversion is a complex and articulated manufacturing chain, shown schematically in Figure 10.11 [609].

In brief, the quartz is first reduced with coke to obtain metallurgic-grade silicon (MG-Si) with a purity of 99%. MG-Si is then reacted with HCl and transformed into trichlorosilane ($SiHCl_3$), a very corrosive liquid that can be purified by multiple distillation processes. The purified $SiHCl_3$ is then reduced via the so-called Siemens process by H_2 to produce high-purity poly-Si, that is, 99.999 99 (EG-Si, electronic grade) or "only" 99.9999 (SG-Si, solar grade). These poly-Si materials can be then transformed into sc-Si ingots by means of the Czochralski process, which involves melting followed by thorough recrystallization of the molten silicon over a seed of sc-Si, also in the presence of the needed doping elements. The cylindrical Si ingot is usually cut into a quasi-square shape, resulting in ~25% material loss; the square ingot is then sliced into wafers. The slicing process takes many hours and it is one of the most costly steps in Si wafer production. It also entails a significant material loss as sawdust of up to ~30%. Finally, the Si wafers are polished to remove any damage caused by the sawing step and cleaned for cell fabrication. There are variants to this complex industrial process, such as the use of $SiCl_4$ or SiH_4 as intermediates instead of $SiHCl_3$. In recent years, advances in manufacturing processes have reduced the material loss of the final steps and decreased the overall energy consumption [610, 611].

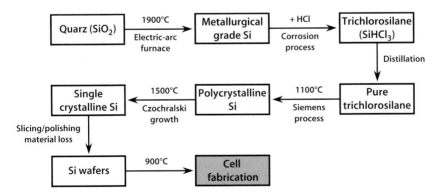

Figure 10.11 Scheme of the industrial process from quartz to single-crystal Si cells.

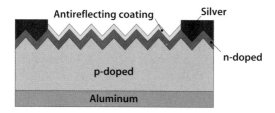

Figure 10.12 Scheme of a section of a crystalline silicon solar cell.

Cell fabrication from doped silicon wafers entails a sequence of complex steps such as chemical etching, high-temperature deposition, diffusion, printing, and annealing. The final product contains the two doped semiconducting materials, an antireflecting coating and the metal contacts on the opposite sides (Figure 10.12).

10.6.2
Material Requirements, Life-cycle Impacts and Cost

The thickness of the wafer used in crystalline silicon technology is currently in the range 150–200 µm and the amount of silicon used is 7–10 g W^{-1}. This means that an average 3 kW$_p$ PV array (~18 m^2), as typically mounted at mid-latitudes to produce the electricity demand of an average household, requires about 25 kg of SG-silicon. For comparison, the dimensions of silicon ingots (see above) are typically 100 kg for monocrystals and 250 kg for multicrystals. It can be reasonably expected that silicon usage will decrease to 5–7 g W^{-1} in the mid-term (2020) and stabilize below 5 g W^{-1} afterwards with wafer thickness <100 µm [612].

The evaluation of the environmental impact of the silicon PV technology has been thoroughly investigated in recent years, by means of cradle-to-grave life-cycle assessments [613–615]. Despite the fact that these analyses can make different assumptions in terms of module manufacturing and final disposal, annual insolation, and conventional technologies displaced, there is wide agreement that silicon PV would bring significant environmental benefits in terms of avoided emission of greenhouse gases, heavy metals, particulate matter, NO$_x$, and SO$_x$, in comparison to current grid-electricity in the US and Europe, mainly produced with fossil fuels [615]. Also energy payback times (EPTs), which used to be above 10 years in the early 1990s, are now within 3–4 years even in the less insolated regions such as northern Europe [613, 615]. Therefore, since operating lifetimes are not less than 25 years, modern solar silicon cells are net energy producers despite their energy-intensive production chain. Their performance will improve as PV production will increase its global share of electricity production, thus starting to breed its own manufacturing [615].

The spectacular growth of the world's PV demand in recent years, sustained by the adoption of feed-in tariffs in many countries and high oil prices, has shaken the silicon market. The price of polysilicon was $40 kg^{-1} in 2007, hit $400 kg^{-1} in mid-2008, then steadily decreased through 2009 and 2010 to less than $60 kg^{-1} [616]. This trend was the consequence of a market oversupply due to increased global polysilicon production sustained by new plants entered in operation all over the world [616, 617]. Solar module manufacturers buy silicon on the open market rather than under contract. This causes a prompt decrease of retail prices because silicon accounts for about 40% of the cell cost and 10% of the final installed panel [618, 619]. Accordingly, the average price of solar modules decreased from $4.88 to 4.15 W^{-1} from early 2007 to mid-2010, whereas throughout this period of time the price of the inverters was unchanged. In August 2010, the lowest retail price for a poly-Si silicon solar module on the US market was below $1.75 W$_p^{-1}$ [620].

Recently the integration of sc-Si solar cell modules on highly flexible and light-weight polymer substrates, by a relatively simple transfer printing technique, has been demonstrated [621]. This opens up new application routes to silicon technology that were previously envisaged only on solution processed organic or nanocrystal films.

10.6.3
Amorphous Thin Film Silicon Cells

Amorphous silicon (a-Si) is the non-crystalline allotropic form of this semiconducting material. It is characterized by a large number of structural and bonding defects, with atoms often lacking neighbors at the bonding distance. These defects can be removed by hydrogenation, so that H atoms combine with many dangling bonds, allowing electrons to be moved through the material without extensive charge trapping [609].

Amorphous silicon absorbs solar radiation more efficiently than crystalline silicon, therefore much thinner layers ($\leq 1\,\mu m$) are needed to harvest sunshine. Deposition can be made at mild temperatures (as low as 75 °C) on glass and plastic, possibly using roll-to-roll processing techniques. Such an easy processing means lower manufacturing costs compared to its more noble crystalline counterparts. Once deposited, a-Si can be doped in a fashion similar to crystalline silicon (c-Si), to form p- or n-type layers. a-Si performs better then c-Si in cloudy sky conditions with a higher share of diffused light; additionally it does not suffer the performance drop of sc-Si and poly-Si at high temperatures experienced on the most sunny days [622].

a-Si is common today in consumer devices with low-power requirements such as calculators and low-cost battery rechargers. More recently, improvements in a-Si manufacturing techniques have made it more attractive for large-area solar cells, with efficiency in the range 6–10%. Flexibility can also make it interesting to manufacture building integrated photovoltaics (BIPV) for roofing, shading systems, and window glazing [623]. All of the above listed advantages of a-Si are counterbalanced by larger space requirements and by lower robustness due to intrinsic photochemical instability.

a-Si remains the most well-developed thin film technology to date and an interesting avenue of further development can be envisaged through the use of "microcrystalline" silicon, which seeks to combine the stable high efficiencies of c-Si with the simpler and cheaper large area deposition technology of a-Si [609].

10.7
Thin Film Solar Cells

Promising advances in non-silicon thin film PV technologies have been made in the last two decades, which might overcome the technical problems associated with a-Si. These technologies might be a cost-effective alternative to crystalline silicon materials [609, 624] even though, starting from 2008, their economic competitiveness has been substantially eroded by the price drop of silicon modules [625].

The currently leading materials are cadmium telluride (CdTe) and copper indium gallium diselenide (CIGS). They are characterized by an optical absorption spectrum very well matched with the solar spectrum, allowing the use of active layers as thin as 1–2 μm. These semiconducting materials are thus amenable to large area deposition on to substrates of up to $1\,m^2$ and allow high volume manufacturing. The semiconductor junctions are formed by using another semiconducting material, typically a thin layer of cadmium sulfide (CdS, n-doped). A transparent conducting oxide layer (e.g. tin oxide) forms the front electrical contact of the cell whereas the rear contact is metallic [626].

CdTe is a polycrystalline material with an optical bandgap of about 1.5 eV, that is, nearly optimally matched to the solar spectrum [609]. It is characterized by a high absorption coefficient ($>5 \times 10^5\,cm^{-1}$) which means that ~99% of photons with energy greater than the bandgap can be absorbed within 2 μm of CdTe film.

The earliest attempt to make solar cells by means of CdTe dates back to the 1950s, but the first relatively large and efficient solar cells were patented by Kodak around 1980. In the following two decades there were several advancements that prompted the birth of several start-up companies, most of which did not survive. The landscape changed in the early twenty-first century, when time became mature for industrial development. Nowadays the largest company in the world manufacturing CdTe systems, First Solar, produces over 1 GW of thin film modules annually and has become one of the largest PV company in the world [627]. The current solar-to-electricity conversion of commercial CdTe modules is 9–11% [628], that is, intermediate relative to a-Si and CIGS. Prototype small-area CdTe cells with efficiencies of almost 17% have been developed [629].

CIGS is a polycrystalline material with physical properties similar to those of CdTe, but with a slightly smaller bandgap of 1.3 eV. CIGS solar cells with conversion efficiencies of 19.5% have been produced, while commercial modules exhibit efficiencies in the range 10–14% [628]. The better performance compared to CdTe is counterbalanced by a more complex and expensive manufacturing technology, due to intrinsic chemical complexity of the material, which is made of four elements, and a higher moisture sensitivity [630]. Another problem can be the availability of indium at an affordable price, due to competition from a variety of electronic applications and modern display technologies. Despite these potential hurdles, the largest CIGS PV company in the world, Nanosolar, claims the high potential of this highly efficient and cheap technology in which the active material is deposited as an ink on flexible aluminum foils [631].

Semiconducting materials for non-silicon solar cells are made of relatively rare elements such as tellurium, indium, germanium, gallium, and selenium. This raises concerns about the possibility of a wide global deployment of thin film PV technologies, due to limited material availability of elements that are obtained essentially as by-products of major metals [632]. For instance, tellurium is an element not currently used for many applications and global annual production is around 150–200 t [633], most of it coming as a by-product of copper extraction, with smaller amounts from lead and gold [632]. It can be roughly estimated that, at current efficiencies and thicknesses (10%, 3 µm), 1 GW of CdTe PV modules would require about 90 t of tellurium, clearly posing supply concerns at current production rates.

The availability of Cd [632] is much less of a concern, taking into account that 1 GW of CdTe requires about 80 t of Cd, whereas annual world production of this metal is around 20 000 t [633]. The main problem with cadmium is toxicity and the possibility that it can be released into the environment over the life-cycle of the PV module. This topic has been investigated in recent years and the results obtained so far are somewhat surprising. CdTe PV requires the least amount of energy during the module production among all of the currently commercially available silicon and thin film technologies [615]. By taking into account the embedded energy in all types of PV modules, which continues to be largely of fossil origin and thus entails sizeable emissions of cadmium upon combustion, the replacement of fossil fuel power plants with CdTe electricity would result in lower emissions of all pollutants, including cadmium itself [615]. Notably, the release of cadmium seems to be negligible even in case of fire, due to robust encasement and sealing of PV modules [634].

A wide investigation considering 23 promising semiconducting materials, most of them still to be developed for solar energy conversion, has recently been undertaken [635]. It demonstrates that CIGS and, even more, CdTe would not have the capacity to meet the current global electricity demand, but there are about a dozen largely unexplored composite materials that, even alone, could make enough PV modules to cover or even exceed today's global electricity demand [635]. This stimulating exercise suggests that thin film technology is extremely versatile in terms of choice of materials and its huge potential is still far from being fully disclosed.

10.8
Organic Solar Cells

One of the main problems with the manufacturing of inorganic solar cells is the deposition of the semiconductor layers, which requires costly and energy intensive procedures [609]. Using materials that can be deposited through cheap solution-based methods would offer an attractive alternative to the established inorganic systems described so far, with substantial reduction of production costs [636]. This approach, however, can be successfully implemented only by leaving the realm of inorganic semiconductors, to focus on organic chemistry and materials. Indeed, the field of organic solar cells (OSCs) (also termed "plastic" or "polymer" solar cells), a relatively marginal research activity until the early 1990s [637], has undergone an impressive growth that has made it an important player in the current PV scenario [638, 639]. OSCs are particularly attractive not only for ease of processing, but also for mechanical flexibility, which may open the way to low cost printing of large "plastic" PV surfaces.

Typical organic solar cells are based on a charge transfer occurring at the interface between two distinct materials, namely an electron donor (D) and an electron acceptor (A), which, at their contact surfaces, constitute the analog of the "inorganic" semiconductor p–n junctions described in Section 10.5. The physical contact between the donor and the acceptor was initially made as a simple bilayer stack of donor–acceptor films about 100 nm thick. Currently, state-of-the-art OSCs are made with a disordered bulk heterojunction (BHJ), in which donor and acceptor are intimately mixed to form an interpenetrating phase network at the nanoscale level (Figure 10.13). The performance of a solar cell manufactured with this strategy is dramatically dependent on the nanomorphology of the donor–acceptor heterojunction, which can be controlled by engineering the molecular structures of D and A, the specific deposition protocol, or the use of thermal treatments [640]. "Molecular" heterojunction concepts were also attempted both covalently, by linking donor and acceptor moieties in molecular dyads or polymeric structures, and via supramolecular interactions [640, 641]. However, these strategies resulted in substantially lower conversion efficiencies than those obtained by blending the donor and the acceptor materials.

It has to be emphasized that the generation of electric current in OSCs is conceptually different to that in "classic" inorganic solar cells. In the latter, light absorption leads directly to the generation of free electrons and holes in the conduction and hole bands, respectively. By contrast, in OSCs, the excited state gener-

b)

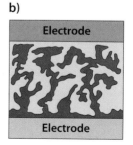

a)

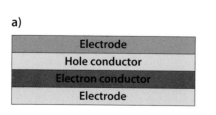

Figure 10.13 Schematic representation of an organic photovoltaic cell with (a) a bilayer morphology and (b) bulk heterojunction morphology.

ated by light absorption is a molecular, and therefore bound, excited state or "exciton." This is essentially an electron–hole pair which either can recombine to emit light or heat or, more usefully, can dissociate and eventually move to the respective electrodes, generating a PV current.

In organic PV devices, excitons have a high binding energy (0.3–0.5 eV) and, moreover, the average distance over which an exciton can diffuse between its generation and its recombination (diffusion distance) is rather short, typically less than 10 nm [642]. Therefore, in order to take advantage of enhanced dissociation at the donor–acceptor interface, excitons should be generated as close as possible to the heterojunction. Practically, a heterojunction should occur within the exciton diffusion distance to facilitate exciton breaking and, additionally, a continuous conducting pathway should exist for electron and holes to reach the electrodes.

Among others [637], the most popular electron acceptor molecules utilized so far in OSCs are C_{60} and C_{70} fullerene derivatives [636, 640, 642]. These carbon spheres are extremely photostable and combine a variety of spectral and electronic properties [643] that make them ideal candidates for PV technologies [642]. A particularly successful fullerene derivative is [6,6]-phenyl-C_{61}-butyric acid methyl ester (C_{60}-PCBM), due to its ease of preparation, high solubility, and capability to pack efficiently in crystalline organic structures with high charge conductivity. Even better, but more expensive, is the analogous C_{70} derivative (C_{70}-PCBM), which is characterized by enhanced absorption in the visible spectral region, which allows greater PV efficiency compared to the same cell prepared from C_{60}-PCBM. Both compounds are depicted in Figure 10.14.

Electron donors that have been widely utilized in OSCs are conjugated organic polymers such as MDMO-PPV {poly[2-methoxy-5-(3,7-dimethyloctyloxy)]-1,4-phenylenevinylene} and P3HT [poly(3-hexylthiophene-2,5-diyl] (Figure 10.14)

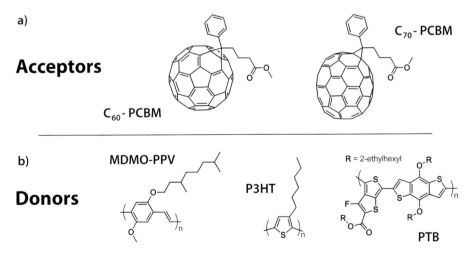

Figure 10.14 Some of the most successful acceptor (top) and donor (bottom) molecules and materials utilized in organic photovoltaic cells.

[636]. Solar cells using these polymers and C_{60}-PCBM reached efficiencies above 4% [644].

Most of the organic materials utilized so far are hole conductors and exhibit a HOMO–LUMO optical bandgap around 2 eV, which is considerably higher than that of silicon and inorganic thin film cells; this limits the harvesting of the solar spectrum to a great extent [636]. In recent times, more sophisticated and stable low band organic semiconductors have been synthesized, such as those based on alternating ester substituted thieno[3,4-*b*]thiophene and benzodithiophene units (PTB, Figure 10.14). They have reached record conversion efficiencies as high as 7.4% when coupled with C_{70}-PCBM [645–647].

Organic materials have relatively strong absorption coefficients (usually $\geq 10^5$ cm^{-1}), thus considerable light absorption is achieved even with ultrathin devices (<100 nm); this partially offsets losses due to low charge mobilities and mismatch with the solar spectrum. The optical absorption intensity and range can be finely tuned via chemical modification of the PV material, providing wider versatility compared to inorganic PV technologies. Enhancement of OSC efficiency is also being pursued with tandem strategies entailing two or even three sub-cells with complementary bandgaps. The record efficiency for such organic multijunction cells is now set at 6.5%, but values as high as 15% are considered attainable [638].

Over the last 15 years dramatic improvements have been made in fundamental understanding, materials processing, device fabrication, and operating performance of OSCs. Meanwhile the cost of PV materials (e.g. fullerenes) has undergone a substantial decrease and wet thin film deposition processes have been improved. The major challenge that OSC technology now has to face is stability [648]. Organic materials are by nature more susceptible than inorganic materials to chemical and physical degradation, for instance by the action of oxygen and moisture, which also endangers the integrity of the electrodes. Current strategies to limit degradation include the design of more robust polymer materials, encapsulation, or application of getter materials and UV filters to limit undesired photochemistry [648]. The rise of the first companies offering OSCs on the market suggests that reasonable stability, at least for niche applications, has been achieved [649, 650].

10.9
Concentrated Photovoltaics and Other Innovative Concepts

Besides the most developed fields of inorganic and organic PV described above, there are a few novel emerging approaches to solar-to-electricity conversion [556]. Among these, the most technologically mature is concentrated photovoltaics (CPV).

CPV systems use a concentrating optic, for example, mirrors or lenses, to harvest sunlight and focus it on a very small PV cell. Concentration levels, depending on active materials and optics, can be up to hundreds of times those of regular insolation. Heat dissipation is crucial for CPV, in order to preserve device performance or even avoid thermal destruction [651]. CPV cells are often mounted on a solar tracker to keep the focal point upon the cell as the Sun moves across the sky. The obvious advantage of CPV is higher light-to-electricity conversion efficiency (theoretically up to 50% for multijunction systems [652]) and a smaller amount of photoactive material employed. To date there exist only a few megawatts of CPV systems operating worldwide, made of single crystal silicon or multijunction cells of group III–V semiconductors (Ga, As, In, P, Sb, Ge), but several hundred megawatts of installations should be deployed after 2010 [652]. A CPV demonstrator cell with efficiency of 40.7% has been produced, which constitutes the first solar cell of any type to surpass the 40% efficiency milestone [653]. Systems with 25% efficiency are now commercially available, but more ambitious 35% targets are set within a few years [654].

Another concept for more efficient sunlight harvesting is pursued by using luminescent solar concentrators (LSCs), which are thin flat plates of highly luminescent materials that absorbs sunlight and concentrate most of the resulting emission to a juxtaposed PV active layer through total internal reflection. Hence a concentrated luminescence input, which can also be optically tuned by a thorough choice of the plate material to match the PV bandgap, is converted into electricity [655, 656].

Several innovative designs of PV cells have been proposed in recent years, but they are still at a research stage. Hybrid solar cells with polymers and semiconductor nanorods (e.g. CdSe) were engineered in which, by controlling the nanorod length, one can tune the distance over which electrons are transported directly through the thin film device. The material bandgap can be altered by varying the nanorod radius, allowing optimization of sunlight harvesting [657]. Novel ultrathin, solution processable, and chemically stable solar cells entirely based on colloidal semiconductor nanocrystals have also been prepared [658]. A similar but more versatile approach entails the use of quantum dots as PV materials in solid state cells or even as sensitizers in liquid dye-sensitized solar cell (DSSC)-type systems (Section 11.6) [659]. Moreover, a hybrid exciton–DSSC concept has been developed which involves donor–acceptor porphyrin–fullerene aggregates deposited electrophoretically on semiconducting electrodes [660].

10.10
Photovoltaics: Global Installation and Market Trends

Due to its delocalized nature, it is difficult to know with high precision the PV capacity installed globally, but it can be safely assumed that about 25 GW, including grid-connected and off-grid, were installed at the end of 2009 [544, 587]. This amounts to about 0.55% of the global 4500 GW installed electric generation capacity [102]. In the near future, PV is expected to consolidate as the fastest growing alternative energy market, with a yearly growth around 40% [587]. In 2010 the largest PV power stations in the world were located in Spain, Germany, and Italy, with a peak power in the range 40–70 MW, all of them adopting crystal silicon technologies. A larger 550 MW thin film project, Solar Topaz Farm, should be deployed in the California Valley by 2013, covering 25 km^2 of ranchland and producing 1100 GWh of electricity annually [627]. The largest PV station in the world, with a peak power of 2 GW, should be built in Ordos City, Inner Mongolia (China), by 2019 at a cost of about \$5 billion [661]. It should cover an area of about 65 km^2 with thin film panels.

In 2007 c-SI technologies still made up almost 90% of the world PV market [611]. This dominance has been constantly eroded in the last few years due to the impressive growth of the thin film industry, which now accounts for over 20% of the total. In the near future, the substantial price decrease of silicon products that occurred recently could slow, in relative terms, the growth of the thin film market. However, making predictions in the PV market, which is quite diversified and possesses a huge potential for growth even in the near term, is extremely daunting. It is also difficult to predict which technology will prevail in the mid to long term, but probably all of the approaches that are established or are under development will find their own position even if, in some cases, in market niches only.

As far as the price of PV electricity is concerned, it has to be emphasized that economic comparisons of electricity production with different technologies is not straightforward [662]. First, base load, peak load, and intermediate load electricity is generated with different technologies and has substantially different costs (see Chapter 13). PV electricity production is bound to solar availability, which is naturally correlated with the electricity demand patterns of modern societies. In other words, intrinsically, PV is a technology to provide peak and intermediate power electricity demand, which are the highest when people are awake and productive, that is, mainly when the Sun is shining. Hence cost comparisons for PV electricity have to be made primarily with the most expensive peak generation options such as natural gas stations. There is wide agreement that by 2020, and probably even before, PV will be cost-effective in providing peak electricity to the grid. In the subsequent decade its cost will drop further, challenging base load utility prices [662] and further fostering, inexorably, the transition to solar energy.

10.11
Solar Energy: Sustainable and Affordable

The theoretical potential of solar energy is immense, about 6000 times the current world primary energy demand of 470 EJ per year. However, taking into account technical boundary conditions that are difficult to estimate precisely, the exploitable potential still remains tens to hundreds of times the current level of energy consumption. For renewable resources, this technical potential must be seen in a dynamic perspective since R&D should improve conversion technologies over time.

Making long-term projections of solar energy deployment on the global scale for any sort of end-use is practically impossible, given the variety of peculiar features of individual geographic areas and of specific technologies. More reliable assessments can be made focusing on electricity only and analyzing homogeneous geographic regions; these studies consistently show that the transition to a solar-powered world can be accomplished within the twenty-first century, if resolute energy policies are embraced worldwide [516, 591, 663, 664]. In the case of the US, based on expected improvements of established and commercially available PV and CSP technologies, it has been argued that solar energy has the technical, geographic, and economic potential to supply 70% of the total electricity demand and 35% of the total energy supply of the country by 2050 [665].

A recurrent concern regarding the wide development of solar collectors is the extension of the Earth's surface that should be mobilized and eventually made unavailable for other essential services such as food production, wildlife habitats, and housing. The Joint Research Center of the European Commission has recently undertaken a detailed study addressing this issue for PV technology, with reference to each single European country and taking into account specific solar availability [666]. In the totally imaginary hypothesis of producing 100% of the EU electricity by means of standard PV modules mounted at the optimum angle and producing $1\,kW_p$ over $9.5\,m^2$, the percentage of land area needed has been calculated for each country and is reported in Figure 10.15. On average, covering ~0.6% of the European territory by PV modules would theoretically satisfy its entire electricity demand. This estimate is somewhat conservative because the latest PV

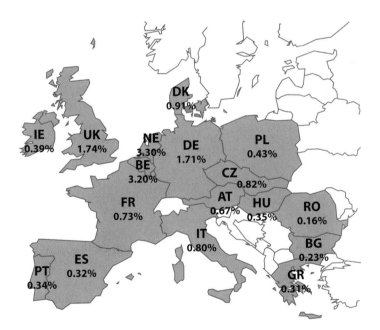

Figure 10.15 Theoretical PV potential: surface of PV modules mounted at the optimum angle that would be needed to satisfy completely the electricity consumption of some selected European countries, expressed as a percentage of the country's area (European average 0.6%) [666].

module technologies have higher efficiency than considered in this appraisal, so the area covered per kW_p is smaller and will certainly decrease in the future. Furthermore, this exercise ignores cross-border electricity trade: no allowance is made for Dutch or German electricity consumption being covered by PV in southern Europe. On the other hand, the PV area does not translate directly into land area covered because (i) the PV modules are not placed horizontally and (ii) a large share of PV panels (currently about 90%) are normally placed on rooftops, which would exist regardless of whether solar panels are installed. Interestingly, for comparison, in countries such as Estonia and Bulgaria the theoretical land requirement for full coverage of electricity demand with PV modules would correspond to the current extent of mineral extraction sites and landfills.

It is important to point out that another question has to be posed before concluding that a wide deployment of PV panels, perhaps in the order of terawatts, is a desirable goal. Would extensive areas of dark-colored, sunlight-absorbing panels, reduce the Earth's reflectivity (albedo), alter the solar radiation balance, and ultimately affect climate stability? A recent study has shown that the climatic benefit of replacing fossil fuels by PV would far outweigh the unfavorable effects generated by the artificial albedo modification: the avoided radiative forcing would be approximately 30 times larger than that of a PV powered world [667].

But what about the economic burden for the citizens? Will people ever want or could simply afford to buy PV devices and become energy entrepreneurs selling electricity to local utilities? To give an answer, we can make some simple calculations analyzing the case of Italy. Current peak demand in Italy is 55 GW. Installing $100\,GW_p$ of c-Si silicon panels all over the country, at the average current turnkey price of about €3000/kW_p, would result in a huge distributed bill of €300 billion, apparently unaffordable. But let us turn our attention to cars. Every year Italians buy an average of 1 000 000 new cars at a (conservative) price of €15 000, thus spending €15 billion. There are 35 million cars circulating in Italy, each of which costs on average of €3000 per year for maintenance (fuel, insurance, taxes, tires, etc.), totaling €105 billion annually. So Italians, *every year*, spend more that €120 billion to run their cars. If we compare this number with the above theoretical investment for PV (€300 billion) and consider that the latter will work with free fuel and very little maintenance cost for at least 25 years, we can probably understand that, even with today's high price, PV is already an affordable choice for affluent countries. In fact people regularly spend similar amounts of or even more money for what they believe to be indispensable, for instance a car. Sure, a Maserati or even a tiny Fiat 500 is definitely more sexy than a boring chunk of silicon encased in a gray aluminum box, but that is another story.

Throughout the previous and the present chapter, we have first analyzed how solar radiation is generated and distributed over the planet and then we have described the wealth of technologies that human ingenuity has conceived to exploit this immense natural resource and produce useful heat or electricity. The great deal of scientific and technological achievements accomplished in the last 50 years, accompanied by a refined ability to analyze complex systems, has provided overwhelming evidence that solar resources can cover in principle any conceivable heat or electricity demand many times over. This is, of course, excellent good news, but it is only one part of the story because our civilization also needs fuels, that is, storable energy to be used on demand. But is it conceivable that we can manufacture huge amounts of fuels using sunshine? This topic of utmost importance is the subject of the next chapter.

11
Solar Fuels

"A tree is essentially made of air and Sun.
When it is burned, it goes back to air,
and in the flaming heat is released
the flaming heat of the Sun
which was bound in
to convert the air into tree."

Richard P. Feynman

11.1
Introduction

Photochemistry, the interaction between light and matter, is perhaps the oldest and certainly the most important natural phenomenon. All forms of life on this planet, all our food, and all fossil fuels are the result of the light–matter interaction as developed by plants in the photosynthetic process where sunlight is converted into chemical energy [88]. Up until now, to cope with food and energy needs, we have essentially relied upon the natural miracle of photosynthesis [536]. In the last few decades, however, we have realized, particularly as far as energy is concerned, that what Nature does (and did) for us is no longer sufficient [57]. One of the grand challenges of twenty-first century chemistry is to accomplish an artificial miracle, namely to produce "solar" fuels, energy rich compounds that can be stored and transported and that, when used, reversibly convert into the starting materials. If we are successful, we will begin to pay back to Nature what we have up until now taken from it [668]. A successive step could be using sunlight to convert abundant, energy poor, low value molecules to any kind of energy rich, high value molecules useful to mankind.

Energy for a Sustainable World: From the Oil Age to a Sun-Powered Future. Nicola Armaroli and Vincenzo Balzani
© 2011 WILEY-VCH Verlag GmbH & Co. KGaA, Weinheim
ISBN: 978-3-527-32540-5

11.2
Natural Photosynthesis

11.2.1
A Complex Process

Photosynthetic organisms are ubiquitous in Nature; they are responsible for the development and sustenance of all life on Earth [536]. It is estimated that currently photosynthesis produces more that 100 billion tons of dry biomass annually [669]. Oil, gas, and coal are also derived from millions of years of photosynthetic activity. The success of photosynthesis as an energy conversion and storage system stems from the fact that the power (sunlight) and the raw materials (water and carbon dioxide) needed for the synthesis of biomass are available in almost unlimited amounts. The photosynthetic process essentially consists in the splitting of water by sunlight into oxygen, which is released into the atmosphere, and "hydrogen," which is not released into the atmosphere but instead is combined with carbon dioxide to make organic compounds of various types (Figure 11.1). The burning of these compounds with oxygen either by respiration (food) or combustion (fossil fuels, wood, biomass) forms the original compounds (water and carbon dioxide) and releases the stored energy that originated from sunlight.

The entire photosynthetic process involves more that 50 distinct chemical transformations that can occur over short or long times [670]. Photosynthetic organisms may be quite different, but all of them use the same basic strategy, in which light is initially absorbed by antenna proteins containing many chromophores, followed by energy transfer to a specialized reaction center protein, in which the captured energy is converted into chemical energy by means of electron-transfer reactions [671].

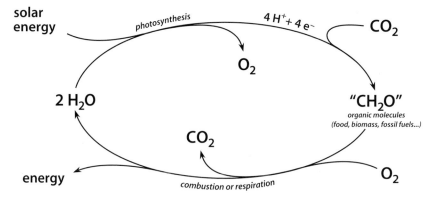

Figure 11.1 Schematic representation of the results of natural photosynthesis, combustion, and respiration processes.

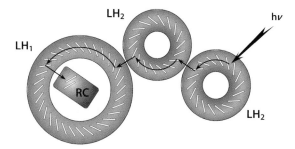

Figure 11.2 Schematic representation of the overall light-harvesting process by LH2 and LH1 antenna complexes in bacterial photosynthesis. RC is the reaction center [672].

11.2.2
Natural Antenna Systems

The better known natural antennas are the light harvesting complexes of photosynthetic purple bacteria. A schematic view of the overall light harvesting process is shown in Figure 11.2. A major breakthrough in the field was the high resolution X-ray crystal structure of the light-harvesting antenna complex LH2 of the photosynthetic unit of *Rhodopseudomonas acidophila* [672]. The complex is composed of rings of bacteriochlorophyll (BChl) molecules that have different absorption and photophysical properties. Energy migration occurs on the picosecond time scale.

The energy collected by the LH2 antenna is then transferred within a few picoseconds to another antenna complex, LH1, which surrounds the reaction center (RC). The RC is the final destination of the collected energy, and it is the site where charge separation takes place. The rate of the successive energy transfer steps from LH1 to the embedded RC is more than 10 times slower (35 ps). The light harvesting complexes of green plants are not well known and, likely, they are more complicated than those of bacterial photosynthesis. There are good reasons to believe, however, that the governing principles of operation are similar to those discussed above.

11.2.3
Natural Reaction Centers

- **Bacterial photosynthesis:** The structures of several bacterial reaction centers are known precisely by X-ray crystallographic investigations [673]. They consist mainly of protein, which is embedded in and spans a lipid bilayer membrane. A simplified view of the structure of the reaction center of *Rhodopseudomonas viridis* is sketched in Figure 11.3a. Detailed photophysical studies have led to a precise picture of the sequence of events participating in photoinduced charge separation [674]. The key molecular components are a bacteriochlorophyll "special pair" (P), a bacteriochlorophyll monomer (BC), a bacteriopheophytin

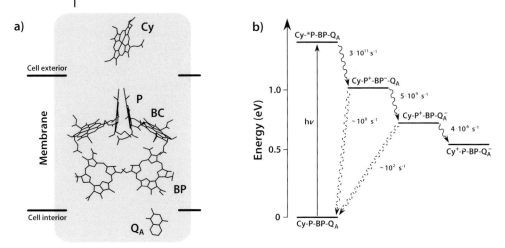

Figure 11.3 (a) A simplified view of the structure of the reaction center of *Rhodopseudomonas viridis*. (b) Energy-level diagram and rate constants of the electron-transfer steps involved in the charge-separation process [674].

(BP), a quinone (Q_A) and a four-heme *c*-type cytochrome (Cy). These molecules are held in a fixed geometry by surrounding proteins, so that the twofold axis of P is perpendicular to the membrane, the periplasmic face lies approximately between P and Cy, and the cytoplasmic face at the level of Q_A. In the RC, excitation of P by absorption of light or, more commonly, by singlet–singlet energy transfer from various antenna systems, is followed by very fast (~3 ps) electron transfer to the BP acceptor. The next step is a fast (~200 ps) electron transfer from BP to Q_A, followed by a slower (~270 ns) reduction of the oxidized P by the nearest heme group of Cy. At that stage, trans-membrane charge separation has been achieved with an efficiency approaching unity and an extremely long lifetime with respect to charge recombination. The rate constants of the various electron transfer steps involved in the charge separation process are summarized in the approximate energy level diagram of Figure 11.3b, together with those of the non-occurring $BP^- \rightarrow P^+$ and $Q_A^- \rightarrow P^+$ charge recombination steps, as determined from experiments with modified reaction centers. Figure 11.3 points out the importance of the supramolecular structure of the RC. The achievement of efficient photoinduced charge separation over a large distance is made possible by optimization of several aspects of this photochemical device: (i) the organization of the molecular components in space, (ii) the thermodynamic driving force of the various electron transfer steps and (iii) the kinetic competition between forward (useful) and back (dissipative) electron transfer processes. How this occurs can be reasonably well understood in terms of electron transfer theory [675]. In the process described above the ultimate electron acceptor is a quinone Q_A. Then the process continues with many other steps leading to the synthesis of adenosine triphosphate (ATP), which fills the majority of the energy needs of the bacterium.

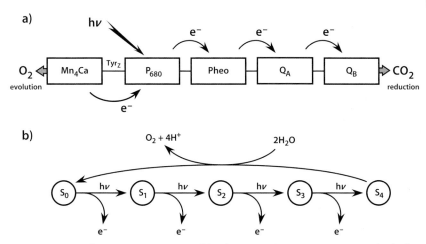

Figure 11.4 (a) Schematic representation of the charge-separation process in PSII; (b) the five redox states ($S_0 \rightarrow S_4$) of the oxygen-evolving Mn_4Ca cluster [674].

- **Photosystem II:** The most important solar energy conversion process is that occurring in green plants where the RC called Photosystem II (PSII) is quite similar to that of the bacterial RC except for a very peculiar donor side which can use water as an electron source and produce oxygen as a "waste" product [676]. In order to do that, PSII must (i) reach potentials high enough to oxidize water (>+0.9 V relative to NHE), (ii) handle such a high oxidation potential in fragile biological structures, and (iii) couple the one-photon–one-electron charge-separation process to the four-electron water oxidation process. The water oxidation moiety of PSII (Figure 11.4a) consists of a triad composed of a multimer of chlorophylls (named P_{680}), a redox active tyrosine amino acid (Tyr$_Z$), and the so-called oxygen evolving complex (OEC), a cluster containing four manganese atoms and one calcium atom (Mn_4Ca). The specific protein environment and one chloride ion are also essential for the water-splitting activity. PSII spans the thylakoid membrane in the chloroplasts and the water oxidizing triad is located closely to one side of the membrane. On direct absorption of a photon or energy transfer from the antenna units, P_{680} is excited and becomes a strong reductant. An electron is then transferred from excited P_{680} to pheophytin (Pheo), which in its reduced state is sufficiently negative to drive the formation of hydrogen. Instead, the reducing equivalent is transported to Photosystem I where it is excited by the energy of a second photon absorbed by a chlorophyll molecule, so as to have enough energy to drive the reduction of carbon dioxide.

  The oxidized primary donor, P_{680}^+, is rapidly reduced by Tyr$_Z$, which, in its reduced state, is hydrogen bonded to a nearby histidine residue. Such a hydrogen bond facilitates oxidation of Tyr$_Z$ which occurs with concomitant deprotonation [proton-coupled electron transfer (PCET)] [677]. Oxidation of water to dioxygen is a four-electron process, so the results of four charge-separation events must be accumulated. This role is played by the Mn_4Ca cluster, which is close to Tyr_Z and is oxidized stepwise by the Tyr_Z radical to a series of states S_i (where i ranges from 0 to 4), as shown in Figure 11.4b. Oxygen evolution occurs when the most oxidized cluster state, S_4, returns, in a four-electron reduction process, to the most reduced state, S_0. This process involves the oxidation of two water molecules accompanied by the transfer of four protons. Although the geometry of the Mn_4Ca cluster is not yet known precisely, the available models do provide a basis for developing chemical mechanisms for water oxidation and oxygen formation [669, 678].

11.2.4
Efficiency of Photosynthesis

Natural photosynthesis converts light energy into chemical energy contained in the organic molecules of biomass. The solar energy conversion efficiency of photosynthesis, η_{solar}, can be evaluated from the following considerations [669]:

- The antenna system absorb most of the incident photons of visible light.

- The excitation energy corresponds to photons of about 680 nm (1.83 eV; 42 kcal per quantum mole); therefore, part of the energy of the shorter wavelength photons is degraded to heat.

- Since the two photosystems work in series, two photons are needed to move an electron/proton from water to carbon dioxide.

- Formation of dioxygen and reduction of carbon dioxide require four electrons.

It follows that photosynthesis uses at least eight photons, some of which have more energy than needed, per O_2 molecule generated and CO_2 molecule "fixed." Therefore, formation of a molecule of glucose, $C_6H_{12}O_6$, requires at least 48 photons. With threshold 680 nm (42 kcal mol^{-1}), the light energy needed to produce glucose is at least 2016 kcal mol^{-1}. Since the energy content of glucose when it burns with oxygen in a calorimeter is 672 kcal mol^{-1}, the efficiency of the photosynthetic process, performed by 680 nm photons, is about 30%. This value drops to about 10%, since only 34% of the incident solar energy can create the photochemistry-driving excited state $*P_{680}$ [679]. In fact, the efficiency is even lower because of the above threshold energy losses, saturation effects (the system takes time to process an energy input and it is not ready to use a second input), and the energy needed to maintain the organization, metabolism, reproduction, and survival of the photosynthetic organisms. A more accurate assessment shows that the maximum efficiency for the conversion of light into chemical energy in the natural photosynthetic processes is around 4.5%, whereas experience shows that, even under the most favorable conditions, crops produce yield of biomass at efficiencies less than 1%, and often as low as 0.1%. Despite these low values, the gigantic amount of solar energy reaching Earth (Chapter 9), makes it possible to cover the planet with terrestrial phytomass and plankton.

11.3
Biomass and Biofuels

11.3.1
Biomass

Biomass, the end product of photosynthesis, can be used to satisfy three fundamental needs of mankind: food, energy, and valuable materials. Traditional biomass remains the dominant contribution to the energy supply of a large number of developing countries.

As we have seen above, biomass is formed through CO_2 fixation and, when burned with oxygen either by combustion (fossil fuels, wood, etc.) or respiration (food), produces the same amount of CO_2 fixed in it (Figure 11.1). Therefore, from the viewpoint of atmospheric CO_2 concentration and related effects (global warming), biomass is said to be carbon neutral, whereas the use of fossil fuels releases into the atmosphere the CO_2 which was fixed eons ago. For this reason, the conversion of biomass in biofuels is considered in several countries, including the EU and US, to be an attractive option to replace fossil fuel.

The advantages and disadvantages of the conversion of biomass into biofuels have been extensively discussed and continue to be at the center of much debate [680–683]. Perspectives can change very rapidly because of technological breakthroughs and changes in government policies, financial incentives, marked prices, environmental concerns, and social opinions.

On a global scale, a future fossil fuel consumption of 20 TW (current consumption ~12 TW) would correspond to almost three times all cultivable land currently used for agriculture globally, even using the best known plants for energy production [669]. This comparison shows that biofuels can at most make a relatively small contribution to replace fossil fuels, unless plant breeding and genetic engineering strategies succeed in increasing substantially the efficiency of biomass production, requiring minimum input of fertilizers, water, and pesticides [684, 685]. Perhaps more important, this shows that if we try to solve our energy problem by relying on biofuels, we will endanger our planet.

11.3.2
Biofuels Today

To be a viable alternative, a biofuel should provide a net energy gain, have environmental benefits, be economically competitive, and be producible in large quantities without reducing food supplies. Figure 11.5 shows schematically the routes that lead from biomass to biofuels.

At present, the production of first-generation bioethanol mainly utilizes conventional agriculture resources (edible feedstocks) rich in carbohydrates such as sugarcane, corn, wheat, and sugar beet. The production of bioethanol is an energy intensive and complex process that involves two key steps, hydrolysis and fermentation, followed by distillation and dehydration. Currently, the US is the world's

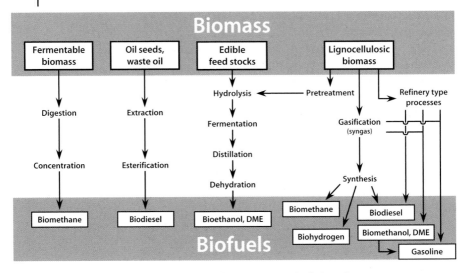

Figure 11.5 Schematic representation of the biomass to biofuels pathways.

largest producer and Brazil the larger exporter, accounting together for 70% of the world's production and 90% of ethanol used for fuel. Brazil produces ethanol from sugarcane and the US uses corn starch [686]. Ethanol has 66% of gasoline's energy content.

The first-generation biodiesel [fatty acid methyl ester (FAME)] is obtained by a variety of chemical processes causing transesterification of triglycerides from vegetable oils (soybean, sunflower, palm, corn) or animal fats, with glycerol as a by-product. Usually biodiesel is produced in slightly modified chemical plants already in use. More recent technologies allow incorporation of glycerol into the biodiesel.

Fuels with small amounts of admixed biofuel (up to 5% by volume, E5) can be used in common gasoline or diesel engines without any form of adaptation. Fuels with higher levels of biofuel are currently available in Brazil (E25) and the US (E10). In Sweden, the UK and the US, E85 is used in flexible fuel vehicles with modified engines. These cars and trucks have sensors to adjust the timing of spark plugs and fuel injectors so the engine runs smoothly no matter what percentage of the fuel might be ethanol. Another type of fuel used in Brazil is hydrated bioethanol, which is bioethanol (E100) that contains small amounts of water. Diesel containing 5, 10, and 30% of FAME are also used (B5, B10, B30). The infrastructure of renewable fuel pumps is often inadequate and needs to be developed [687].

The EU has planned to reach a 20% share of energy from renewable sources by 2020 and a 10% share of renewable energy specifically in the transport sector [688]. The 2005 US Energy Bill [689] requires biofuels to cut the country's gasoline consumption by 20% by 2017 and by 30% by 2030. But it has been estimated that even dedicating all US corn and soybean production to biofuels would meet only

12% of the gasoline demand and 6% of the diesel demand of the country [690]. As part of the ongoing effort to support the development of domestic renewable fuels, in August 2009 the DOE announced the availability of up to $5.5 billion from the American Recovery and Reinvestment Act to increase the use of higher ethanol blends up to E85 by expanding refueling infrastructure and launching targeted outreach to promote public awareness [691].

Biobutanol, that can be produced from sugar beet, is reported to have some advantages over ethanol, including low vapor pressure and tolerance to water contamination; it can be blended into gasoline at higher concentration without having to retrofit vehicles.

Biogas is a cheap and versatile fuel composed of a mixture of CH_4 and CO_2 that is usually generated from anaerobic bacterial digestion of biomass. Almost any type of organic matter (e.g. sewage sludge, animal wastes, industrial effluents) is suitable for the production of biogas, which can be directly utilized in cooking and heating systems. Biogas can avoid wood burning and smoke hazards in underdeveloped countries and can be delivered with natural gas in industrialized nations.

Production of biofuels entails several problems which can be summarized as follows.

- **Biofuel-food competition:** Generation of biofuels from edible feedstocks is in competition with food production. Approximately 500 kg of grain, needed to feed a person for 1 year, yield 200 l of ethanol. In some countries widespread biofuel production could cause relatively extended changes in land use and consequently have environmental, economic, and social impacts not easy to predict and to cope with. The use of conventional crops for the manufacture of biofuels has been criticized in the report OECD-FAO 2007–2016 [692], which pointed out that the growing use of cereals, sugars, and vegetable oils to produce fossil fuel substitutes is underpinning crop prices and, indirectly, through higher costs of animal feed, also the prices of livestock products. Several studies have indeed recognized biofuel production as a major driver of food prices. This relationship raises an important policy issue, since EU and US government policies provide incentives to biofuel production; it has been suggested that such policies should be reconsidered in light of their impact on food prices [693, 694].

- **Environmental impact:** One of the reasons for replacing fossil fuels with biofuels is the need to reduce CO_2 production. Biofuels are, by themselves, carbon neutral because burning them just returns to the air the CO_2 that plants use to grow. However, even in the case of spontaneous biomass, the material has to be harvested, transported to a central location, and converted into a usable form. These operations need energy, and whether the net energy and ecological balance is positive cannot be taken for granted. In the US, the remnants of the nitrogen-based fertilizers used to grow corn are swept into the Gulf of Mexico creating dead zones so devoid of oxygen that most sea life cannot exist [695]. Furthermore, extensive use of ethanol in Brazil (40% of the fuel used) has been found to cause air pollution in the largest cities [696].

According to some experts, growing biofuels is probably of no benefit and in fact is actually making the climate issue worse. One reason is that food crops produced by intensive farming with nitrogen-containing fertilizers release N_2O (Section 7.4.3), which has about a 300 times greater greenhouse effect than that of CO_2, thus outbalancing the benefit of CO_2 production [697]. The sensitivity of the process to the nature of the raw material is underlined by the following example: ethanol from corn in the US and from sugarcane in Brazil is estimated to emit 18% and 91%, respectively, less CO_2 than gasoline [698]. The reason for this difference is that bagasse from sugar cane is used as a source of heat (and electricity) in the preparation of ethanol, including crushing and distillation. In contrast, ethanol from corn requires external energy which in the US comes mostly from coal. If emissions from land use changes resulting in massive deforestation to produce food crop-based biofuels are considered (the so-called carbon debt), the use of biofuels has a negative effect on climate [699, 700]. Other elements of concern are the effect of monoculture practice on biodiversity, excessive water use, eutrophication, and soil erosion, particularly in developing countries. In conclusion, first-generation biofuels are not at all carbon neutral and, depending on specific circumstances, they may also cause extensive damage.

- **Energy, economic, and social issues:** From the energy viewpoint, the benefit of biofuels, and also of any other energy resource, should be judged from the energy return on energy invested (EROI) index (Section 2.4). Some studies [689] estimate that in the US ethanol from corn or grain provides 25% more energy than is used to produce it (EROI 1.25), and biodiesel returns 93% more energy than that used to produce soybeans (EROI 1.93), but others believe that these figures are too optimistic and that the energy inputs to process the corn into ethanol amount to 99% of the energy produced by using ethanol, which means that the EROI index is very close to one [701]. After 30 years of optimization of productivity and scale economy, the cost of ethanol obtained from sugarcane in Brazil is fully competitive with gasoline without any subsidy. The production cost of ethanol from corn in the US and from wheat and sugar beet in Europe, however, is two and four times higher, respectively, and becomes competitive only because it is heavily subsidized.

In general, production and use of biofuel are conditioned by several global and local factors, such as availability of land and water, climate, type of feedstock employed, technology used for transformation, subsidies, cost of fossil fuels and tax on them, market conditions, and public perception. Each factor should be carefully evaluated within the real context, making reference to fossil fuels. The impact that biofuels could have on economies, especially of developing countries, should also be considered. A generalized concept, based on land availability and production costs, is that biofuel implementation will promote the economies of developing countries because they will produce and sell biofuels to the industrialized countries. However, potential damage caused by biofuel production should also be considered (Section 11.3.4). For example, in several Asian countries palm oil production is linked to extensive deforestation [702]. According to a report released by the United Nations Development Programme [703], 98% of rain forest in Indonesia will be destroyed by 2022 if the current rate of destruction is not stopped.

11.3.3
Second-generation Biofuels

To avoid the biofuel–food competition, efforts are currently directed towards second-generation biofuels, that is, produced from lignocellulosic materials, organic wastes, and terrestrial or marine biomass not involved in food chains (Figure 11.5). The production of second-generation biofuels utilizing non-edible feedstocks, however, is still in the stage of research and development [704–706]. A major doubt about the second generation of biofuels is whether there will be sufficient land available to grow the biomass needed for making them. Supporters emphasize that non-food crops can be cultivated in alternative lands including the so-called "wastelands," tropical zones, or even arid regions. Many of the biomass feedstocks are self-seeding crops (they do not need to be planted and re-seeded after being harvested) and require little or virtually no fertilizer input. These approaches can therefore be applied to "marginal lands" where the soil cannot support food crops. In this way, they will not interfere with the land dedicated to food crops, but some experts say that marginal lands should be used to grow trees, which would absorb amounts of CO_2 larger than emission avoided by the use of biofuels [707]. Second-generation biofuels could also be made from forests and from any kind of wood [708], but the carbon sequestered by restoring forests is often estimated to be greater than the emissions avoided by the use of liquid biofuels [709]. The complex and partly not yet known role of forests on climate change should be carefully investigated [710].

Second-generation biofuels are classified in various groups depending on the technologies employed for their preparation, which include chemical, biological and thermochemical conversion [58]. All of these technologies need to be improved. Second-generation biodiesel can be prepared by chemical conversion from non-foods crops like *Miscanthus*, a perennial grass, and *Jatropha*, a plant resistant to drought and pests, that thrives on poor soil and land unsuitable for food crops and produces seeds containing 27–40% oil. After oil extraction, the residues of the crops can be processed into biomass to power electricity plants. Currently the oil from *Jatropha* seeds is used for making biodiesel fuel in the Philippines and in Brazil, where *Jatropha* grows naturally. Hundreds of projects to produce biodiesel from *Jatropha* are being developed in India and in many developing countries, with controversial results [711]. Biodiesel from plants like *Jatropha* is reported to be a cheaper and more environmentally friendly alternative to fossil fuel, particularly for the commercial airline industry. The power per unit area of *Jatropha* on good land in hot countries is $0.18 \, W \, m^{-2}$, which reduces to $0.065 \, W \, m^{-2}$ on wasteland. It has been estimated that even if all Africa were covered with *Jatropha* plantations, the power produced would be only one-third of world oil consumption [701].

Biodiesel could also be obtained from aquatic biomass (microalgae, macroalgae, any other vegetable biomass). Algae are cell organisms that convert via photosynthesis water and atmospheric CO_2 into a variety of chemicals including methane and hydrogen [683, 712]. The efficiency of the process increases in water heavily enriched with carbon dioxide, achieving a power of $4\,W\,m^{-2}$. Very important companies have recently decided to enter this field [713, 714]. One goal is genetic engineering of photosynthetic algae to secrete hydrocarbons or hydrogen from their cells; this achievement would allow continuous production of the compounds in bioreactors. An inherent advantage of algae over other biofuels is that they can be grown on non-agricultural land. Biofuels and a variety of chemicals can be economically produced by catalytic cracking of biomass in oil refineries, requiring little capital investment [715]. Fuel production by biomass is discussed in Section 11.4.

One of the problems with second-generation biofuels is that is it not easy to find enzymes capable of breaking down cellulose. Microorganisms evolved to take care of themselves, not of our energy needs. Therefore, their metabolism should be re-engineered to excrete fuels and, at the same time, to be reasonably happy with this, otherwise they will rapidly evolve away from what we want them to do [716–718]. Production of biofuel is usually accompanied by the formation of numerous by-products, including chemical intermediates, proteins, and dyes. Use of these by-products may facilitate commercialization of biofuels.

11.3.4
Biofuel Perspectives

In general, biofuels are too costly in terms of land, raw materials, and energy, and their benefits have often been oversold by their proponents before vital environmental assessments of the technology were completed. As usually happens, sustained hype is followed by a backlash. Since the background for biofuel production can be very different depending on natural, technological, environmental, social, and political conditions, it would be quite useful to have a set of procedures, standards, and metrics capable of predicting whether it is convenient to develop biofuel production in any particular situation. An attempt in this direction is the life-cycle analysis (LCA), an important principle in understanding energy efficiency, pollution prevention, and sustainability. The LCA is obtained by measuring a balance of energy inputs and outputs, and material used and waste generated, usually in comparison with a reference system. In the case of biofuels, LCA is "field-to-wheel" and the reference system is fossil fuel ("well-to-wheel"). Relevant features, for example, for ethanol, are [719]: (i) the *net energy value* ($MJ\,l^{-1}$), defined as the energy content in 1 l of ethanol minus the fossil fuel energy used to produce it; (ii) the *net petroleum offset*, which is the number of liters of gasoline displaced by 1 l of ethanol; (iii) the *net carbon reduction*, which is the carbon emission reduction resulting from the consumption of 1 l of biofuel. Of course, other parameters would also be important (water consumption, erosion, eutrophication, loss of biodiversity, etc.), but they are difficult to quantify. In the US, the Environmental

Protection Agency (EPA) works on LCA as required by a law aimed at reducing the carbon footprint for different biofuels [720]. The available results show that biofuels can hardly help in reducing CO_2 emissions. Therefore, governments and supporters of biofuels should state clearly that the main objective of using domestic agriculture is not combating climate change, but having an internal source of energy to reduce the dependence on foreign sources.

In general, the development of biofuels, of both the first and second generations, will only be possible if it is supported by incentives and favorable regulations [721]. In several countries there is increasing concern regarding the use of generous public subsidies for promotion of biofuels, but it should be recalled that fossil fuels also receive a variety of known and hidden subsidies [591].

11.4
Future Options for Transportation Fuels

- **Synthetic gasoline:** The above discussed fuel production from biomass should be framed within the more general issue of synthetic fuels [208, 722]}. In July 2009, worldwide commercial synthetic fuel production capacity was about 240 000 barrels (38 000 m^3) per day, with numerous new projects under construction or development. Reforming of natural gas (Section 5.6) and gasification of coal (Section 6.4), biomass, and coal–biomass mixtures all produce syngas, which can be converted to diesel and gasoline or to methanol. Methanol can then be used as a transportation fuel or converted into gasoline (Figure 11.5). According to a recent report [58], the most promising approach to near future fuel transportation in the US is to convert syngas to methanol and then use methanol-to-gasoline technology to produce gasoline, which fits directly into the existing US fuel-delivery infrastructure. The economics of synthetic fuel manufacture vary greatly depending on the feedstock used, the technology employed, site characteristics such as feedstock and transportation costs, and the cost of equipment required to control emissions [723]. Concerning the ecological impact, one cannot expect any advantage in using synthetic fuels compared with the direct use of the original material (natural gas, carbon, or biomass). Therefore, this option should not be supported with incentives by responsible governments unless synthetic fuel production is coupled with carbon sequestration (Section 6.6).

- **Natural gas:** See Section 5.8.1.

- **Hydrogen:** The potential of hydrogen as a transportation fuel will be discussed in detail in the frame of the so-called hydrogen economy (Chapter 14).

- **Methanol:** Methanol (CH_3OH) is the simplest, safest, and easiest to store and transport liquid oxygenated hydrocarbon. Currently, methanol is a commodity chemical, but it can profitably be used as a transportation fuel. Indeed, its importance in the energy scene is increasing, and it has been proposed that methanol could be the basis of an entirely new economy: the methanol economy [111]. At present, methanol is prepared almost exclusively from syngas, a mixture of CO and H_2 obtained from the incomplete combustion of fossil fuels (mainly natural gas and coal). Methanol can also be prepared from biomass (wood, agricultural by-products, municipal waste, etc.), by oxidative conversion of methane, avoiding the initial preparation of syngas, or by reductive hydrogenative conversion of CO_2.

Methanol is a liquid that can be easily handled, stored, distributed, and carried on-board vehicles. Methanol has an octane number of 100, higher than gasoline, and can be blended with gasoline as an oxygenated additive. Alternatively, it can be used in today's internal combustion engines with minor modifications. Dehydration of methanol leads to dimethyl ether (DME), a relatively clean fuel which is considered one of the most promising candidates for substituting conventional diesel oil. Methanol can also be used to generate electricity in fuel cells (Section 14.7).

Methanol, however, also has some significant drawbacks. It is harmful to human health, in particular if ingested, absorbed through the skin, or inhaled. Methanol has about half the energy density of gasoline and, with pure methanol as a fuel, cold start problems can occur because it lacks the highly volatile compounds generally found in gasoline. Acidic products are formed during combustion of methanol, which cause the wearing of valves, valve seats, and cylinders much more than with hydrocarbon burning. Certain additives may be added to motor oil in order to neutralize these acids. Some of the materials used in gasoline distribution, storage devices, and connectors are incompatible with methanol, which corrodes aluminum, zinc, some plastics, and rubbers. This means that introduction of methanol on large scale would require the construction of a new distribution system and the use of flexible fuel vehicles that could run on a mixture of gasoline and methanol. These infrastructural barriers would be avoided by converting methanol to gasoline [58].

Concerning the issue of climate change mitigation, using as a fuel methanol prepared from fossil fuels does not help compared with direct use of the originating fossil fuel. However, recycling of CO_2 into methanol, would provide useful fuels and would also mitigate human-caused climate changes [111].

11.5
Artificial Photosynthesis

11.5.1
The Need for Solar Fuels

Mankind's technical evolution advances by observing and mimicking Nature, but it develops along unpredictable routes. In fact, we have often discovered that mankind's problems can be solved in different, usually simpler, ways than those exploited by natural evolution: Nature moves by walking, mankind has invented the wheel; Nature flies by moving wings, mankind by jet propulsion; Nature uses the brain for memory and computation, mankind has invented the computer. To exploit sunlight, Nature has developed over billions of years photosynthesis, a very complex process that produces high energy compounds through an intricate net of chemical reactions; mankind, instead, has invented a very simple ploy, the photovoltaic process, to convert sunlight into electricity with an energy efficiency at least an order of magnitude higher than that of the natural photosynthetic process. Nature has also transformed over billions of years the products of photosynthesis in fossil solar fuels – oil, gas, and coal – that are far more useful than electricity [536]. They are concentrated forms of energy that can be easily stored and transported. Since oil, gas, and coal reserves will eventually be exhausted and their use causes severe damage to human health and the environment, scientists are now engaged in finding ways to use sunlight for producing *clean* fuels. With this aim, they have carefully investigated the complex mechanism of natural photosynthesis and are now trying to create artificial photosynthetic processes much simpler than the natural one and capable of producing the simplest solar fuels: hydrogen and methanol.

The need for and the possibility of achieving the production of fuels by non-biological photochemical reactions were first anticipated by the Italian chemist Giacomo Ciamician about a century ago (see Box) [724]. At Ciamician's time, most of the energy was supplied by coal, a dirty fuel not easy to extract, transport, and use, and responsible for heavy pollution and many health problems. In the following years, coal was progressively replaced by oil, a fuel much easier to extract, transport, and use and, apparently, not dangerous for health and the environment. With the availability of such a friendly, cheap, and abundant fuel, there was no need to think about solar energy conversion. Later, exploitation of natural gas and, most important, the discovery of nuclear energy that was expected to generate electricity "too cheap to meter" induced people to believe that the energy problem was solved forever. The idea of artificial photosynthesis was reconsidered by a few scientists after the energy crisis of the 1970s [725, 726], but only in recent years, after full realization of the damage caused by burning fossil fuels (Chapter 7) and of the intrinsic difficulties with nuclear energy (Chapter 8), has it become the object of extensive investigations [727].

The Prophet of Solar Energy

Giacomo Ciamician (Figure 11.6) was professor of Chemistry at the University of Bologna from 1889 to 1922. Inspired by the ability of plants to make use of solar energy, he was the first scientist to investigate photochemical reactions in a systematic way. Most of Ciamician's considerations and predictions on the utilization of solar energy are contained in a famous address presented before the VIIIth International Congress of Applied Chemistry, held in New York in 1912 [724]. The address, entitled "The Photochemistry of the Future," was presented in English and, later, it was also printed in journals and booklets in

German, English, and Italian. After about a century, it is worthwhile reading this fascinating paper because Ciamician's intuitions appear to have been inspired by a prophetic spirit.

Ciamician began his address as follows: "Modern civilization is the daughter of coal, for this offers to mankind the solar energy in its most concentrated form; that is, in a form in which it has been accumulated in a long series of centuries. Modern man uses it with increasing eagerness and thoughtless prodigality for the conquest of the world and, like the mythical gold of the Rhine, coal is today the greatest source of energy and wealth. The Earth still holds enormous quantities of it, but coal is not inexhaustible. The problem of the future begins to interest us. Is fossil solar energy the only one that may be used in modern life and civilization? That is the question." If the word "coal", which was by far the most commonly used fuel at that time, is replaced by "fossil fuels," such a statement holds good even today.

Ciamician estimated that "The solar energy that reaches a small tropical country is equal annually to the energy produced by the entire amount of coal mined in the world," and noticed that "The enormous quantity of energy that the Earth receives from the Sun, in comparison with which the part which has been stored by the plants in the geological periods is almost negligible, is largely wasted." Then, Ciamician entered the core of the energy problem, wondering "whether there are not other methods of production which may rival the photochemical processes of the plants." He realized that "the fundamental problem from the technical point of view is how to fix the solar energy through suitable photochemical reactions" and forecast that "by using suitable catalysts, it should be possible to transform the mixture of water and carbon dioxide into oxygen and methane, or to cause other endo-energetic processes," which is indeed the aim of today's artificial photosynthesis.

The final part of the Ciamician's address is wonderful and is worth reporting almost completely: "Where vegetation is rich, photochemistry may be left to the plants and, by rational cultivation, solar radiation may be used for industrial purposes. In the desert regions, unsuitable for any kind of cultivation, photochemistry will artificially put their solar energy to practical uses. On the arid lands there will spring up industrial colonies without smoke and without smokestacks; forests of glass tubes will extend over the plants and glass buildings will rise everywhere; inside these will take place the photochemical processes that hitherto have been the guarded secret of the plants, but that will have been mastered by human industry which will know how to make them bear even more abundant fruit than Nature, for Nature is not in a hurry and mankind is. And if in a distant future the supply of coal becomes completely exhausted, civilization will not be checked by that, for life and civilization will continue as long as the Sun shines! If our black and nervous civilization, based on coal, shall be followed by a quieter civilization based on the utilization of solar energy, that will not be harmful to progress and to human happiness."

Exactly as we continue to believe today.

Figure 11.6 Giacomo Ciamician, pioneer of photochemistry and prophet of solar energy conversion, while watching flasks under solar irradiation on the roof of his laboratory at the University of Bologna, Italy, ca 1910.

11.5.2
Choosing the Right Type of Photoreaction

It has long been known that photoexcitation can bring about a variety of chemical reactions in which light energy is converted and stored into chemical bonds [728]. After the energy crisis of the 1970s, several types of endoergonic reversible photochemical reactions (e.g. photodissociation, valence photoisomerization) were proposed for artificial conversion and storage of solar energy, but the results have been disappointing [726]. The photoreaction chosen by Nature in the natural photosynthetic process is photoinduced electron transfer (Section 11.2.3). This process creates a charge-separated state that is the starting point to generate high-energy molecules. Since the mechanism of natural photosynthesis has been at least in part elucidated, the photoinduced electron transfer approach has been chosen by almost all scientists in the attempt to produce artificial solar fuels.

Plants and other types of photosynthetic organisms (algae) have an overall efficiency of solar energy conversion usually much less then 1%, but the efficiency of the early photochemical and chemical reactions of photosynthesis is very high (Section 11.2). This discovery opens up the possibility of developing high efficiency, artificial, molecular-based solar energy conversion technologies which exploit the principles of the front-end processes of natural photosynthesis and produce with high yield a simple fuel like hydrogen instead of complex organic molecules whose synthesis requires an intricate series of chemical steps. Today, our knowledge of the natural process is indeed sufficient to provide a blueprint for the design of an artificial photosynthetic system.

Currently, several research groups are trying to mimic closely the early events of the natural photosynthetic process with the aim of arriving at solar fuel production by photoinduced electron transfer reactions performed in solution. Other scientists are trying to produce fuels or electricity by photoinduced electron transfer processes taking place at solid state–solution interphases. Finally, as discussed in Chapter 10, photoinduced electron transfer is the basic event that leads to electricity generation in photovoltaic devices.

11.5.3
Choosing the Right Chemical Substrate

For solar fuel production to be economically and environmentally attractive, the fuels must be formed from abundant, inexpensive raw materials such as water and carbon dioxide. Liquid water should be split into molecular hydrogen and molecular oxygen [674, 729] and carbon dioxide in aqueous solution should be reduced to ethanol with the concomitant generation of dioxygen [730, 731]. Carbon dioxide reduction, however, is a very difficult process from a kinetic viewpoint. Therefore, the attention of most scientists operating in the field of solar fuels is focused on the photochemical water splitting reaction [674, 732]:

$$H_2O + \text{sunlight} \rightarrow H_2 + \frac{1}{2}O_2 \tag{11.1}$$

As discussed in detail in Chapter 14, if hydrogen could be generated from photochemical water splitting by using solar energy, both the energy and the environmental problems of our planet would have largely been solved. It should be noted, however, that storing the equivalent of the current energy demand would require splitting more than 10^{15} mol per year of water, which is roughly 100 times the scale of the most important chemical reaction performed in the chemical industry today, namely nitrogen fixation by the Haber–Bosch process [733]. This comparison tells us once again that perhaps we are asking too much of ourselves and of our planet.

Water splitting by sunlight, of course, is not a straightforward process, otherwise it would already happen in Nature. The electronic absorption spectrum of water, in fact, does not overlap the emission spectrum of the Sun, so that direct water dissociation by sunlight cannot take place (Figure 11.7) [725]. On the other hand, the free energy for the dissociation of liquid water into molecular hydrogen and molecular oxygen is 1.23 eV (29 kcal mol^{-1}). As shown in Figure 11.7, the excess energy needed for direct dissociation is due to the formation of radicals that, besides requiring more energy to be produced, cannot be stored. Therefore, the problem with conversion of solar energy using water as a substrate is to find systems in which water dissociation takes place by using lower energy photons without formation of intermediate radical species. This means that the water splitting reaction by sunlight must be sensitized by species absorbing visible light and capable of being involved in cycles in which the absorbed energy can be profitably

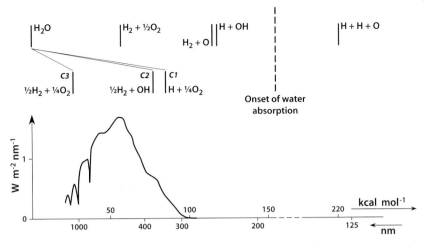

Figure 11.7 Spectral distribution of solar radiation at sea level with AM1, and energy needed for the various water splitting processes [725].

used. For liquid water, three different cycles can be devised (C1–C3, Figure 11.7), which correspond to the following threshold energies: C1, 332 nm; C2, 367 nm; C3, 1010 nm. The advantage of the C1 and C2 systems over direct photodissociation of water in the gas phase is that only one radical is formed. From the thermodynamic viewpoint, the most convenient process is C3, the evolution of molecular oxygen and molecular hydrogen from liquid water (Equation 11.2), whose low energy thermodynamic threshold (1.23 eV) in principle allows conversion of about 40% of the solar energy. This cycle, however, requires two multielectron processes (Equations 11.3 and 11.4), including the very complex four-electron oxidation of two water molecules coupled to the removal of four protons (Equation 11.4):

$$\frac{1}{2}H_2O \rightarrow \frac{1}{2}H_2 + \frac{1}{4}O_2 \qquad \Delta G° = +1.23\,eV \tag{11.2}$$

$$2H_2O + 2e^- \rightarrow H_2 + 2OH^- \qquad E° (pH\,7) = -0.41\,V\ vs\ NHE \tag{11.3}$$

$$2H_2O \rightarrow O_2 + 4H^+ + 4e^- \qquad E° (pH\,7) = -0.82\,V\ vs\ NHE \tag{11.4}$$

Since in a photochemical process each photon can transfer only one electron, it follows that two catalysts must be present in the system: one should collect electrons to produce molecular hydrogen, which is a two-electron reaction, and the other should collect holes to produce dioxygen, which is a four-electron process. This means that oxygen evolution requires four successive photoinduced electron transfer steps (threshold +1.23 eV), coupled with proton transfer (PCET processes), as happens in the natural photosynthetic system.

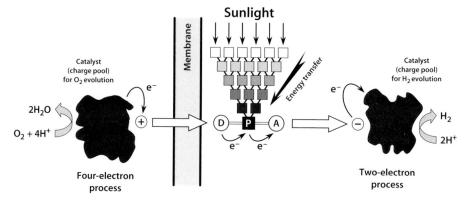

Figure 11.8 Schematic representation of the strategy for photochemical water splitting (artificial photosynthesis). Five fundamental components can be recognized: an antenna for light harvesting, a charge separation triad D–P–A, a catalyst for hydrogen evolution, a catalyst for oxygen evolution, and a membrane separating the reductive and the oxidative processes [674].

11.5.4
Components of an Artificial Photosynthetic System

The best way to construct artificial photosynthetic systems for water splitting (or, more generally, for solar fuel production) is that of mimicking the molecular and supramolecular organization of the natural photosynthetic process: light harvesting should lead to charge separation, that must be followed by charge transport to deliver the oxidizing and reducing equivalents to catalytic sites where oxygen and hydrogen evolutions should occur separately. Therefore, a plausible artificial photosynthetic system should include the following basic features (Figure 11.8) [674]: (i) an antenna for light harvesting, (ii) a reaction center for charge separation, (iii) catalysts as one-to-multielectron interfaces between the charge-separated state and the products, and (iv) a membrane to provide physical separation of the products.

11.5.5
Coupling Artificial Antenna and Reaction Center

The complexity of the natural photosynthetic systems, largely related to their living nature, is clearly out of reach for the synthetic chemist. Single photosynthetic functions, however, such as electronic energy transfer and photoinduced electron transfer, can be duplicated by simple artificial systems. The important lesson from Nature is that efficient conversion of light into a long-lived charge-separated state requires the involvement of supramolecular structures with very precise organization in the dimensions of space (location of the components), energy (excited state levels and redox potentials), and time (rates of competing processes). Such an

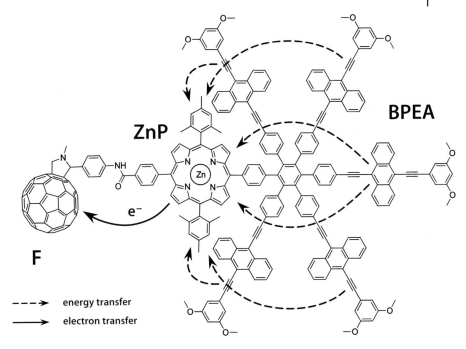

Figure 11.9 Energy- and electron-transfer processes occurring in a heptad consisting of five bis(phenylethynyl)anthracene (BPEA) antenna units and a zinc porphyrin–fullerene (ZnP-F) electron donor–acceptor module organized around a hexaphenylbenzene scaffold [734].

organization, which in natural systems comes as a result of evolution and is dictated by intricate intermolecular interactions, can be imposed in artificial systems by choosing selected building blocks and fixing them in a suitable structure via covalent or non-covalent bonding [77].

A successful example of this strategy is shown in Figure 11.9 [734]. In this supramolecular system, the hexaphenylbenzene scaffold provides a rigid and versatile core for organizing antenna chromophores and coupling them efficiently with a charge-separation moiety. The compound contains five bis(phenylethynyl) anthracene (BPEA) antenna units and a zinc porphyrin–fullerene (ZnP-F) electron donor–acceptor module. The BPEA antenna chromophore was chosen because it absorbs strongly in the visible region, as do carotenoid polyenes in photosynthetic organisms. Energy transfer from the five antennas to the zinc porphyrin occurs on the picosecond time scale with unitary efficiency, as happens in natural antenna systems. After light harvesting, the first singlet excited state of the zinc porphyrin P donates an electron to the attached fullerene to yield a ZnP$^+$–F$^-$ charge-separated state, again with almost unit efficiency. This clever molecular device, however, is useless alone: after charge separation has been achieved, a fast back electron transfer reaction (nanosecond time scale) dissipates the converted light energy.

11.5.6
The Problem of Multi-electron Redox Processes

The main problem with artificial photosynthesis is to succeed in taking advantage of the energy of the charge separated state, created by light absorption, to generate high energy stable species, that is, a fuel. In the water splitting reaction (Figure 11.8) this means coupling photoinduced charge separation, which is a one-photon–one-electron process, with hydrogen and oxygen evolution, which are multi-electron transfer processes (Equations 11.3 and 11.4). As discussed above, from the standard redox potentials of the two corresponding half-reactions, the free energy demand for water splitting (Equation 11.2) is 1.23 eV. For the charge-separated states of several investigated D–P–A triad systems, the difference in redox potentials of the oxidized and reduced molecular components is larger than 1.23 V [674]. Thus, with such systems, water splitting by means of photoinduced charge separation (Equation 11.5), followed by Equations 11.6 and 11.7, is thermodynamically feasible:

$$D–P–A + h\nu \rightarrow D^+–P–A^- \tag{11.5}$$

$$D^+–P–A^- + H_2O \rightarrow D^+–P–A + \frac{1}{2}H_2 + OH^- \tag{11.6}$$

$$D^+–P–A + \frac{1}{2}H_2O \rightarrow D–P–A + \frac{1}{4}O_2 + H^+ \tag{11.7}$$

None of those systems, however, would evolve hydrogen and oxygen upon simple irradiation in aqueous solution. There is, in fact, a fundamental kinetic problem. The photoinduced charge separation is a one-electron process (i.e. A^- and D^+ are one-electron reductants and oxidants). On the other hand, the reactions depicted by Equations 11.6 and 11.7, although written in one-electron terms for stoichiometric purposes, are inherently multielectron processes (two electrons for Equation 11.6 and four electrons for Equation 11.7). Therefore, although relatively long-lived charge separation can be achieved with supramolecular systems, the reactions depicted by Equations 11.6 and 11.7 are hopelessly slow to compete with charge recombination:

$$D^+–P–A^- \rightarrow D–P–A \tag{11.8}$$

This problem is common to any conceivable fuel-generating processes. The answer to this general problem lies in the possibility of accelerating multielectron redox reactions by use of artificial catalysts (Figure 11.8) [674, 725]. A catalyst for multi-electron redox processes is essentially a "charge pool," that is, a species capable of (i) acquiring electrons (or holes) from a one-electron reducing (or oxidizing) species in a stepwise manner at constant potential, and (ii) delivering these electrons (or holes) to the substrate in a "concerted" manner to avoid the formation of high-energy intermediates.

From the field of heterogeneous catalysis, metals and metal oxides are known to be good candidates for this type of process. For example, colloidal platinum is a good catalyst for photochemical hydrogen evolution and colloidal RuO_2 is a moderately efficient catalyst for photochemical oxygen formation. Both platinum and ruthenium, however, are rare and very expensive. More recently, a catalyst for oxygen evolution which is formed from Earth-abundant cobalt and phosphate ions has been reported [733]. Much research is needed in this field.

Apart from solid-state materials, discrete supramolecular species can also be conceived as catalysts for multielectron redox reactions. As we have seen in Section 11.2.3, Nature has solved this problem by creating the Mn_4Ca oxygen evolving complex [669]. Several investigations have been performed in the field of oxygen generating catalysts [735–738]. In order to understand the mechanisms of the reactions involved and to optimize one of the two sides of the process (Figure 11.8), experiments can be performed in the presence of sacrificial reductants or oxidants that prevent fast charge recombination and allow the occurrence of hydrogen or oxygen evolution.

An artificial supramolecular catalyst for multielectron redox processes must contain several equivalent redox centers (at least as many as the electrons to be exchanged), with the appropriate redox properties to mediate between the charge-separated state and the substrate. The electronic coupling between such centers should be not too strong, otherwise the "charging" process (stepwise one-electron transfer to or from the catalyst) could not occur at a reasonably constant potential. The centers should, on the other hand, be sufficiently close to cooperate in binding and reducing (or oxidizing) the substrate. Apparently, electron transfer alone cannot satisfy these requirements. The lesson coming from Nature (Section 11.2.3) is that multiple electron-transfer processes can be profitably accomplished when accompanied by proton transfer to maintain electric neutrality. This can occur by a sequence of two distinct reactions, or by a concerted process (PCET). PCET processes are advantageous from the thermodynamic viewpoint, but they are inevitably more complex than either electron or proton transfer because both electrons and protons must be transferred simultaneously [739]. The potential of biologically templated photocatalytic nanostructures is also actively investigated [740].

Much has to be learnt in the field of artificial photosynthesis concerning the choice of the best chromophoric groups, electron acceptors, electron donors, and homogeneous and heterogeneous catalysts for hydrogen and oxygen evolution.

11.5.7
Water Splitting by Semiconductor Photocatalysis

The basic principles of water splitting by irradiation of a semiconductor photocatalyst are illustrated in Figure 11.10 [741]. Irradiation with photons having energy equal to or greater than the bandgap of the semiconductor promotes an electron (e^-) to the conduction band (CB), with concomitant formation of a hole (h^+) in the valence band (VB). Electrons and holes that migrate to the surface of the semiconductor (in competition with charge recombination) can react with water generating H_2 and O_2, respectively. At pH 0, the reduction reaction that leads to H_2 generation

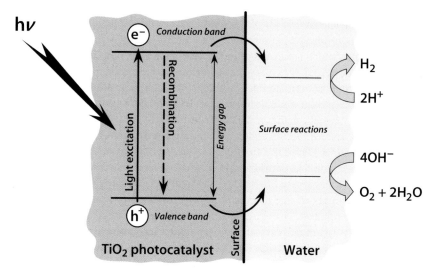

Figure 11.10 Fundamental steps for water splitting by solid-state photocatalysis. For details, see text.

takes place when the bottom of the CB is located at more negative potential than the reduction potential of the H^+/H_2 couple (0 V vs NHE) at pH 0, and the oxidation reaction that leads to O_2 formation occurs when the top of the VB is positioned more positively than the oxidation potential of the H_2O/O_2 couple (1.23 V vs NHE at pH 0). Therefore, excitation can lead to water splitting only if the photon energy is equal to or greater than 1.23 eV, which corresponds to about 1010 nm. It follows that visible light is energetically sufficient to carry out water splitting, but it should be pointed out again that 1.23 eV constitutes the thermodynamic requirement of the process and that in practice the photonic energy needed is larger. The pioneering experiment in this field was performed in 1972, using TiO_2 as a semiconductor and UV light excitation because TiO_2 has a very high energy gap (3.2 eV) [742]. In order to perform water splitting by sunlight, semiconductors should have a bandgap small enough to be excited with visible light. Several hundred semiconductor systems, including species containing dopants to modify the energy gap and/or co-catalysts on the surface to facilitate the reaction with water, have been investigated [741, 743]. Dye-sensitized photocatalysts (see the next section) and two-step photoexcitation have been studied [744]. In many cases the experiments were performed in the presence of sacrificial reductants or oxidants to test and optimize the occurrence of hydrogen and oxygen evolution, respectively. So far, the maximum quantum efficiency for overall water splitting upon visible light irradiation achieves only a few percent. The roles played by several factors such as composition, structure, particle size, defect density, and surface structure have not yet been elucidated.

11.6
Dye-sensitized Solar Cells

Solar power can be converted directly into electrical power not only in solid state photovoltaic devices (Chapter 10), but also in photoelectrochemical cells, often called Grätzel cells from the name of the Swiss scientist who has developed them [745–747]. These cells are based on sensitization of wide gap semiconductors by dyes capable of exploiting sunlight, that is, visible light.

The working principle of a dye-sensitized solar cell (DSSC) is shown schematically in Figure 11.11a. The system comprises a nanocrystalline semiconductor TiO_2 electrode of very high surface area and a conductive and transparent glass counter electrode, immersed in an electrolyte solution containing a redox mediator (R). A photosensitizer (P) is linked in some way [usually by $-COOH$, $-PO_3H_2$, or $-B(OH)_2$ functional groups] to the semiconductor surface. Light is absorbed by the sensitizer that is promoted to an excited state. The excited sensitizer then injects, on the femto- and picosecond time scale, an electron into the conduction band of the semiconductor (step 1 in Figure 11.11a). The oxidized sensitizer is reduced by a relay molecule (step 2), which then diffuses to discharge at the counter electrode (step 3). As a result, a photopotential is generated between the two electrodes under open-circuit conditions, and a corresponding photocurrent can be obtained on closing the external circuit by use of an appropriate load. A great number of photosensitizers of the Ru–oligopyridine family have been employed. The most efficient ones are those bearing two NCS^- and two substituted 2,2'-bipyridine (bpy) ligands (see, e.g., the compound shown in Figure 11.11b), which exhibit intense absorption bands in the visible region. A variety of solvents of different viscosity and of redox mediators have been used, the most common being the I_3^-/I^- couple in acetonitrile solution.

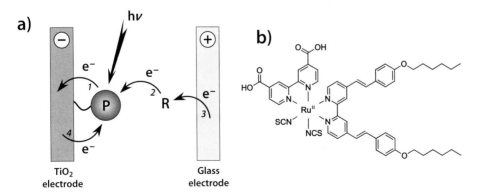

Figure 11.11 (a) Working principle of a dye-sensitized solar cell. P is a photosensitizer linked to semiconductor electrode and R is an electron relay molecule. (b) Examples of an Ru–oligopyridine-type photosensitizer.

In the scheme of DSSCs, the absorption spectrum of the cell is separated from the bandgap of the semiconductor, so the cell sensitivity is more easily tuned to match the solar spectrum. The cell efficiency depends on the dynamics of several kinds of processes, such as the way in which the electrons move across the dye–TiO_2 and the dye–relay interfaces, and the way in which charges move through the dye, the TiO_2 particles, and the electrolyte. So far, a maximum global efficiency of 11% has been reported [747]. DSSCs are already available from commercial providers.

DSSCs might look at first sight quite different from the photosynthetic systems discussed in Sections 11.2 and 11.5. On closer inspection, however, analogies do appear. DSSCs are clearly based on photoinduced charge separation. From this viewpoint, they can be regarded as heterogeneous "pseudo-triads" in which the semiconductor surface acts as the primary acceptor and the relay as the secondary donor. As in any triad, the efficiency of charge separation and energy conversion depend critically on the kinetic competition between the various forward processes and charge-recombination steps. The main difference from photosynthetic systems is simply that the redox potential energy of the charge-separated state is not stored in products of subsequent reactions, but rather is used directly to produce a photocurrent.

Grätzel cells can be modified to perform water splitting (artificial photosynthesis) instead of producing electricity [735, 748, 749]. This result has been obtained by the following array [748]. A nanoparticulate TiO_2 photoanode is sensitized by a ruthenium–polypyridine complex in which one of the bipyridine ligands bears phosphonates that bind selectively to TiO_2, while a second ligand bears a malonate moiety that links a nanoparticulate $IrO_2 \cdot nH_2O$ oxygen-evolving catalyst. In this system, solar production of hydrogen and oxygen was obtained with the help of a small potential electric bias. Although the process is inefficient and the dye is progressively bleached, this system shows that artificial photosynthesis can be achieved.

11.7
The Solar Fuel Challenge

To overcome the diurnal cycle and the intrinsic terrestrial intermittency of sunlight, efficient exploitation of solar energy requires (i) sunlight capture, (ii) conversion of the captured light into a usable form of energy, and (iii) storage of the converted energy. Photochemistry can accomplish this task directly by producing solar fuels, that is, converting the absorbed light into high-energy, but kinetically stable chemical bonds. Natural photosynthesis has long since performed this fundamental function, producing biomass that gave rise to the solar fossil fuels (oil, gas, and coal) that currently sustain civilization. Since these resources will eventually be exhausted and their use causes severe damage to human health and the environment, there is an urgent need to produce new, clean solar fuels. Photosynthesis itself is relatively inefficient (<0.5%) when measured on a yearly average basis per unit area of insolation. If cellulosic conversion technology can be successfully developed and deployed economically [750], biofuels will provide a potentially significant contribution to liquid fuels for transportation uses. But even if all the land not currently used for food crops were to be dedicated to biofuel production, only a small fraction of mankind's energy needs would be covered. Furthermore, biofuels cannot give important help in reducing CO_2 emissions. Clearly, mankind needs to find a way to convert and store solar energy in chemical bonds of clean fuels with a significantly higher efficiency than plants or algae.

As described above, an important solar energy conversion and storage approach involves water splitting with formation of molecular hydrogen. On paper, this process is a very simple chemical reaction (Equation 11.2), but if it has to occur by visible light excitation and store sunlight into the chemical energy of hydrogen (Equation 11.1), it becomes a complex process in which one-photon–one-electron excitation must be coupled with a multielectron overall transformation (Figure 11.8). Moreover, to ensure charge neutrality in the system, proton transfer must accompany electron transfer (PCET). Because of its complexity, this photoinduced process needs the assistance of mediators (catalysts) which, whether molecular or solid, will invariably be involved in bond-making and bond-breaking processes. The design of specific water splitting strategies is a fascinating and challenging problem of modern chemistry [674, 751]. Significant advances in basic science are needed for this technology to attain its full potential that, hopefully, will allow mankind to achieve a definitive solution to the energy and climate problems.

12
Other Renewables

"The country that harnesses the power of clean,
renewable energy will lead the 21st century."

Barack Obama

12.1
Hydroelectric Energy

12.1.1
The Rise of Hydropower

The use of falling water to provide energy has a long history in many parts of the
world (Chapter 3), mainly with the use of waterwheels. At the end of the nineteenth century, the water turbine slowly replaced the waterwheel due to its higher
efficiency and power. In the first half of the twentieth century, hydropower fostered
the industrialization process in many countries [752]. Coal-rich states such as the
UK, US, and Germany used hydropower and coal powered thermal plants, but
coal-poor states such as Italy and France turned essentially to their "white coal,"
that is, the hydropower resource. Today hydropower is by far the largest source of
renewable energy, covering about 16% of the global electricity demand and 2.2%
of the total primary energy supply (TPES). Interestingly, nuclear energy represents
a much higher amount of TPES (5.9%), but provides a lower global share of electricity with respect to hydropower (13.8%) [4]. This apparent discrepancy reflects
the about threefold higher generation efficiency of hydroelectric technology (~90%)
versus thermonuclear (~33%).

Hydroelectricity generation is based on turbines that convert the energy of
falling water into mechanical power (Figure 12.1). At 100% conversion efficiency,
it takes 100 kg of water falling a distance of about 6 m to keep a 100 W light bulb
going for 1 min. Two major approaches are utilized to harness and exploit the
potential energy of water flows: (i) mountain lakes or reservoirs held back by dams;
these systems have large generating capacities, in the range from ca 5 MW to
several gigawatts, and are operated by electricity utilities that feed electricity into
the grid; (ii) flows of rivers or streams, normally without involving reservoirs and

Energy for a Sustainable World: From the Oil Age to a Sun-Powered Future. Nicola Armaroli and
Vincenzo Balzani
© 2011 WILEY-VCH Verlag GmbH & Co. KGaA, Weinheim
ISBN: 978-3-527-32540-5

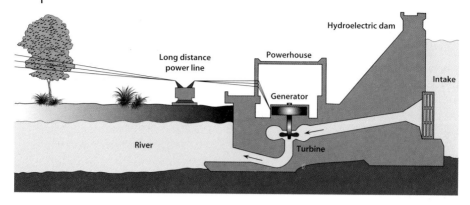

Figure 12.1 Schematic section of a hydroelectric power plant with water reservoir.

dams. These small facilities, that produce from a few kilowatts up to a several megawatts, may be privately owned and operated; they can provide power to villages, isolated communities, or industrial processes requiring a relatively large and steady supply of electricity.

In principle, hydropower could also be obtained by allowing ocean water to flow through turbines into nearby areas below sea level, from where water would be removed by evaporation. Possible locations could be the Dead Sea, the Qattara Depression in Egypt, and Death Valley in California.

12.1.2
Potential, Current Deployment, and Use

Surface water runoff is one of the mightiest natural forces on Earth and, over eons, a key agent in sculpting the morphology of our continents through erosion. The estimated global surface runoff is $44\,500\,km^3$ per year [3]. The Amazon alone carries about 16% of the water that flows through the planet's rivers. Waterfalls are the sites with highest concentration of released hydropower. Inga Falls in the Congo release a steady average of 16.3 GW, Iguassú in South America 4.9 GW, and Niagara Falls 3.4 GW [3]. Even just the latter, alone, might in principle provide a substantial fraction of the baseload electricity for an industrial country of the size of Germany.

According to recent assessments, hydropower could theoretically provide about 60 EJ of energy [516], approximately 12% of the current TPES, to be compared with the actual 2.2% currently provided [4]. However, taking into account technical, environmental, and legal restrictions, only a fraction of this potential can be exploited, suggesting that there is still room for expansion.

As a source of electricity, hydropower is second only to fossil fuels. It is exploited in at least 160 countries and in some of them (e.g. Brazil, Canada, Austria, Norway, Venezuela and many African countries) it accounts for the large majority of the internal electric energy production [80, 753]. The overall hydroelectricity power corresponds to a generating capacity of 850 GW, with an annual production of about 3 PWh [754]. An additional 120 GW of capacity is under development, mainly in China and India, but most of the existing facilities worldwide will need to be modernized or decommissioned by 2030, since they have been in service for almost a century [80]. Indeed, in some of the first countries that built dams for hydropower, such as the US, the rate of dam removal is now exceeding the rate of construction [755]. There are about 800 000 dams operating worldwide; 45 000 are higher than 15 m ("large"), with major dams over 150–250 m in height. It is estimated that almost half of the world's rivers contain at least one large hydroelectric dam. By far the largest electric generation facility in the world is fed by the Three Gorges Dam in China, with a rated capacity of 22.5 GW and a surface area of over 1000 km².

Initially, most of the hydropower projects were built to provide a primary baseload to the power system and this still occurs in countries where the technology occupies a significant share in the power generation mix. However, in recent decades, hydroelectricity emerged primarily as a peak power technology, buffering the fluctuating electricity demand throughout the day, thanks to the fact that water turbines can get started at any time within a few seconds. For this reason hydropower is ideal to smooth the production of inherently intermittent sources such as wind or solar energy, and this role is expected to increase substantially in the near future [756].

12.1.3
Advantages, Disadvantages, and Environmental Impact

The main advantage of hydropower is that it produces electricity at the cheapest price, with a supply that can range from tiny to huge. The reservoirs created by dams are also normally conceived as a tool to control devastating floods that, in the past, have plagued wide geographic areas with innumerable casualties. The main disadvantage of hydropower is invasiveness. As an average, 100 MW of hydroelectric power claims an area of about 50 km², 200 times more than geothermal and thousands of times more than wind [516]. World artificial water reservoirs occupy an area approximately half the size of Italy [52]. Interestingly, it has been argued that the 10 800 km³ of water artificially trapped on continents has decreased the sea level rise due to climate change by about 3 cm over the last 50 years [275].

In recent decades, concerns have cooled interest in dam construction in most industrialized countries, because hydroelectric projects may suffer from hidden costs that make them economically and ecologically questionable [757, 758]. Dams have physically displaced millions of people worldwide, and most of them have never regained their former livelihood. In many cases, dams have led to the disappearance of several valuable sacred, archeological, and cultural sites and to a significant and irreversible loss of species and ecosystems. Many dammed rivers, such as the Colorado River in the US and the Yellow River in China, no longer consistently reach the sea. Water exiting a turbine usually contains very little suspended sediment, which can lead to scouring of river beds and loss of river banks;

attempts to restore ecosystems through silt released with forced floods into the Colorado river were unsuccessful [759]. On the other hand, silt is slowly deposited behind dams and progressively reduces their power capacity. Another serious problem is that hydropower facilities can damage fish, restrict or delay migration, increase predation, and affect water quality [760].

The ecological consequences of the gigantic Three Gorges Dam in China are just starting to appear and only time will tell if an acceptable ecological equilibrium will be achieved for the local population and biota [761]. Meanwhile it has been conjectured that the 2008 catastrophic earthquake in China's Sichuan Province, that killed some 80 000 people, could have been caused by stress from the water piled behind the new gigantic Zipingpu dam, which might have triggered the failure of a nearby geological fault [17, 762]. Dam failures are generally cata-strophic; a recent preliminary estimate has concluded that in the period 1907–2007 there were more than 180 000 direct deaths caused by accidents at energy infra-structures, 94% of which are attributable to hydropower facilities [763].

Hydroelectric generation is expected to be vulnerable to climate change (Chapter 7), due to local or regional variations of rainfall and surface water levels associated with reduced glacial melt [764]. Furthermore, recent research on climate change has shown that decaying plant matter from flooded areas releases methane and carbon dioxide [765]. Nevertheless life-cycle emission of greenhouse gases of hydropower remains one of the lowest among electrical technologies, 17–22 g of CO_2 equivalent (CO_{2eq}) per kWh, above only those of wind, CSP, and tidal [516].

A hydropower facility is estimated to produce 200 times more energy than that required to be built [52]. The high land requirement (see above) is somewhat moderated if a life-cycle analysis approach is considered [576]. Hydropower has the highest electricity conversion yield, the lowest operating costs, the longest plant life compared with other large-scale generating technologies, and a widely geo-graphically spread availability. Finally, hydropower stations are unique multipur-pose energy infrastructures which can provide additional vital services such as irrigation water, navigation improvements, and amenities for recreational use.

12.1.4
Hydropower Future

Hydroelectricity is a mature technology, with little room for expansion in the most widely exploited regions of the past, such as Europe and North America. In con-trast, there is considerable opportunity for new capacity in Africa, Asia, and South America which could expand hydropower up to ten-, three-, and two-fold relative to the current level, respectively [80]. Although the world energy system is increas-ingly interconnected and transnational, hydropower maintains attractiveness, for a given country, as an intrinsically domestic electricity resource. Difficulties may sometimes arise because the exploitation of this "domestic" resource actually affects the water control and availability of another country downstream, showing once again the extent to which we are interconnected on our small and fragile planet.

It is conceivable that small hydropower (SHP) projects, up to 10–20 MW, will emerge as a more sustainable alternative to the traditional hydroelectric dams [766]. SHP will become a relatively important player in the frame of a distributed electric system managed through a smart grid (Chapter 13). SHP has now reached a global installed capacity of 60 GW worldwide; most of it is installed in China, where SHP grew by 4–6 GW annually during 2004–2008 [544].

12.2
Wind Energy

12.2.1
Brief Historical Notes

For centuries, wind has been utilized not only for sailing ships but also for grinding grain and pumping water [71] (Chapter 3). Windmills were used by American farmers to mine groundwater for crop irrigation and livestock and personal use; windmills were also utilized to fill water tanks for steam locomotives and produce electricity for residential use, until the electric grid was implemented during the 1930s. Practically, windmills helped to transform the arid Great Plains into a breadbasket.

The development of wind turbines for power generation started at the end of the nineteenth century and remained substantially unchanged for several decades even when, after World War II, the first relatively large wind turbines (100–300 kW) became available. Interest in wind power finally rose again in the aftermath of the oil shock of the early 1970s; the modern wind power industry began in 1979 with the serial production of wind turbines by a few Danish manufacturers [767]. Starting from the early 1990s, the wind power sector has become one of the fastest growing industries, generating tens of thousands of jobs worldwide. Wind power has grown to such an extent (over 30 times in the period 1995–2009) that it now can hardly be defined as "alternative" technology.

12.2.2
Wind Power Technology

Modern wind turbines (or aerogenerators) can be categorized into two main classes based on the orientation of the rotor relative to the ground, that is, vertical and horizontal (Figure 12.2). Vertical axis wind turbines exhibit some benefits such as a lighter structure and the possibility of accepting wind from any direction. These features can make them interesting for small wind catching systems mounted on the top of large buildings. Nowadays, however, almost all of the commercial wind power systems have a horizontal axis, with the electric generator mounted atop behind the blades.

The key components of a classic horizontal axis wind turbine in its simplest design are depicted in Figure 12.3. Externally, five components are usually distinguishable:

Figure 12.2 Examples of aerogenerators with rotating axis vertical (a) and horizontal (b).

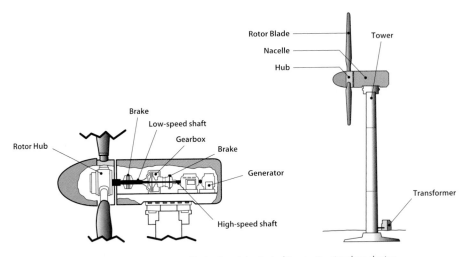

Figure 12.3 Main components of a horizontal axis turbine in its simplest design.

1) **Blades:** These are made of fiber-glass or carbon fiber materials or, when small, of lightweight metals such as aluminum. The average rotation diameter is about 80 m, with the largest systems as large as 100–120 m. Three-blade machines operate at lower speed than two-blade systems, and produce less acoustic disturbance (aerodynamic noise). In noise-sensitive environments a

tip speed limit can be set. A basic feature of wind technology is that the larger the width of the blades the more air friction they experience, thus lowering the wind speed that can start the windmill, but also the overall efficiency [768].

2) **Hub:** This is the unit that holds the blades in place and attaches it to the shaft.

3) **Nacelle:** The nacelle is the main body of a horizontal axis wind turbine and contains several components: a low-speed and a high-speed shaft connected through a gearbox; an AC electricity generator; brakes which stop rotation in case of power overload or system failure; and other key ancillary components (not shown) including an anemometer and several electronic controllers such as those to align the rotor with the direction of the wind.

4) **Tower:** Wind blows faster at higher altitudes because of the drag of the sea or land surface. Doubling the altitude of a turbine increases the expected wind speeds by about 10% and the expected power by 34% (see below). Currently wind turbine towers support hubs placed as high as 125 m above the ground [768].

5) **Transformer:** The voltage generated by a wind turbine is generally below 1 kV, even if some larger turbines may generate up to 3 kV. Individual turbines of a wind farm are clustered into another transformer substation, where the voltage is stepped up to higher voltage for connection to the existing electric grid.

The power produced by a wind machine is directly proportional to the swept area and proportional to the cube of the wind speed. Therefore, doubling the size of the machine doubles the power obtained, but doubling the wind speed increases the power eightfold. Modern wind turbines are designed to work most efficiently at wind speeds in the range 7–15 m s^{-1}. At high wind speeds, normally above 25 m s^{-1}, blades are feathered to prevent damage. There are two basic approaches used to control and protect a wind turbine from excessive winds: pitch control and stall control [768].

A key factor to characterize an energy-producing facility is the so-called *capacity factor*, which represents the amount of energy produced in one year divided by the total amount it could generate if it ran at full capacity. A typical wind farm operates with a capacity factor approaching 30%. Of the 58 projects installed from 2004 to 2006, over 25% had capacity factors greater than 40% [516]. This substantial progress is due to improved site selection and constant enhancement of the reliability and efficiency of wind power machines.

Modern wind turbines have been proven to operate effectively in a temperature range between −20 and 40 °C; specific cold and hot weather versions are available. The operating lifetime is normally set at 20 years but, since many turbines from the 1970s are still operating today, a 30 year lifetime seems to be more realistic [516]. Construction times of wind farms are just a few months, that is, incomparably faster than for any other electric power facility. Repowering, dismantling, or even relocation of wind turbines can be accomplished with a simplicity and speed that cannot be rivaled by any other electric technology.

12.2.3
The Huge Potential of Wind Power

About 0.87 PW of the overall 90 PW of solar radiation absorbed by the Earth is converted into the kinetic energy of winds in the atmosphere, which are dissipated ultimately by friction at the Earth's surface [3, 769].

Theoretically, only about 60% of the kinetic energy of the wind can be converted into mechanical energy turning a rotor (Betz limit). This limit has nothing to do with inefficiencies in the generator, but is due to the very nature of wind turbines themselves [770]. Windmill efficiency, in fact, is limited because the wind must keep on moving after passing through the spinning blades. Practically, when the wind blows steadily at the optimum speed, the best modern wind turbines can reach an efficiency of 40%, that is, about two-thirds of the theoretical limit. Estimations of the global potential of wind generated electricity have been accomplished in recent years, taking advantage of an increasingly accurate resolution of the wind fields across the planet provided by satellites, aircraft, balloons, ships, buoys, and dropsonde measurements [769, 771].

A recent study indicates that a network of land-based 2.5 MW turbines restricted to non-forested, ice-free, non-urban areas operating at as little as 20% of their rated capacity could supply more than 40 times the current worldwide consumption of electricity and over five times the total global use of energy in all forms. Estimates were also made for quantities of electricity that could be obtained by using a network of 3.6 MW turbines deployed in ocean waters at depths of <200 m within 50 nautical miles (92.6 km) of the closest coastlines. Data for some top energy consuming countries are reported in Table 12.1. These data unambiguously show that the potential of wind energy is indeed huge, but it has to be emphasized that

Table 12.1 CO_2 emissions, current electricity consumption, and annual wind energy potential onshore and offshore [769] for some selected top CO_2-emitting countries (CO_2 emission and electricity consumption are from the US Energy Information Administration, Country Energy Profiles, http://tonto.eia.doe.gov/country/index.cfm).

Country	CO_2 emission (Mton)	Electricity consumption (TWh)	Potential wind energy (TWh)	
			Onshore	Offshore
U.S.	5832.8	3923.8	74000	14000
China	6533.5	2835.0	39000	4600
Japan	1214.2	1007.1	570	2700
Russia	1729.4	840.4	120000	23000
India	1494.9	568.0	2900	1100
Germany	828.8	547.3	3200	940
Canada	573.5	536.1	78000	21000
U.K.	571.8	345.8	4400	6200
South Korea	542.1	386.2	130	990
Italy	454.9	315.0	250	160

the fraction of this resource that can actually be developed will be subject to economic, geographic, and environmental constraints. For instance, exploitation of the large off-shore potential of Japan would need significant investments in transmission systems.

Another challenge is the matching of supply with consumption, taking into account that the availability of wind cannot follow the daily pattern of electricity demand (Chapter 13). This problem can be solved with more widely dispersed wind facilities which belong to the same electric grid system but are placed in regions which tend to be uncorrelated meteorologically [769]. The escalation of wind energy production is spontaneously creating these conditions, which will limit the oscillating supply problem and make this technology an increasingly attractive option also for base load generation in the near future [772].

12.2.4
Current Deployment and Trends

Wind is the world's fastest growing electric technology. The growth of the wind sector in the period 1991–2008 was even stronger than that of the nuclear industry in its golden age (1961–1978) [773]. The amount of global wind capacity installed in 2009 alone (38 GW) in much higher than the new nuclear power installed in the decade 1999–2009 [773]. The global cumulative capacity of wind energy in the period 1990–2009 is depicted in Figure 12.4.

In 2009, the EU new wind installed capacity of 10160 MW was by far the first among all technologies (37%) with gas stations ranking second at 24% and PV third at 21%, whereas for coal and nuclear the new installed capacity were largely outpaced by the decommissioned one [774]. The European Commission 1997 White Paper target of 40 GW by 2010 was reached by 2005 and the actual installed

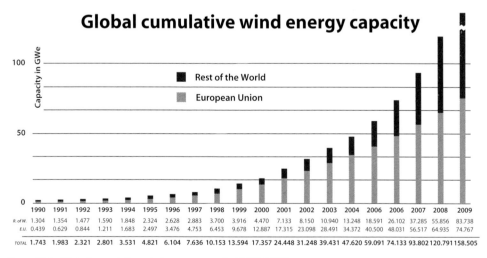

Figure 12.4 Global cumulative wind energy capacity (1990–2009).

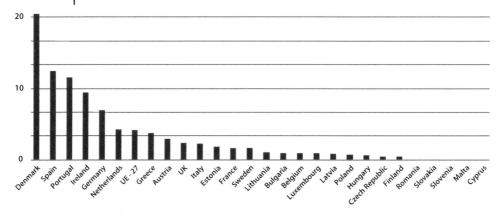

Figure 12.5 Percentage wind power share of national electricity demand in the EU (2008) [773].

capacity at the end of 2010 will be more than twice as much [774]. In 2008, the wind electricity production of the EU amounted to 143 TWh and is forecast to grow to 580 TWh by 2020 [773]; for comparison, the declining production of nuclear energy in the EU amounted to 940 TWh in 2007 [775].

The wind power share of national electricity demand in EU countries is depicted in Figure 12.5. The median value is a notable 4.1%, with three countries exceeding 10% and Denmark above 20%. Germany is slightly below 7% but, as the largest economy in the area, has the highest absolute production with 43 TWh in 2008. Spain was the second wind energy producer of the continent in 2009 [776], with an output of 36.2 TWh, not far from nuclear (52.7) TWh. Notably, in more mature markets such as Germany and Denmark the replacement of first-generation wind turbines with modern multi-megawatt machines (repowering) is ongoing.

In the US wind accounted for 39% of all new electrical capacity in 2009, up from less than 2% in 2004. Nearly 10 GW of new wind turbines were installed in 2009 thanks to the Recovery Act incentives, bringing the cumulative US value to 35 GW, ahead of Germany (25.8 GW), which had been the world leader for 15 years [777]. China is another booming market for wind power, mainly as a consequence of the 2005 Renewable Energy Law which aims to produce 8% of the country's electricity by 2020 from renewable sources other than hydropower [778]. China's installed wind energy was 0.34 GW in 2000, 1.3 GW in 2005 and 25.1 GW in 2009 [779]. A recent study has argued that wind could in principle accommodate all of the Chinese demand for electricity projected for 2030, about twice the current consumption, in an economically sustainable fashion [780]. In 2009 there were more than 80 countries around the world with commercial wind power installations, Mongolia and Pakistan being among the latest newcomers [544]. Projects are under development in several African countries, such as Kenya and Tanzania, whereas in North Africa, particularly Morocco and Egypt, several hundred megawatts of capacity are already installed.

It has to be emphasized that in 2010 only 1.5% of wind energy was installed offshore, with the UK as the leading country [768]. The huge potential of this untapped resource is most likely to be the next frontier for further expansion of an industrial sector which is already providing about 110 000 jobs in EU alone [768].

12.2.5
Environmental Impact

Wind power has definitely one of the lowest environmental footprints among all the electricity generation technologies: no land to disturb for extracting the resource, no emission of carbon or atmospheric pollutants, no residues of any sort to stockpile somewhere, no dams to build, no water for cooling, minimal exploitation of natural resources. Life-cycle assessments indicate that wind has the lowest emission of CO_{2eq} per kWh produced (≤ 10 g), with hydroelectric at ~20 g and nuclear at 40–110 g [58, 514–516]. Due to the intrinsic simplicity in terms of materials employed and design, energy payback times of wind turbines, regardless of size, are normally estimated as 1–6 months [516, 781]. The most pessimistic analysis indicates 12 months, which is still the smallest among modern energy conversion technologies [782].

The footprint on the ground or ocean floor of one large 3–5 MW wind turbine with tubular tower is 10–20 m^2 [516]. Depending on the resource availability and terrain, wind turbines use only 1–10% of the wind farm areas [576] and there are countless examples of turbines installed on agricultural land on which cultivation, grazing, or recreation can continue substantially undisturbed. The spacing between turbines is crucial to limit turbulence and maintain optimal operating conditions and efficiency; normally, turbines have to be kept three to five rotor diameters apart to avoid such problems.

Negative environmental consequences related to wind energy production are essentially visual impact, noise, and fatalities of birds through collision with the blades. The last is the major concern because visual intrusion is subjective and noise in modern turbines is basically that of the wind interacting with the rotor blades, which is quieter than other types of modern-day equipment. Even in quiet rural areas, the sound of the blowing wind is often louder than that of the turbines. It has been estimated that wind turbines in the US currently kill about 10 000– 40 000 birds annually [783]. For comparison, 5–50 million birds are killed every year by the thousands of communication towers and hundreds of millions by collision with windows and moving vehicles [516].

The extensive use of wind energy could have multiple interrelations with climate change. On the one hand, climate change might alter the wind fields over the planet, affecting (not clear in which direction) the overall useful energy that can be extracted [784]. On the other hand, the extensive exploitation of wind resources might generate some climate change. In fact it has been argued that very large amounts of wind power might produce non-negligible climatic change at continental scales, but with unimportant effects on global mean surface temperatures.

At any rate, the enormous global benefits produced by reducing emissions of CO_2 and air pollutants would likely show, once again, that the wind energy pros far exceed the related cons [785].

12.2.6
The Cost of Wind Power

Cost evaluation of energy services is an extremely tricky business. As discussed in Section 7.1, there are no market prices so unrelated to "real" costs such as those of coal, oil, and gas due to externalities and subsidies. Furthermore, as argued in Chapter 8, assessing the cost of nuclear energy is pure fiction because the nuclear chain has never been closed and generous public funding is never considered for a technology that is intrinsically incapable of surviving in a free market competition.

The price at which wind energy is currently sold on the market is comparable to, and in some cases even lower than, that of conventional technologies [768]. In perspective, this economic advantage can only improve because the price of the energy feedstock (wind) will be $0 forever and the technology, being simple and established, will get cheaper and cheaper.

Last but not least, the availability of the wind resource can be known with basically millimeter precision on every corner of the planet. This is simply delightful for operators wishing to invest money in the energy sector, beleaguered by the uncertainties of future supply of fossils and fissionable materials. No surprise that they are increasingly reluctant to invest, for instance, in the nuclear sector, preferring to address money where investments are much more secured. This is the very reason why wind has been so successful in the last decade. This is why other renewables will follow suit.

12.3
Ocean Energies

The powerful vastness of the ocean stores gargantuan amounts of energy in the form of heat, currents, waves, and tides, which have solar or gravitational origin. Only a minuscule amount of the ocean energy is currently harvested and utilized. In the long term and for some geographic locations, ocean energy might represent a viable option within a diversified renewable portfolio. The three main ocean energy technologies will be briefly described, in decreasing order of development and deployment.

12.3.1
Tidal Energy

Tidal energy is the energy dissipated by tidal movements, which are the regular rise and fall of the surface of the ocean originated by the gravitational force of the

High Altitude Winds: The Next Frontier?

Wind is stronger at high altitudes. In particular, between the narrow slice of the atmosphere at altitude 7–16 km, there are strong west-to-east "jet streams." The first global mapping of jet streams [786] shows that their power density can be as high as $10\,000\,W\,m^{-2}$, that is, about 1000 times higher than on the ground. Accordingly, some visionaries, entrepreneurs, and inventors believe that high altitude winds will be the most revolutionary renewable of the twenty-first century [787].

Two basic design concepts for the exploitation of relatively high altitude jet streams have been proposed. The first entails the transmission of mechanical energy from altitude to the Earth's surface, where conventional generators produce electricity at the ground. An example is the so-called KiteGen consisting of light airfoils (kites) placed at moderate altitude (800–1000 m) and tethered to ground-based heavy machinery for power generation [788]. Tethering is accomplished through high resistance lines which both transmit the traction of the kites and, at the same time, control their direction and angle to the wind [789]. The second basic approach aims at capturing wind energy closer to jet streams with the generators placed aloft and the electricity transmitted to the ground with a tether [790, 791].

Conversion of high altitude winds would offer minimal intrusion on the Earth's surface and production not limited to selected regions. Potential drawbacks would be risks of crash and, in case of exploitation on a vast scale, possible climate alterations due to the depowering of high winds [786].

Moon and the Sun on the Earth and by the centrifugal force due to the rotation of the Earth and Moon about each other [792]. Tidal phenomena occur with great precision, that is, twice every 24 h, 50 min, 28 s [756], which is an unrivaled advantage over most of other renewable energies.

Tidal fluctuations have an estimated global power of about 3 TW [3], of which only 0.8 TW could theoretically be exploited without interfering significantly with cycles of the biosphere [516]. The availability of tidal energy is rather site specific and a tidal range is amplified by factors such as shelving of the sea bottom, funneling in estuaries, reflections by large peninsulas, and resonance effects when the tidal wave length is about four times the estuary length. The Earth's highest tidal ranges combined with ideal geographic and morphologic conditions are found in the Bay of Fundy (Nova Scotia, Canada). Very favorable sites are also located in Alaska, southern Argentina, northern France, Britain and White Sea bays. The theoretical power of the 28 best sites is 360 GW [3], similar to the current global capacity of nuclear reactors.

Tidal energy is made of potential and kinetic components, hence tidal power facilities can be classified into two main categories: tidal barrages and tidal current turbines, which use the potential and kinetic energy of the tides, respectively.

In principle, the technology for tidal barrages is identical to that for hydroelectric generation (except that the water flow runs in both directions); several types of barrages have been designed [793]. At present there are only four tidal barrage power plants in operation. The largest one, with a capacity of 240 MW, entails a 720 m long barrage enclosing 22 km^2 of the estuary of the Rance river in Brittany, France. There are many sites undergoing feasibility studies, the largest one being on the river Severn estuary in southwestern UK, which might provide 5% of the country's electricity, but is facing fierce opposition from several environmental groups [794] because tidal barrages pose several threats, especially through alteration of the flow of saltwater into and out of estuaries.

The tidal current turbine technology is still at a preliminary stage of development but can be environmentally less impacting because it does not block channels. In principle it is similar to wind energy technology, but to operate turbines fully immersed in salty water is technically more demanding because they experience greater forces and must withstand chemical degradation and biofouling [795]. Several tidal turbine prototypes have been proposed [793] and the first megawatt-scale facilities will be opened soon [796].

The life-cycle greenhouse emissions of a 100 MW tidal turbine farm has been estimated at around 14 g CO_{2eq} per kWh, lower than for hydroelectric energy, with an energy payback time of 3–5 months [516]. Tidal energy technologies are expected to undergo substantial growth in the next two decades, providing a relatively limited, but no longer negligible, contribution to the world's electricity supply [793].

12.3.2
Wave Energy

Ocean waves are generated by transfer of kinetic energy from winds to surface waters and their global power is estimated at around 90–100 TW [3, 797], of which 1–10 TW is practically accessible along the world's 800 000 km of coastlines and a much smaller fraction (~0.1–0.5 TW) is likely to be economically recoverable [516, 797, 798], with an upper estimate of 4.2 PWh per year of electricity production [516] (about 25% of the current global consumption of 17.5 PWh). The greatest potential for wave energy exists in strongly windy locations at temperate latitudes between 40° and 60° north and south, on the west coast of continents of the Northern Hemisphere, and on the east coast in the Southern Hemisphere. One of the richest nations in terms of wave energy resources is the UK.

At least six main types of wave energy converters (WEC) devices are currently developed, which can be either floating or submerged (partially or completely); the three most common schemes are shown schematically in Figure 12.6. Waves are characterized by low frequencies and irregular velocities, hence efficient conversion and transmission to the electric grid are difficult. A further issue is the capability of resisting the harsh conditions of the open ocean, withstanding potentially destructive storms and chemical corrosion by salty water [795]. As far as environmental impact is concerned, life-cycle greenhouse emissions of WEC

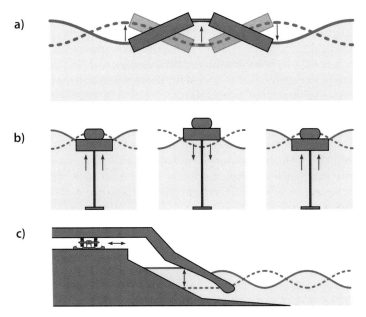

Figure 12.6 Three technologies for the exploitation of wave energy. (a) Attenuator: a floating device oriented along the propagation direction of the waves, which flexes as they pass. (b) Point absorber: a buoyant float which, excited by the heave motion of the waves, moves relative to a spar anchored to the ocean floor. (c) Oscillating water column (OWC): it is one of the various "terminator devices" which extend perpendicular to the direction of wave and capture or reflect its power; they are typically located onshore or near the shore. In an OWC water enters through a subsurface opening into a chamber with air trapped above it. The wave action moves up and down the captured water column like a piston, forcing the air passing through an opening connected to a turbine.

devices are estimated at around $22\,\mathrm{g}$ CO_{2eq} per kWh, very similar to hydroelectric energy, with an energy payback time of 1 year and an estimated lifetime of 15 years [516].

The first commercial WEC appeared only in recent years off the coasts of Portugal, Spain, and the US, with a combined capacity of a few megawatts [799]. In the mid to long term, wave energy will likely constitute a viable option, but perhaps not a major one, within a diversified renewable energy portfolio. Deployment of WECs on a large scale will require a deeper evaluation of potential environmental impacts, which is now lacking due to limited experience in the field [800].

12.3.3
Ocean Thermal Energy

The sunlight that hits the ocean's surface is strongly absorbed, raising the temperature of the top layer of tropical oceans up to about $28\,°C$ all year round. In contrast, most of the deep portions of the oceans are at temperatures of about $4\,°C$.

In order to consider seriously the possibility of ocean thermal energy conversion (OTEC) technologies [801] there must be a difference of at least 22 °C between the ocean surface and a depth of 1000 m. This limits exploitable sites to the oceanic equatorial belts, particularly in the western Pacific area [800].

The production of electricity by OTEC entails the use of the heat stored in warm surface water to make steam and drive a turbine, while cold water is pumped upwards from the deeper ocean layers to recondense the steam. The process can be accomplished by means of a closed cycle using a working fluid such as ammonia, or through an open cycle in which the warm seawater is evaporated inside a vacuum chamber [801]. Operations connected with OTEC technologies could offer some ancillary benefits, such as the production of desalinated warm water after vaporization, or the availability of an abundant cold water flow that could be used for air conditioning or aquaculture [800].

The potential of OTEC is huge, with the latest assessment estimating about 5 TW of electricity that could be produced theoretically [802]. However, the environmental and climatic impact of the extensive deployment of OTEC technologies would probably be large, due to the disruption of the vertical thermal structure of the oceanic water column [802]. OTEC technologies are still in the infancy stage of development and perhaps will never be a major renewable energy technology, despite their considerable potential.

12.4
Geothermal Energy

12.4.1
The Geothermal Resource

The Earth's surface is illuminated and heated from above by an average solar energy input of about 170 W m^{-2} (Chapter 10) and is also heated from beneath by a minuscule endogen heating flux, of about 90 mW m^{-2} [3]. Its distribution over the planet is far from being homogeneous, since there are hot spots, signaled by the presence of geysers and fumaroles, in which the Earth's heat is spectacularly released with high intensity. The global heat power output of spaceship Earth, integrated over its entire surface, is about 42 TW, that is, about three times the current global rate of energy consumption [3].

The gargantuan power of geothermal heat, over billions of years, has literally forged the Earth by breaking up and reassembling continents, shaping mountain chains and ocean ridges, and generating earthquakes, tsunamis and volcanic eruptions [3]. This resource originates from the primordial heat trapped during the Earth's formation, from gravitational pressure, and from the decay of long-lived radioactive isotopes with half-lives comparable to the age of the Earth.

The average temperature gradient within a few kilometers beneath the Earth's surface is about 30 °C km^{-1}. Values may range from about 10 °C km^{-1} in the ancient "cold" continental crust to over 100 °C km^{-1} in areas of active volcanism. Our planet is cooling, but very slowly: the temperature of the mantle has decreased no more than 300–350 °C in three billion years, remaining at about 4000 °C at its base. Hence the geothermal energy resource base as a whole can be definitely considered a renewable source on the timescale of human civilization although, at the local level, this might not be true.

12.4.2
Electricity Production

Geothermal power capacity reached over 10.7 GW in 2009 [544]. The US is now developing more than 120 projects, representing at least 5 GW, and 3 GW are in the pipeline in other countries. In 2009, the world geothermal electricity production exceeded 67 TWh [544], an amount higher than the annual electricity consumption of Austria [102] and at least 10 times smaller than the technical potential [516]. Conversion of the energy contained in hot rock into electricity is accomplished by using water as a heat transport medium to the Earth's surface, where vapor is produced and then injected into turbine generators, like any other conventional electricity technology (Chapter 13). The steam temperature in geothermal power plants is lower than that of conventional fossil fuel facilities, therefore the electricity conversion efficiency (10–17%) is smaller due to the Carnot limit [803]. Nonetheless, they deliver electricity at a price comparable to that of coal-fired power plants [804].

All the geothermal energy produced today relies on naturally occurring hydrothermal systems shown schematically in Figure 12.7, where a continuous circulation of heat and water occurs [803]: the fluid enters the reservoir from the recharge zones and leaves it through discharge areas (hot springs, wells). It is thus fairly clear that industrial exploitation of a geothermal field is sustainable over time only if fluid recharge can keep up with extraction. In this sense geothermal energy, locally speaking, may not be a renewable resource and, for this reason, reinjection of the extracted fluid has become a common practice in industrial geothermal facilities [805].

The most valuable and economic fields are vapor dominated because the "product," extracted at high pressure (5–10 bar) and at temperatures as high as 200 °C, can be injected directly into the turbine. The two most famous geothermal fields in world, The Geysers in California and Larderello in Italy, are vapor dominated. The first feeds the largest geothermoelectric power plant in the world (725 MW), and the second generates 550 MW of electricity and was initially developed by the pioneer of geothermal energy, Piero Ginori Conti, who tested the first geothermal power generator in July 1904 [803].

Water dominated fields require a separator between the well and the power plant to split steam and water; the latter can be injected back into the reservoir or utilized to produce electricity through a so-called binary cycle plant. In these facilities a secondary, low boiling point working fluid (e.g. ammonia or isobutene), vaporized by hot water (>85 °C) through a heat exchanger, drives the turbine. These systems are typically small (1–3 MW) and their conversion efficiency is poor (2.8–5.5%),

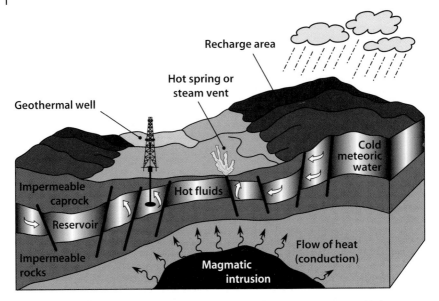

Figure 12.7 Scheme of a hydrothermal system that enables the exploitation of geothermal heat for energy production. The fluid reservoir is placed between two impermeable rock layers but, at the same time, is somehow connected with the Earth's water cycle; the upper impermeable layer is relatively thin. Intrusion of hot molten rock (magma) at relatively shallow depth induces convective circulation of groundwater inside the reservoir permeable layer.

but this strategy has emerged as a cost-effective and reliable way to convert large amounts of low temperature geothermal resources into useful electricity [803, 806]. Today about 15% of geothermal plants operate in the binary mode.

12.4.3
Heat for Direct Use

Use of geothermal heat worldwide is estimated to be around $50\,\mathrm{GW_t}$, roughly five times the installed electric capacity, with a growth of 30% during the last decade [544, 807]. In available global inventories the largest share (>30%) is attributed to geothermal heat pumps which, as discussed in Section 10.2.3, should be more properly regarded as solar energy technologies [809]. Space heating and industrial uses claim about 50% of demand, and a variety of minor applications (greenhouse heating, aquaculture, agricultural drying, bathing, and swimming) account for about 20% of direct uses [807]. Clearly, since heat transportation is costly and inefficient over long distances, production and exploitation sites have to be very close.

Heating and cooling in the industrial, commercial, and domestic sectors constitute around 40–50% of total global final energy demand. Therefore, the potential for geothermal energy in the direct use sector is very large and, in the years to come, is expected to be developed all over the world [80]. Iceland increased its installed geothermoelectric capacity 2.5-fold between 2005 and 2008 [754].

12.4.4
Advantages, Disadvantages, and Perspectives

The greatest advantage of geothermal energy is that, unlike other renewable electricity technologies, it can provide 24/7 base load electricity. The capacity factor averaged worldwide is 67% but it can be as high as 90% (e.g. in Iceland) [754]; these values are fairly comparable with conventional fossil or nuclear technologies. Additionally, the exploitation of geothermal energy does not involve land use far from the production site (e.g. mining) or a large transportation infrastructure to carry the energy feedstock (vapor or water) to the transformation facility. A 100 MW geothermoelectric plant requires about $0.3 \, km^2$, 75% less than a coal power station [516]. Geothermal technologies are relatively simple and can normally be adopted also in developing countries [810]; in some cases, for instance El Salvador, Kenya, and the Philippines, they cover a relevant part (>15%) of the electric demand [810].

The main drawback in handling geothermal heat is that it brings into the biosphere contaminants of various sorts which are normally confined underground. Typical gaseous contaminants are CO_2, H_2S, NH_3, CH_4, N_2, and H_2 [803].

Without abatement, the specific emissions of sulfur from geothermal power plants are about half of those from coal-fired plants. There is some potential risk of water pollution by a variety of toxic substances if the condensed water is released into the environment, but this problem is limited by the practice of reinjection. Notably, in binary plants, where the geothermal fluid is passed through a heat exchanger and reinjected underground without exposure to the atmosphere, environmental pollution is negligible. Other problems that may arise from exploitation of geothermal fields are land subsidence [803] and induction and microearthquakes [811, 812]; continuous monitoring is undertaken at geothermal facilities and in their surroundings to control such events and take countermeasures.

A life-cycle balance shows that geothermal electricity has an emission of 15–55 g CO_{2eq} per kWh, comparable to photovoltaics and nuclear [515, 516].

12.4.5
The Next Frontier: Going Deeper

In order to achieve a more widespread availability of geothermal resources, a different approach is being pursued, which is based on the concept that geothermal energy is obtainable not only through the naturally occurring (and rather rare) hydrothermal systems as depicted in Figure 12.7, but also "artificially" [803]. This approach can be implemented through an enhanced geothermal systems (EGSs), which aims at exploiting fields characterized by hot dry rocks (HDRs) with low permeability that do not allow spontaneous circulation of water or vapor. In such systems, permeability is stimulated artificially by hydraulic fracturing, high-rate water injection, and/or chemical treatment [803, 813]. EGS/HDR technologies are at an advanced test stage in many European countries as well as in the US and Japan. The largest EGS project in the world is a 25 MW demonstration plant currently being developed in central Australia. A report published by MIT states that EGSs have the potential to supply about 10% of the US base-load electricity by 2050 [813].

There is some concern about the seismicity that EGS technologies may trigger [814], but their successful development would likely constitute a key step towards energy transition because they might make available an abundant and relentless renewable energy flow almost everywhere on the planet. To achieve this result, substantial engineering progress has to be accomplished over the next decade [813, 815]; the technological capacity rather than the availability of geothermal resources will define the future share of this energy source. Indeed, we might be just at the very beginning of the exploitation of such an immense underground energy resource.

Part Five
Energy Carriers

Energy for a Sustainable World: From the Oil Age to a Sun-Powered Future. Nicola Armaroli and Vincenzo Balzani
© 2011 WILEY-VCH Verlag GmbH & Co. KGaA, Weinheim
ISBN: 978-3-527-32540-5

13
Electricity

"We will make electricity so cheap
that only the rich will burn candles."

Thomas Alva Edison

13.1
Basic Concepts

Electricity is not a primary energy, but it is the most flexible and convenient form of energy. It can be converted without any losses to useful heat, even to generate temperatures higher than combustion of any fossil fuel; it can be turned with high efficiency into mechanical energy; it is used to generate light; it can be easily adjusted with very high precision and is clean and silent at the point of consumption. There are indeed good reasons to believe that we are moving towards an electricity-based economy.

The role of electricity in shaping the modern world relies on the widest man-made entity ever made, the electricity grid [816]. The distribution of electric energy through the grid sustains the way of life of developed countries, which today is totally electricity dependent. When for any reasons the grid fails, all activities are paralyzed. Electricity makes the difference between poor and rich countries; in fact, more than one-third of people in the developing world, some 1.6 billion people, do not have access to this basic energy service [32]. Since the electricity grid represents a highly capital-intensive investment, the most realistic solution to provide electricity for poor and dispersed communities is to develop distributed generation based on renewable energy sources.

For an electricity generator, it is important to distinguish the concepts of electric power (sometimes termed capacity) and electricity generated (energy) for many discussions. Electric power is the rate at which electric energy is transferred, usually measured in watts (W). Electricity generated is the amount of electric energy actually produced by a generator and it is usually measured in watt-hours (Wh). Switching between measurements of power and electricity generated is easily done by multiplying the power by the number of hours it is used. Therefore,

Energy for a Sustainable World: From the Oil Age to a Sun-Powered Future. Nicola Armaroli and Vincenzo Balzani
© 2011 WILEY-VCH Verlag GmbH & Co. KGaA, Weinheim
ISBN: 978-3-527-32540-5

a 500 MW generator that is used at its peak power for 6 months (4380 h) would generate 2.19 GWh of electricity.

The energy power of the most common electric devices is shown in Table 13.1, where an estimation of the electric energy consumption per day is also given. Electricity can be generated by different sources as exemplified in Table 13.2.

Table 13.1 Energy power of electric devices and estimated energy consumption per day (for the most efficient appliances available on the market).

Device	Power	Time per day (h)	Energy per day
Quartz watch	1 μW	24	0.000024 Wh
Clock radio	1 W	24	24 Wh
Table fan	25 W	4	100 Wh
Compact fluorescent lamp	21 W	5	105 Wh
Laptop computer	20 W	3	60 Wh
Refrigerator	20 W	24	480 Wh
Freezer	90 W	24	2 300 Wh
Incandescent lamp	100 W	5	500 Wh
Desktop computer	150 W	2	300 Wh
Air conditioning	600 W	3	1 800 Wh
Toaster	800 W	0.1	80 Wh
Microwave oven	1400 W	0.5	700 Wh
Electric heater	2000 W	2	4 000 Wh
Washing machine	900 W	1	900 Wh
Dishwasher	800 W	0.5	400 Wh
Electric oven	3000 W	0.5	1500 Wh
Clothes dryer	4000 W	1	4 000 Wh
Electric cars:			
G-Wiz	13 kW	2	26 kWh
Tesla Roadster	185 kW	2	370 kWh
Electric trains:			
ETR 200 (Italy)	1 MW	10	10 MWh
Frecciarossa (Italy)	8.8 MW	10	88 MWh
TGV (France)	8.8 MW	10	88 MWh
Shinkansen (Japan)	13.2 MW	10	132 MWh

Table 13.2 Installed capacity of the largest electric facilities in the world belonging to the main current technologies.

The largest wind turbine (Emden, Germany)	7 MW
The largest concentrated solar power plant (Andasol 1, Spain)	50 MW
The largest photovoltaic power plant (Fratta Polesine, Italy)	72 MW
The largest group of geothermal power plants (California, US)	725 MW
The largest nuclear reactor (AREVA EPR)	1.65 GW
The largest oil-fired electricity station (Riyadh, Saudi Arabia)	3 GW
The largest wind turbine power plant (Texas, US)	4 GW
The largest coal-fired electricity station (Kendal, South Africa)	4.1 GW
The largest nuclear power plant (Kashiwazaki-Kariwa, Japan)	8.2 GW
The largest hydropower plant (Three Gorges Dam, China)	18 GW

13.2
Illumination

At the beginning of the twentieth century most of the electricity was used for residential lighting, but now illumination takes a relatively small share (about 19%) of global electricity consumption [817]. This is due not only to the great disparities in typical power ratings of lights and common appliances (Table 13.1), but also to the increasing luminous efficacy. This is defined as the ratio of luminous flux, emitted from a light source, which is usable for vision [in lumens (lm)] to the electric power (in watts) consumed by the source. In 1882 carbonized fibers in Edison's first lamp produced about 1.4 lm W^{-1}, five times that of a candle [3]. Standard incandescent lights generate about 15 lm W^{-1}, which corresponds to about 2.2% of conversion efficiency of electricity into light. It has been estimated that a moderate affluent person uses for lighting about 4 kWh per day, including a small fraction of public lighting (0.1 kWh per day for street-lights and 0.005 kWh per day for traffic lights) [701]. Unfortunately, because of their very low efficiency, billions of bulbs waste billions of kWh, contributing a substantial fraction of greenhouse gases that are changing the climate [817].

Substantial progress in illumination is being obtained with compact fluorescent lamps (CFLs), which have much higher efficiency (20%, ~70 lm W^{-1} efficacy) and longer lifetime (8000–20 000 h). Although CFLs are about 3–5 times more expensive than incandescent lamps, replacing incandescent bulbs with CFLs reduces the cost of lighting about fivefold. In the US, where there are 4 billion light bulbs installed, the DOE estimates that using CFLs would save an amount of electricity equivalent to that produced by twenty 1000 MW power plants [818]. Incandescent bulbs and conventional halogen bulbs will be phased out gradually from many markets worldwide by 2015.

CFLs do not rely on heat to produce visible light, but on fluorescence. To do that, they must contain a small amount of mercury vapor which, upon excitation with a beam of electrons, emits UV light. Excitation by UV light of a phosphor deposited on the inside surface of the glass tube causes visible fluorescence. If broken during use or improper disposal, the 5 mg of mercury contained in CFLs can be dispersed into the environment [819]. Hence in countries where there is a small percentage of coal-based power generation and little or no CFL disposal (e.g. Sweden and Switzerland), the use of fluorescent lamps may increase mercury emissions. On the other hand, significant mercury emission reductions are expected in areas with massive electricity production by coal (e.g. China and Romania) [442].

A new generation of lights is based on light-emitting diodes (LEDs), electronic devices made from a variety of inorganic materials, which operate on the principle of electroluminescence [820]. They are based on semiconductor diodes in which electric flow generates electrons and holes which recombine, releasing energy in the form of light [821, 822]. The color of the emitted light is determined by the intrinsic properties (energy gap) of the semiconductor employed. The LED is usually small in area (less than 1 mm^2) and needs optical components to shape its radiation pattern. LEDs are about 10 times more expensive than CFLs, but present many advantages including lower energy consumption (efficacy over $100\,lm\,W^{-1}$), longer lifetime (up to 50000–100000 h), improved robustness, smaller size, and faster switching. LEDs of many different colors are available and are extensively used as low-energy indicators. White light can be produced by mixing the colors of red, green, and blue LEDs or using phosphor material to convert blue or UV LED light to broad-spectrum white light [820]. In this field, technology is moving fast and LEDs are beginning to replace traditional light sources in automotive and general lighting.

LEDs based on organic compounds are called organic light-emitting diodes (OLEDs) [823, 824]. At present, OLEDs have lower efficiency and shorter lifetime than inorganic LEDs. However, polymer-based OLED may be flexible, a property that can be exploited for a variety of applications. For outdoor illumination, the most usual solutions are low pressure sodium lamps (50–150 lm W^{-1}, lifetime 10000–40000 h). They are the most efficient light sources now available, although rather poor in rendering colors [817].

13.3
Traditional Power Generation

13.3.1
Demand and Supply

Electricity represents about 40% of the primary energy consumption and less than 20% of the final energy demand. Electricity production rose steadily at an average rate of 3% per year from 1990 to 2007, but by only 1.3% in 2008, and decreased in 2009 due to the economic crisis. Its share in the total energy consumption is also rising. By 2030 this share is expected to increase above 20% in the end-use energy budget [825].

Electricity can be generated by using a variety of primary energy sources and more than 60% of the production is obtained from fossil fuels (Figure 2.5). The amount of electricity generated (TWh) and the per capita consumption (kWh) of the top 10 producing countries are collected in Table 2.3.

The demand for electricity fluctuates widely from summer to winter, from working days to holiday, and throughout the day from hour to hour and even minute to minute (Figures 13.1 and 13.2) [826]. The seasonal fluctuations differ from country to country. In northern Europe, the peak demand is in winter for lighting and heating. In warmer regions, for example, southern Europe and the southern US, the peak demand is in summer for air conditioning. Such a strongly fluctuating demand represents a big problem for power utilities because electricity cannot be stored. Usually, this problem is solved by using different types of generators, taking into account costs as well as other parameters. In principle, a fluctuation like that shown in Figure 13.1 would be best met by dividing the conventional supply into four areas: (i) base load (usually about 30–50% load of installed capacity), intermediate load (20–30%), peaking load (10–20%), and

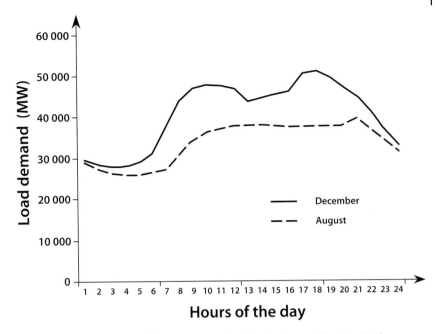

Figure 13.1 Electricity demand fluctuation in Italy during the hours of the day (third Wednesday of August and December 2009).

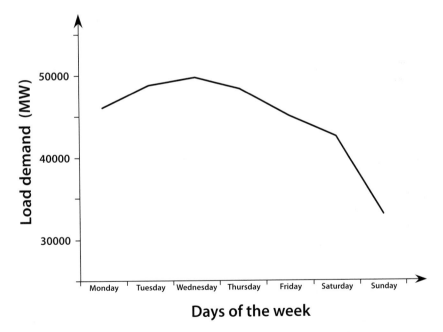

Figure 13.2 Electricity demand fluctuation in Italy along the days of the week (at 11 a.m. throughout the second week of March 2010).

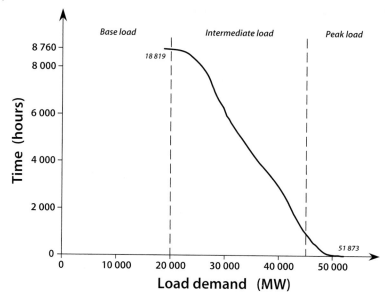

Figure 13.3 Hourly load demand on the Italian grid in 2009. Base load, intermediate load and peak load are indicated schematically.

reserve load (the remaining 10–20%). Figure 13.3 shows an example, taken from Italy [826], of the number of hours for which a specific power is required during a year. In the case shown, an appropriate base load would have been about 20 GW, the intermediate load 20–25 GW, and the peak load 10–15 GW. Since the installed capacity in Italy in 2009 was around 105 GW, an ample reserve (about 50 GW) was available.

The base load is usually supplied by large plants that are best kept running for long periods and have the lowest generating cost. Intermediate demand is usually met by older, smaller plants with higher generating costs, while peak demand is provided by reserve gas- and oil-fired plants, or by hydroelectric power. Superimposed on the daily demand curve are short-term peaks associated with the personal habits of consumers or with external inputs. Controllers at power stations can often predict demand spikes by studying television program schedules. The response time to meet sudden demands should be of the order of seconds to minutes. Intermittent renewable energies must find a place in such a generation scheme that, however, can be substantially modified by controlling demand (Section 13.8).

The production cost of electricity depends on the cost of the primary energy source used, the efficiency of its transformation into electricity, and the costs of power plant, operation, maintenance, and transmission to the final users. The real cost, however, is very difficult to evaluate because it also depends on several other parameters, including the life-cycle of the overall system, explicit and implicit subsidies, the premium associated with political, social and environmental risks, potential environmental tax regimes, and calculation of externalities [827]. In some cases, the social and environmental externalities associated with our present energy economy are even larger than the market prices of the dispatched electricity.

13.3.2
Thermal Power Plants Based on Fossil Fuels

A thermal power plant consists of a furnace that burns the fossil fuel to boil water and generate steam. The dynamic pressure generated by expanding steam turns the blades of a turbine that rotates a generator, a machine capable of converting mechanical energy into AC electric energy, by creating relative motion between a magnetic field and a conductor. The steam that leaves the final (low pressure) stage of a turbine is condensed in cooling towers or low temperature heat sinks. Where economically and environmentally possible, electric companies prefer to use cooling water from a river, a lake, or the ocean, instead of cooling towers. This type of cooling can reduce the building expenses and may have lower energy costs for pumping cooling water through the plant's heat exchangers. However, the wasted heat can cause the temperature of the water to rise detectably, which may be dangerous for fish and aquatic organisms.

13.3.2.1   Coal-fired Power Plants
Coal is the most abundant and dirtiest of fossil fuels (Chapter 6). About two-thirds of the coal mined in the world is used in electricity generation [828]. Just to give a visual impression, a standard 1 GW plant consumes 100 train loads of coal per day. Since coal is not expensive, coal-fired power plants are usually employed to cover base load (Figure 13.3). Modern coal-fired power stations are often large in size (1–4 GW) and generally operate with an efficiency in the range 30–40%.

The greatest producer of electricity from coal-fired plants is the US, followed by China and India. In China, electricity is obtained mainly by carbon-fired power plants, ~80% [825], and for the period 2004–2015 China has planned to build about 800 1 GW coal-fired plants, an average of about one plant every 5 days [220]. As large as this number may appear, it should be recalled that it corresponds to one plant per 1.7 million inhabitants, and it should also be noted that in the same period the US, where the energy consumption per person is almost four times higher than in China (Table 2.3), has planned the construction of almost 50 coal-fired power stations.

13.3.2.2   Oil or Gas Power Plants
Burning oil produced about 25% of world's electricity in 1973 and 5.6% in 2007, whereas burning gas increased form 12.1% in 1973 to 20.9% in 2007 [4]. The progressive shift from oil- to gas-fired power plants is due to the discovery and exploitation of major gas fields, the cleanliness of natural gas as a fuel, the construction of pipelines and LNG carriers to convey the gas to the market, and the development of more efficient simple- and combined-cycle gas turbines.

A gas turbine, also called a combustion turbine, is a rotary engine that extracts energy from a flow of combustion gas. A typical large simple-cycle gas turbine may produce 100–400 MW of power and usually has 35–40% efficiency, with the remaining part of the energy being lost as low temperature heat. In a combined-cycle gas turbines (CCGT) plant, a gas turbine generator produces electricity and the waste heat is used to make steam to generate additional electricity via a steam turbine. For large-scale power generation a typical set would be a 400 MW gas turbine coupled to a 200 MW steam turbine giving 600 MW. If the plant produces only electricity, efficiencies close to 60% can in principle be achieved.

At first sight, a simple- or combined-cycle power station appears quite inefficient because most of the chemical energy of the fuel is eventually "wasted" as heat that, rather than being dissipated in cooling towers, rivers, or the sea, could be delivered to heat buildings. This idea is even more appealing in the case of small power stations that could be placed within town districts, providing both electricity and heat. In such co-generation or combined heat and power (CHP) systems, steam circulates around the district in well-lagged pipes and customers may be metered for the heat received as well as for the electricity [701]. CHP power stations have indeed expanded in recent decades in Europe, particularly in Denmark, Finland, and The Netherlands, and microturbines are now becoming the preferred choice of power and heat co-generation not only for town districts, but also for hospitals, hotels, industries, farms, and even off-shore oil rigs.

While the CHP technology is certainly beneficial, it is worthwhile pointing out its limits. For CHP plants, it is common practice to lump together the efficiency of electricity production and heat production into a single total efficiency. For example, if the efficiency of electricity generation is 15% and the efficiency of heat generation is 60%, the CHP power plant is said to be 75% efficient, and this number is favorably compared with the less than 60% electrical efficiency of an electricity-only gas power station. This comparison, however, is misleading since electricity is more valuable than heat. Furthermore, while it is assumed that using the waste heat does not affect the efficiency of electricity production, this in fact cannot be true for thermodynamic reasons because the efficiency of heat engines is related to the temperature difference between furnace and cold sink, and the latter is warmer when heat has to be delivered to buildings.

A gas-fueled power plant produces about 310 g of CO_2 per kWh, compared with 750 g kWh^{-1} for a coal-fired power plant [829]. It should be considered, however, that natural gas leaks out during the journey from the well to the power station (Chapter 5), and methane has a nearly eight times greater global warming effect than CO_2.

13.3.3
Hydroelectric Power Plants

Hydroelectric power (Section 12.1) is presently the world's largest renewable source of electricity. In terms of electricity produced it is second only to fossil fuel and generates 10 times more power than geothermal, solar, and wind power combined [797], Figure 2.5.

An important advantage of hydropower is that turbines can be designed to run either forwards or backwards, either generating power from falling water or using power to pump water up to a high-level reservoir. When demand rises, the upper reservoir can be drained back through the turbine into the lower reservoir. The overall efficiency of this process can be in excess of 70% and maximum generation can be reached in only a few seconds. This method is used to store excess electricity produced overnight by coal or nuclear power stations, which run best when operated steadily. Hydroelectric systems can also be used to compensate for the fluctuating energy generated by renewable resources [756]. The fact that hydroelectric systems require no fuel means that they also require non-fuel-extracting infrastructures and no fuel transport.

Large hydroelectric power generation facilities have many drawbacks (Section 12.1) but they will continue to expand, presumably adding a terawatt capacity [797].

13.3.4
Nuclear Power Plants

A full discussion of nuclear power is given in Chapter 8. Nuclear power plants have relatively low fuel costs and can run at full blast almost constantly, delivering up to 90% of their capacity. In fact, they cannot be switched on/off or operated at lower power without losing efficiency. This makes them well suited to providing base load power to national grids (Figure 13.3). In 2009, Lithuania produced 76% of its electricity from nuclear energy, followed by France (75%), Slovakia (53.5%), and Belgium (52%). This share is 20% for the US and about 2% for China and India [469, 830]. Commercial exploitation of nuclear energy to obtain electricity dates back to 1957. In 2009, nuclear power plants generated 2558 TWh, 13.5% of the world's commercial electricity (down from 15% in 2006, 16% in 2005, and 18% in 1993) [471].

The reasons why nuclear power is not expanding are discussed in detail in Chapter 8; cost, fuel availability, and waste disposal are the most important hurdles. Despite the optimism of nuclear agencies [831, 832], a nuclear renaissance is very unlikely.

13.3.5
Contribution by Other Energy Sources

Minor amounts of electric energy have been traditionally produced by *burning biomass* in the form of wood, crop residues, and waste incineration (Section 11.3), as well as by exploiting geothermal (Section 12.4) and tidal energy (Section 12.3).

13.4
Traditional Electricity Grid

The electricity grid, through which all modern economies are powered [816], was dubbed the greatest invention of the twentieth century by the American Academy of Engineering in 2000, surpassing in importance even the automobile, the airplane, and the computer. The grid can deliver energy to the user more cheaply, and in far more versatile form, than coal or other forms of fuel hauled to each user's locations and consumed on site. Electrification has long since become the embodiment of political programs all around the world.

A complete grid system comprises three components: generation, transmission, and distribution of electrical power. Electricity generation can be obtained as described above. Transmission can take place by alternating current (AC) or direct current (DC). High voltage AC transmission (230–765 kV in North America, 100–400 kV in Europe) by aluminum lines is the method most commonly used. Transmission networks are linked to distribution networks which supply electric energy to the end-users at a voltage that can vary according to the type of use.

The grid is a complex and fragile system since each of its components must operate regularly and, in any case, must be protected (which means disconnected) to avoid damage. Even a minor local problem can trigger a cascade of events that sometimes becomes uncontrollable. Since electric energy cannot be stored, instantaneous balancing differences between power supply and demand must be achieved in the grid. To cope with variable and sometimes unpredictable demand, grid utilities must bring generation assets online to ensure reliability (peaking plants). In the US, 10% of all generation assets and 25% of distribution infrastructure are required for roughly 5% of the time. As we will see later, a more effective way to match supply and demand is likely to be turning on and off *demand* for power rather than *supply*, a method seldom used at present.

Today, large interconnected electricity grid infrastructures are the norm for all industrial countries. Even if a utility can be completely self-sufficient, it is often advantageous to have the opportunity to buy and sell power to and from neighboring utilities. In North America, two major and three minor interconnected grids, not synchronous with each other, are operative. In Europe, 42 electricity companies in 34 nations are integrated in the European Network of Transmission System Operators for Electricity (ENTSO-E) [833].

Traditionally, many countries have had a state monopoly for the generation and distribution of electricity. A monopoly status makes sense at the distribution level, but not so much for the generation of electric power that can use a variety of primary energy resources. Therefore, starting from the 1990s, many countries began to deregulate their electricity system in an attempt to stimulate free market competition. Deregulation usually involved the separation (unbundling) of the generation, transmission, and distribution operations into distinct financial entities and constrained electric utility companies to buy electricity from independent producers. In theory, deregulation would promote efficiency, lower the price of electricity, and encourage private capital to invest in new power stations and transmission lines. In practice, however, deregulation has originated new problems. Private companies tend to focus on short-term returns and have little interest in making investments that will not show a profit for many years. In fact, deregulation has contributed to make the electric system more fragile. North America and other countries have experienced major blackouts in the last decade. On August 14, 2003, the largest blackout in history affected some 50 million people in the eastern US and Canada and caused an estimated $7 billion economic loss. Aside from underinvestment in the energy infrastructure, poorly designed deregulation can cause other damage, including market manipulation, as happened during the 2000–2001 energy crisis in California [834, 835] (see Box).

The 2000–2001 California Electricity Crisis

Although the origin of the crisis is attributable to a rapid growth of energy demand caused by the increase in population and the mushrooming of computers and industrial equipment, the shortage of energy was substantially created by an unstable commercial situation that arose following partial deregulation of the electricity market. In order to increase competition, in March 1998 a substantial amount of installed capacity was sold to independent power producers, putting the California utilities that distributed electricity into a tough situation. Energy deregulation policy froze or capped the existing price of energy that the distributors could charge to customers, but deregulating the producers of energy did not lower its wholesale cost. While demand for electricity was increasing, new producers, instead of creating more power and driving down prices, charged more for electricity; they also used moments of spike energy production to inflate price. In January 2001 producers began shutting down plants to increase prices. On March 19–20, 2001, blackouts affected 1.5 million customers. The California energy market allowed energy companies to charge higher prices for electricity produced out-of-state. It was therefore advantageous to make it appear that electricity was imported from other states, an operation described as *megawatt laundering* [834]. Furthermore, the capacity available for the transportation of electricity along power lines was deliberately manipulated to create the appearance that the power lines were congested (*overscheduling*). As a result, many retail utility providers went bankrupt, and eventually California had to re-regulate the electric supply.

A major company responsible for market manipulation was Enron. The Chairman of the California Power Authority made the following testimonial statements about deregulation and Enron's involvement in the electricity crisis [835]: "There is one fundamental lesson we must learn from this experience: electricity is really different from everything else. It cannot be stored, it cannot be seen, and we cannot do without it, which makes opportunities to take advantage of a deregulated market endless. It is a public good that must be protected from private abuse. If Murphy's Law were written for a market approach to electricity, then the law would state 'any system that can be gamed, will be gamed, and at the worst possible time.' And a market approach for electricity is inherently gameable. Never again can we allow private interests to create artificial or even real shortages and to be in control. Enron stood for secrecy and a lack of responsibility. In electric power, we must have openness and companies that are responsible for keeping the lights on. There is no place for companies like Enron that own the equivalent of an electronic telephone book and game the system to extract an unnecessary middleman's profit."

According to a 2007 study by the DOE, retail electricity prices rose much more from 1999 to 2007 in states that adopted deregulation than in those that did not [836]. In Europe, too, waves of electricity outages and increases in electricity costs have led to widespread discussions about the wisdom and value of deregulation.

13.5
Power Generation from New Renewables

13.5.1
Intermittency and Fluctuation

Renewable energies like hydro, geothermal, and tidal power (Chapter 12) have long been used to generate electricity and their increasing exploitation would not change the fundamental features of the traditional electric infrastructure which is based on the following premises: (i) electric energy is produced in large amounts in a few places (centralized generation), usually located at sites that are not too remote from large consumer markets; (ii) the generated energy is continuously adjusted to a fluctuating demand; and (iii) power is distributed by the grid throughout the country to the final customers (decentralized use). Exploitation of the two most important renewable energy sources, wind and solar energy, implies profound changes in the electricity system because such resources are, by their own nature, strongly delocalized over the territory and their availability is intermittent and often unpredictable.

Usually, no more than 20% of intermittent energy can be connected to the grid without disturbing the grid operativity. Extrapolation of this argument leads the opponents of wind and solar energy to state that such resources cannot be exploited unless parallel back-up conventional plants are provided that should be turned on and off to compensate for fluctuations of supply. In such a case, of course, the cost of wind (and solar) energy would be prohibitive. It should be pointed out, however, that besides back-up supply, there is another equally effective way to match supply and demand, namely maneuvering demand as a function of the fluctuating supply. While this option looks difficult on a small scale and on the basis of traditional grid management, it will be a key feature of smart electricity grids (Section 13.8). Indeed, controlling energy demand is a strategy already used in traditional, cleverly managed grids [816].

As discussed below, the utilization of intermittent renewable energies, the increasing number of customers, the continuous differentiation of their requirements, and the increasing use of electricity in the transport sector call not only for a smart grid, but also for systems capable of storing electric energy and for the full involvement of customers as main players in dealing with the electricity problem.

13.5.2
Electricity from Wind

Use of wind power to generate electricity (Section 12.2) is expanding faster than even its fiercest advocates could have wished a few years ago. At the end of 2009 the installed capacity was 160 GW, with an increase of 38.3 GW (32%) compared to 2008 [773]. The first five nations for total capacity installed are the US, Germany, Spain, China, and India. In 2009, wind power accounted for 39% of all new electrical capacity installed in the US and 37% of that installed in Europe. The development is accompanied by a general diversification process, as shown by 17 nations having installations of more than 1 GW compared with 13 countries in 2007.

The prospects for wind electricity development are indeed very bright. For 2020 China has set a national target for wind at more than double that of nuclear [837]. According to the Energy Watch Group, in 2030 electricity production from wind energy will come close to, or even outpace, hydropower energy [838].

Installation costs for wind power are around $1.8 million per megawatt for onshore developments and between $2.4 million and $3.0 million for offshore projects [797]. In many countries, subsidies lower the prices below that of coal. The market is growing rapidly, especially in China, where at the end of 2008 at least 15 companies were commercially producing wind turbines and several dozen more were producing components.

Wind power differs from solar power in that it is available, in principle, 24 h per day, but in practice its intensity is highly variable. Between October 2006 and February 2007 there were 17 days when the output from Britain's 1632 windmills was less than 10% of their capacity and one day when it was only 2% [839]. This shows that on connecting wind electricity to the grid two problems arise: long periods of reduced production and sudden changes in energy supply [701].

13.5.3
Electricity from Solar Energy Conversion

Solar energy can be directly or indirectly converted into electricity by means of several techniques. In practice, only concentrating solar power (CSP) in its various configurations and photovoltaics (PV) with a variety of different materials can produce important amounts of electricity (Chapter 10).

CSP plants produce steam by solar heat and then use it to generate electricity by the same technology as used in fossil fuel and nuclear plants. Solar heat can be accumulated and stored during the periods of intense sunshine and then used overnight or on cloudy days. In order to ensure uninterrupted service during overcast periods or bad weather, the turbines can also be powered by oil, natural gas, or biofuels. Waste heat from the power-generation process may be used to desalinate seawater or to generate cooling. Because the costs of raw materials for solar thermal power stations are rising more slowly than the price of fossil fuels, CSP may become competitive earlier than previously expected. New solar thermal power plants with a total capacity of more than 2 GW are at the planning stage, under construction, or already in operation. In the US, starting from 2008 several utilities have began to invest in solar electricity [840] with 4 GW of CSP plants. CSP plants situated in the Middle East and North Africa are expected to play a fundamental role within the DESERTEC cooperation initiative between European, Middle East and North African countries for market introduction of renewable energy and interconnection of electricity grids by high-voltage DC transmission.

The PV capacity, whether grid connected or off-grid, is extremely delocalized by its own nature. It produces a small share of electric energy, but is increasing at a rate of 40% per year. In the US, in response to expected carbon regulation, utilities have began to invest also in PV systems. In June 2009, 28 US utilities announced a total of more than 2200 MW of new PV projects [841]. Utilities know that nuclear plants are too expensive and, in the long term, natural gas will likely not be widely available at low cost. Therefore, they have begun to look at the benefit of building fuel-free 30-year power plants. Supported by a variety of incentives, a major part of today's solar industry is focused on installations that are owned by or directly benefit the customer. Utilities are also beginning to be interested in aggregating the energy produced by distributed PV systems, thereby creating a virtual PV power plant.

Being bound to sunshine, PV is a technology very useful to provide peak and intermediate power electricity demand, which are highest during the central hours of the day. Most likely, in well insulated regions, PV will be cost-effective in providing peak electricity to the grid in a few years.

13.6
Energy Storage for Electricity Supply Networks

13.6.1
Role of Storage

For wind and solar technologies, there is little correlation between where and when electric energy is produced and where and when it is needed. Since the electricity grid cannot store energy, solutions have to be found in order to cope with the highly variable supply from solar and wind. In the early phase of introducing renewable sources of electricity into areas with well-established electricity grids, the need for storage technologies is not so acute. Fossil fuel stations, particularly gas-powered ones, can serve as standbys for when the wind does not blow or the Sun does not shine. Usually, a grid can accommodate up to 20% of electricity from renewable sources without the need for dedicated storage facilities. It is obvious, however, that a more generalized use of electricity (e.g. in the transport sector) coupled with the extension of the contribution of renewable energies to generate electricity will require storage components as an integral part of the electric systems [539].

Energy storage can be beneficial to both utilities and their customers not only to support the use of renewable energies, but more generally to solve a variety of problems concerning reliability and quality of electricity [842]. For example, energy storage can be used by a customer to accommodate the minute–hour peaks in the daily demand curve, when the price of electric energy is related to peak-power demand; the energy store is replenished during off-peak hours and is discharged during the period of the customer's peak demand.

In principle, there are numerous possible options for electricity storage: (i) directly as electricity in supercapacitors and superconducting magnetic energy storage (SMES); (ii) by reversible conversion of electric energy into chemical energy in batteries or by water electrolysis followed by the use of the generated hydrogen in fuel cells; (iii) by reversible conversion into kinetic energy in flywheel or potential energy in pumped-hydro and compressed air energy storage (CAES). Depending on the specific type of application, it will be necessary to choose the storage option taking into account the amount of energy that has to be stored, the length of the storage period, the rates at which energy can be fed into or extracted from the storage, and several other parameters like efficiency, environmental impact, risk, seating ease, and life-cycle [829].

Large amounts of energy can be stored by electrolytic hydrogen production, or by pumped hydro and CAES, which, however, require special topological or geological formations. The other energy storage systems rank in the following descending order: in terms of power rating, batteries, SMES, flywheels, supercapacitors; and in terms of discharge duration, batteries, flywheels, supercapacitors, SMES. The quality of an electricity storage system is evaluated on the basis of the following set of specifications:

- **Capacity:** amount of energy stored by the system, measured in Ah;
- **Cycle life:** number of charge–discharge cycles that the system can resist without losing a certain amount of its capacity;
- **Specific energy:** energy stored per unit weight, $Wh\,kg^{-1}$;
- **Energy density:** energy stored per unit volume, $Wh\,l^{-1}$;
- **Energy efficiency:** ratio Wh-output/Wh-input;
- **Peak power:** maximum power density, $W\,kg^{-1}$.

Other requirements concern: rate of recharge, voltage plateau reliability, preservation, temperature range of operation, safety, need for maintenance operations, and ecological impact.

13.6.2
Pumped Hydro

These systems consist of two water reservoirs, in the form of lakes, separated by a substantial difference in vertical distance. When there is a surplus generating capacity (usually at night), water is pumped from the lower reservoir to the upper. When extra electricity is needed, the process is reversed and the falling water passes through a turbine to generate electricity. In some schemes, the turbine is used as both a generator and a pump. Pumped hydro is available at almost any scale with discharge times ranging from several hours to a few days. Other advantages of pumped hydro are high power (up to 3 GW), high efficiency (70–85%), long preservation (up to years), and fast response (of the order of seconds). More than 300 pumped-hydro schemes are in operation worldwide, corresponding to 90 GW capacity [842]. This technique is currently the most cost-effective means of storing large amounts of electric energy on an operating basis, but capital costs and the presence of appropriate geography are critical decision factors.

Interesting examples of international cooperation based on pumped hydro are the agreement between France and Italy, which allows pumping up water in Italian hydroelectric plants at night by using nuclear electricity generated in France, and export of intermittent wind electricity from Denmark (wind capacity 3.2 GW) to Norway and Sweden to generate hydroelectric power that is sold back to Denmark when needed.

13.6.3
Compressed Air Energy Storage (CAES)

Energy can be stored in compressed air on a scale similar to pumped-water energy systems. For very large systems there is no practical alternative to underground storage. Off-peak electricity is used to compress air to 40–70 atm pressure in either one or two stages. When extra electricity is required, the compressed air is pre-heated in a heat exchanger, mixed with a small amount of natural gas, and injected into a two-stage turbine [829]. A few compressed air energy storage (CAES) plants have been constructed, based on large salt caverns. That at Huntorf in Germany is a 290 MW system that can generate power for up to 4 h. Another plant, located in McIntosh, Alabama, has a capacity of 110 MW. A larger facility (2700 MW) is under development in Norton, Ohio. It should be noted that the overall efficiency of CAES facilities is lower than that of pumped-water energy storage systems because air heats up under compression and then cools down in the underground reservoir.

13.6.4
Flywheels

Minutes to hours storage of electricity in a flywheel is obtained as follows: electricity to be stored powers an electric motor that causes rotation of a flywheel, thereby converting electric energy into kinetic energy; the kinetic energy is then reconverted into electric energy when the motor, used as a generator, is powered by the kinetic energy of the wheel. The energy stored is proportional to the mass moment of inertia of the wheel and the square of the angular velocity. The amount of energy that can be stored by a flywheel is orders of magnitude less than that stored by pumped-water or compressed air systems. High-speed flywheels made of composite materials have an energy density up to 100 Wh kg^{-1}.

Figure 13.4 A KERS for a formula 1 car (Flybrid Systems).

Flywheels have several advantages compared to batteries: (i) capacity to absorb and release energy at a high rate; (ii) long life; (iii) low life-cycle costs; and (iv) no environmental impact. Recently, flywheels have been introduced to recover as kinetic energy part of the energy that would be dissipated as heat with a conventional brake [kinetic energy recovery systems (KERS)] Figure 13.4. A flywheel system designed for energy storage in a racing car, of weight 25 kg, with a total packaging volume of 13 l, and rotating at over 60 000 rpm in an evacuated chamber, can store 0.1 kWh of energy (4 Wh kg^{-1}) and can accept or deliver 60 kW of power [843]. Electric batteries capable of delivering that much power would weigh about 200 kg. KERS are now under development for road vehicles.

Flywheels are also used in electric rails to capture regenerative braking and to supply voltage support at weak points of the network.

13.6.5
Superconducting Magnetic Energy Storage (SMES)

Superconducting magnetic energy storage (SMES) systems store energy in the magnetic field created by the flow of direct current in a superconducting coil which has been cryogenically cooled to a temperature below its superconducting critical temperature. Once the superconducting coil is charged, the current will not decay and the magnetic energy can be stored indefinitely. The stored energy can be released back to the network by discharging the coil [844].

The round-trip efficiency of SMES systems is greater than 95% and the main parts in an SMES are motionless, which results in high reliability. They are, however, very expensive because of the energy requirements of refrigeration and the high cost of superconducting wires. SMES units in the range 1–10 MW are used to provide grid stability in distribution systems and power quality at manufacturing plants requiring ultra-clean power, such as microchip fabrication facilities.

13.6.6
Electrostatic Energy Storage (Capacitors)

A capacitor is a device made of two conductors separated by an insulator. On the application of a voltage, the device stores energy by the separation of positive and negative electrostatic charge. Years of research have led to systems in which the insulator has been replaced by finely divided porous carbon materials that store electrostatic charge in the form of ions (super-capacitors) or by a combination of large surface area with a material that can be reversibly oxidized and reduced (ultra-capacitors).

Capacitors are used to store small amounts of energy (up to 1 kWh) where many cycles of operation are required and charging must be completed quickly. They can play as "energy smoothing" and momentary-load devices. Their energy density is normally $6\,Wh\,kg^{-1}$, but progress in this field is fast. Capacitors and batteries have quite different properties that make them ideal partners for several applications [829]. Batteries store electric energy, whereas capacitors are power devices that store little energy but can be charged and discharged at much greater rates than batteries.

13.6.7
Batteries

Batteries convert electric energy into chemical energy that can be stored and then back-converted to electric energy. They are based on reversible oxidation/reduction processes of various kinds (see below). Batteries have long been the most practical way of storing electric energy on the small to medium scale.

In order to place the battery storage problem in a real context, we can take as a reference the most important form of stored energy, namely oil, which is characterized by the following properties: (i) specific energy, $11.9\,kWh\,kg^{-1}$; (ii) energy density, $10.3\,MWh\,m^{-3}$; (iii) easy to deliver; (iv) usable at ambient temperatures and pressures; and (v) storable for long times without appreciable loss. These are targets that it will be very difficult to match by any battery.

13.6.7.1 Battery Requirements
Rechargeable batteries come in many different sizes and capacities. The cycle life of a battery is measured by the number of charge–discharge cycles that it can resist without losing more than 20% of its capacity. Degradation is caused by irreversible side reactions. The specific energy for current batteries is between 30 and $160\,Wh\,kg^{-1}$, that is, at least two orders of magnitude lower than the specific energy of oil. The energy efficiency in most cases is around 60–80%. The peak power may

range between 80 and $1000\,W\,kg^{-1}$ [829]. A desired requirement is that the power delivered is stable over a good depth of discharge. Other important factors, beside price, are charge retention on open circuit stand, rate of recharge, temperature range of operation, reliability, need for maintenance operations, safety, benign nature of components and efficient material reclamation at the end of service-life. The relative importance of the various requirements depends of the kind of application and it is thus difficult to make general comparisons among the various types of batteries.

13.6.7.2 Types of Batteries

Lead-acid batteries   Over 90% of the world market for medium-to-large rechargeable batteries concerns lead-acid batteries, invented 150 years ago and repeatedly improved throughout the years. These cheap batteries are based on the overall reaction shown in Equation 13.1. The standard cell voltage is 2.04 V.

$$2PbSO_4(s) + 2H_2O(l) \rightarrow PbO_2(s) + Pb(s) + 2H_2SO_4(aq) \tag{13.1}$$

A typical lead-acid car battery has the following properties: *cycle life*, 500–800 cycles, strongly dependent on temperature (ideal temperature: 10–25 °C); *specific energy*, 30–$40\,Wh\,kg^{-1}$; *energy density*, 65–$75\,Wh\,l^{-1}$; *energy efficiency*, 50–92%; *power peak*: 150–$400\,W\,kg^{-1}$; *voltage plateau*, 2.1 V; *self-discharge rate*, 3–20% per month; *capacity* (Ah), decreasing strongly with decreasing discharge time. Because of their reliability and relatively low price, lead-acid batteries are extensively used for both stationary and transport applications. They can be constructed by using different lead alloys and can be designed according to different schemes so as to satisfy specific uses [829]. There is concern about the environmental consequences of improper disposal of old batteries, although lead-acid battery recycling is one of the most successful recycling programs in the world [845].

Alkaline batteries   The most popular ones are nickel–cadmium (NiCd) batteries that operate according to the overall reaction shown in Equation 13.2 (the alkaline electrolyte is not consumed):

$$Cd(OH)_2 + 2Ni(OH)_2 \rightarrow Cd + 2NiO(OH) + 2H_2O \tag{13.2}$$

For normal NiCd batteries, voltage (1.2 V) and specific energy (30–$40\,Wh\,kg^{-1}$) are fairly similar to those of lead-acid batteries. NiCd batteries offer some advantages over lead-acid batteries, for example, faster charge and discharge, minimal loss of capacity and longer cycle life, but if they are repeatedly recharged after being only partially discharged they gradually lose their maximum energy capacity (memory effect). They are expensive and are a matter of environmental concern because cadmium is a toxic heavy metal and therefore requires special care during battery disposal. Nowadays less toxic nickel–metal hydride (NiMH) batteries are progressively replacing the NiCd batteries in spite of higher cost and a high self-discharge rate. Because of their high power, NiMH batteries are used in hybrid electric vehicles.

Another kind of battery, nickel–zinc (NiZn), is increasingly important; because of their fast recharge they are used in the light electric vehicle sector. Considerable progress has recently been made concerning cost, energy density, and cycle life of *nickel–zinc* [846] and *zinc–air* batteries [847].

Flow batteries A flow battery may be seen as a hybrid of rechargeable battery and fuel cell [829]. In flow batteries an electrolyte containing one or more dissolved electroactive species flows through an electrochemical cell that converts chemical energy to electricity with formation of a spent material. Additional electrolyte is stored externally, generally in tanks, and is usually pumped through the reactor (gravity feeding is also possible). Flow batteries can be rapidly "recharged" by replacing the electrolyte liquid (in a similar way to refilling fuel tanks for internal combustion engines), while the spent material can be recovered for re-energization. In a flow battery, capacity and energy density are independent of each other: the capacity is determined only by the size of the reservoir, whereas the power is determined by the design and size of the cell stack.

Flow batteries are normally considered for relatively large (up to many MWh) stationary applications, for example, (i) for load leveling, where the battery is used to store cheap night-time electricity and provide electricity when it is more costly, (ii) for storing energy from renewable sources such as wind or solar for discharge during periods of peak demand, and (iii) for providing energy when the main power fails. The most common flow batteries are based on the redox reactions between Zn^{2+} and Br^- or between the oxidation states of vanadium.

Sodium batteries Sodium is a very reactive metal that can ignite or explode when treated with water. Therefore, is not an obvious candidate as an electrode in a battery. On the other hand, it has some interesting properties: (i) the electrochemical reduction potential of Na^+ (−2.71 V) is much more negative than that of other metals (e.g. −0.76 V for Zn^{2+}); (ii) it is a light element; (iii) it is abundant and cheap; and (iv) it is non-toxic. Coupling liquid sodium with liquid sulfur at temperatures of 300–350 °C, it is possible to construct a *sodium–sulfur* battery that can be used for storage of wind energy [848]. Replacing sulfur with nickel chloride led the Zeolite Battery Research Africa Project to construct a *sodium-nickel chloride* (Na–$NiCl_2$) battery called ZEBRA. This battery utilizes molten sodium chloroaluminate ($NaAlCl_4$, melting point 157 °C) as the electrolyte [849]. The ZEBRA battery has a nominal voltage of 2.58 V and an attractive specific energy and power (120 Wh kg^{-1} and 180 W kg^{-1}, respectively). The normal operating temperature range of ZEBRA batteries is 270–350 °C, cycle life 3000 cycles, and durability 8 years. These batteries are robust, reliable, safe, and require no maintenance. Therefore, they have been extensively used in the last 15 years for a variety of cars and buses. There is a limitation, however: when not in use, ZEBRA batteries should be left under charge so that they will remain molten and be ready for use when needed.

Lithium-ion batteries Lithium-ion batteries are rechargeable batteries not to be confused with lithium batteries, which are not rechargeable. The most common material for the anode is graphite. The cathode is a lithium–cobalt layered oxide, $LiCoO_2$ (or related compounds), and the electrolyte consists of lithium salts, such as $LiPF_6$ or $LiClO_4$, dissolved in an organic solvent (e.g. diethyl ether). The battery is assembled in the discharge state. During charge, Li^+ ions are withdrawn from the layered structure of $LiCoO_2$, transported through the electrolyte and then intercalated into graphite. As a result, cobalt is oxidized from Co^{3+} to Co^{4+}. This process is reversed on discharge. Specific properties of Li-ion batteries are as follows: cycle life, ~1200 cycles; specific energy, 150–200 Wh kg^{-1}; energy density, 250–530 Wh l^{-1}; energy efficiency, 80–90%; power peak, 300–1500 W kg^{-1}; voltage plateau, 3.7 V; self-discharge rate, 5–10% per month; durability, 1–3 years; capacity (Ah), decreasing strongly with decreasing discharge time.

Because of their high specific energy, Li-ion batteries have been extremely successful: more than 500 million cells are produced per year. Progress with Li-ion batteries is very rapid and involves the use of a variety of new materials [850–852]. Since cobalt is expensive, $LiCoO_2$ is often replaced by $LiNiO_2$ or better by Li–Mn oxides which are not only cheaper, but also less toxic. Using lithium iron phosphate, $LiFePO_4$, ultrahigh discharge rates have been obtained, comparable to those of ultracapacitors. Nanophosphate technology is at the basis of the A123Systems [853] high-power lithium-ion batteries.

Of course, Li-ion batteries are not free from disadvantages. They may be irreversibly damaged if discharged below a certain voltage, may explode (very rarely) if overheated or if charged to an excessively high voltage, and their capacity declines from time of manufacturing at a rate that depends on temperature. An important step towards the electrification of personal transportation is the development of cost-effective, long-lasting, and abuse-tolerant Li-ion batteries [854].

13.6.8
Electrolytic Hydrogen

For storing electric energy over long periods, reversible conversion to chemical fuels is an attractive choice, since chemicals are inexpensive to store and low efficiency is less critical for long storage periods. The most attractive chemical for this process is hydrogen (H_2) generated locally by electrolysis of water using intermittent excess solar or wind power. When combined with air or oxygen (O_2) in engines or fuel cells, H_2 can regenerate electricity on demand (Chapter 14).

13.7
Plugging-in Transportation

13.7.1
Hybrid and Full Electric Vehicles

In 2001 an Indian company launched a small three-door electric car, the G-Wiz [855]. It is powered by eight 6 V lead-acid batteries (peak power of 13 kW), and has a range up to 77 km with energy consumption of 21 kWh per 100 km, about four times better than an average fossil fuel car. In 2006 Tesla Motors presented the Tesla Roadster, an all-electric sport car powered by a lithium-ion battery pack that weighs 450 kg and stores 53 kWh [856]. This car costs around €100 000 and exhibits the following features: overall weight 1220 kg, range 350 km, engine maximum power 185 kW, acceleration from 0 to 100 in 3.9 s, and energy consumption 15 kWh per 100 km, that is, five times less than that of an average fossil fuel car (80 kWh per 100 km).

In the last 10 years, hybrid vehicles [857], like the popular Toyota Prius, have also been introduced in the market. They have dual gasoline and electric engines, the latter powered by a NiMH battery pack that is recharged by the combustion engine and, in part, by energy recovered from braking. The third-generation model of this car consumes 3.9 l of gasoline per 100 km, produces 89 g of CO_2 per kilometer, and costs €25 900. After having used the electric motor for less than 2 km, the battery has to be recharged by the combustion engine.

Most automakers are now engaged in designing and constructing electrified vehicles that can run exclusively, or in part, on electricity from the grid. Shifting to plug-in electric drive is strongly supported by public opinion [858] and by some governments [859]. This could be the biggest change the automobile industry has seen since the beginning of mass production of vehicles about 100 years ago.

Electric vehicles (EVs) have a natural advantage in that they can achieve efficiencies around 40% compared with 15% for those using internal combustion engines. However, if a fossil fuel electric plant is generating the electricity (efficiency about 35%), then the overall efficiency of the electric vehicle will be about 14%, comparable to that of a normal car. The advantage would be the reduction of emissions of air pollutants in congested urban environment, because the power plant is likely to be in a less populated area.

Full development of EVs faces several difficulties [860]. A first necessity is that the EV cars have to be light so as to carry as many batteries as possible to increase mileage performance and to have sufficient power. Lightweight cars have always been produced in Europe and Japan, but not in the US.

Batteries, of course, should also be as light and small as possible, while having high energy density, very high in/out efficiencies, a high level of safety and a life of at least 10 years. Lead-acid batteries (40 Wh kg^{-1}) are too heavy and also NiMH batteries (60–120 Wh kg^{-1}) are not satisfactory. Lithium-ion batteries are promising: with their higher energy density (150–200 Wh kg^{-1}), a range of 500 km is achievable. However, at present they are very expensive, less durable, and prone to overheating. Furthermore, there might be the problem of lithium availability, which is mostly concentrated in South America, and rate of recovery [861].

The motivation to shift from cars based on internal combustion engines to electricity is that of increasing sustainability of the transport sector by decreasing oil consumption and carbon dioxide emissions. This result, of course, could also be substantially pursued simply by reducing the power, speed, and weight of traditional cars. It is questionable, indeed, that hybrid technology is used to build cars like the Mercedes Class S 400 HYBRID, which consumes 8.2 l of gasoline per

100 km and emits 179 g km^{-1} of carbon dioxide. Although the fuel consumption and emissions are much smaller than for the non-hybrid version of the same car (14.9 l per 100 km and 355 g km^{-1} of CO_2), it should be clear that using such powerful, large, and heavy hybrid cars is not the right way to contribute to a sustainable world.

It must be emphasized that the current combined capacity of the internal combustion engines powering the world's road vehicles amounts to approximately 40 TW, while the present capacity of the world's electric power stations is about 4.5 TW [862]. This comparison highlights, once again, that the transformation of our energy system *must* be accompanied by a substantial change in the mobility of people and goods and, more generally, in our lifestyle.

13.7.2
Infrastructure

Mass produced electric vehicles, because of their restricted mileage performance, will require a ubiquitous charging infrastructure of wall sockets which currently is not available. Plug-in should be possible not only in the garage of the house, but also in parking lots of working places, public parking areas, commercial centers, and so on. In constructing such an infrastructure, it should be taken into account that the vehicles should plug into the grid not only for charging batteries, but also for communicating with the grid to control and optimize the charging operation. This implies a strict, smart interconnection between electric utility systems and EVs.

The EVs will also play the role of mobile electricity storage resources since their batteries can be cycled on and off to manage peak electrical demand and help utilities' reliance on variable renewable energies integrated within the grid. Furthermore, when the grid fails to deliver energy for any reason, the power stored in the car batteries could be used for emergency power supply. The components needed to create such a smart infrastructure already exist; the challenge is linking and scaling them.

Development of plug-in cars should be strictly related to increasing production of electricity by renewable energy such as solar and wind, hence EVs represent an important driving force to develop the use of renewable energies. Their development will be determinant in solving the energy and climate problems and in creating entrepreneurial opportunities and new green jobs.

13.8
Smart Grid

The economy of affluent nations depends heavily on electricity, and its role is expected to increase further in the future with the growth of renewables (wind, PV, geothermal). We are indeed moving from a fossil fuel economy to an electricity economy, even if large part of the electricity will continue to be generated for several years by burning gas, oil, and especially coal. In a hypothetical hydrogen economy (Chapter 14), electricity production and distribution will also play a major role because most of the hydrogen would likely be produced through the use of the electric vector. Therefore, the traditional, fragile, and insufficient electricity grids need to be re-engineered to play the role of engines in any developed and developing nation [863].

As we have seen before, the hardest task for a grid is to match supply to demand. Since demand is difficult to predict, grid utilities use to turn power production up and down by means of peaker plants. To face an increasing and fluctuating demand and cope with the deployment of intermittent renewable energies, in the near future the electricity industry will be forced to make a transformation from a centralized, producer-controlled to a distributed, consumer-interactive systems. This requires the creation of automated, widely distributed energy delivery networks, the smart grids [864, 865]; they will be based on the use of digital technology to enable a two-way flow of electricity and information, capable of optimizing supply and demand. A global network of this type could be compared with the Internet and the term "electranet" has been coined to emphasize the similarity [866]. A smart grid should perform several important functions. It should be capable of sensing system overloads and rerouting power to prevent a potential outage. It should be able to accept and seamlessly integrate wind and solar energy, whether from large farms or from individual units scattered across millions of rooftops. It should enable real-time communication between the consumer and the utility to allow consumers to tailor their energy consumptions on the basis of individual preference, such as price and environmental concerns. It should be capable of delivering high quality power, free of sags, spikes, disturbances, and interruptions. It should be increasingly resistant to attack and natural disasters. It should create new opportunities and markets by means of its ability to capitalize on plug-and-play innovation. Last, but not least, it should be increasingly greener to offer a genuine path towards energy saving and significant environmental improvement. Smart grids will help to make our planet smarter [867].

The smart grid is also a necessity for developing plug-in hybrid or all-electric vehicles. In the US a study on the potential for plug-in electric cars estimates that they would make up half of the annual car sales by around 2025 [119]. With the help of a smart grid, electric cars would be recharged when electric demand runs below what base load and other non-peaking plants can produce. For example, fickle wind currents usually generate much power at night. If we can shift auto recharging to night time and synchronize it to when currents are strong, we could save money, help the economic management of the utility, and eventually increase the value of that non-carbon-emitting wind power. The smart grid will give more value to intermittent renewable energy and electric cars can greatly help the reliable behavior of the grid.

In the US, following the Independence and Security Act of 2007, a research, development and demonstration program for a smart grid has been initiated with public/private partnerships under the leadership of the DOE [865]. Serious attention to the newer, greener grid should be a continuing priority for governments and private investors around the world. With effective investment, regulations, and incentives, the enormous task of remodeling the grid for renewable energy will be a benefit to both the environment and the global economy [868].

13.9
Towards an Electricity Powered World

If we wish to reduce substantially the use of fossil fuels, in the future we should mostly rely on nuclear, wind, and solar energy. These primary energies essentially produce electricity. Therefore, most of the human activities, including transportation, will progressively move towards an electricity-based economy. If most of the electricity will be produced by nuclear energy, the electrical network will continue to be based on centralized production and decentralized distribution schemes. If we wish to rely on renewable energies [869], not only smart grids, but also storage of electric energy will be essential.

Storage will play not only a technical, but also an economic role. Since the price of wholesale electricity varies with time, *when* electricity is sold, in fact, is just as important as *how much* electricity is sold. Pumped water and CAES are currently economic storage means for utilities, but they need very specific natural formations. Therefore, their contribution is likely to be insufficient and, in any case, localized in a few regions. Whether flywheels, magnetic storage, and capacitors will play a role is difficult to say. Therefore it appears that, at least in the short term, intermittent renewables will either depend upon the grid for back-up power, or use batteries for energy storage. Batteries are very modular and thus particularly suitable for use with small scale distributed renewables. They are, however, still very expensive and have a relatively short life and environmental drawbacks. Thus, in the more distant future, when much larger greenhouse gas emission reductions will become necessary, electrolytic production of hydrogen could be perhaps the best storage solution. Although H_2 electricity storage is less energy efficient (40–50%) than compressed air storage (60–70%), H_2 has far lower costs of storage capacity.

It is therefore important that hydrogen research and development efforts focus on technologies enabling efficient integration of future carbon-free transportation and electricity generation. It will be very important to develop higher efficiency reversible systems that can produce H_2 from electricity as well as electricity using H_2, potentially in homes or in vehicles. Not only batteries of electric vehicles, but also hydrogen stored in a fuel cell car (Section 14.7) could be used to provide electric energy to a home in case of need. For example, 5 kg of H_2 contained on board a fuel cell car could provide 75 kWh of electricity from the garage to the connected building, an amount of energy sufficient to power a European family home for more than 5 days.

Of course, policy attention should be paid to future regulations covering distributed electricity generation, hydrogen vehicles, fuel stations, and electricity systems for buildings. This will ensure economic and efficient interaction between all these critical components of an energy system that will be predominantly powered by well-integrated intermittent renewables.

14
Hydrogen

"The two most common elements in the universe
are hydrogen and stupidity."

Harlan Ellison

14.1
Introduction

In the past 20 years, a hydrogen economy has often been proposed by the media, and also by some economists and scientists [870, 871], as a means to solve the problems caused by the use of fossil fuels. At first sight, the idea sounds quite simple: (i) hydrogen is one of the most plentiful elements on Earth and in the Cosmos; (ii) some vehicle manufacturers have demonstrated that hydrogen can be used directly in an internal combustion engine, and fuel cell-powered prototype cars have also been constructed; (iii) combustion of molecular hydrogen, H_2, with oxygen produces heat, and combination of molecular hydrogen and oxygen in a fuel cell generates electricity and heat; (iv) the only byproduct is water, whereas burning of fossil fuels generates CO_2 and a variety of pollutants. Therefore, if hydrogen could promptly replace fossil fuels, both the energy and the environmental problems of our planet would have been solved.

Unfortunately, however, there is no molecular hydrogen on Earth. Molecular hydrogen has to be produced by using energy, starting from hydrogen rich compounds. Therefore, hydrogen it is not an alternative fuel, but an energy carrier [729, 872]. This is the central, but not unique, problem toward a hydrogen economy.

The hope on the advent of a clean world powered by hydrogen was further promoted in 2003 by US President Bush with a $1.2 billion plan called Freedom-CAR and The Hydrogen Fuel Initiative, aimed at commercializing hydrogen-powered cars by 2020 [873]. This plan was funded largely with money redirected from other renewable energy and energy efficiency projects. Most of the US scientific community did not agree since the beginning, pointing out that the plan "was deflecting attention from the hard, even painful measures that would be needed to slow our business-as-usual carbon trajectory" [874]. Recently, several questions have been raised about the design and the goals of the Bush initiatives

Energy for a Sustainable World: From the Oil Age to a Sun-Powered Future. Nicola Armaroli and Vincenzo Balzani
© 2011 WILEY-VCH Verlag GmbH & Co. KGaA, Weinheim
ISBN: 978-3-527-32540-5

[873]. In the meantime, Japan, China, Canada, and even Europe [875] launched similar projects. In the past few years most automakers have invested large amounts of money in the development of hydrogen-powered cars. More than 70 prototype cars and trucks as well as dozens of buses have been constructed. Despite all these efforts, the challenges that lie in the way to a hydrogen economy are enormous [729, 876, 877]. Most scientists believe that the shift to a hydrogen economy will not occur soon and might also not occur at all unless a large research effort is set up [3, 878, 879].

14.2
Properties and Industrial Uses

Hydrogen is the most abundant element in the universe, making up 75% of normal matter by mass, and is the third most abundant element on the Earth's surface. Due to its high reactivity, hydrogen in Nature is nearly always combined with other elements. It is present in water (H_2O) and thereby in every living organism. It is also present in hydrocarbons (e.g. methane, CH_4), in organic compounds, and in several other natural as well as artificial compounds.

Hydrogen is an invisible, non toxic, light gas (Table 14.1). The small H_2 molecule can diffuse through most materials and make, for example, steel brittle. It is highly flammable and burns in air at a very wide range of concentrations. At room temperature, uncompressed hydrogen occupies $11\,250\,\text{l}\,\text{kg}^{-1}$; pressurizing it into a high pressure (350 atm) steel tank reduces this to $56\,\text{l}\,\text{kg}^{-1}$. Hydrogen liquefies at $-253\,°C$. Liquefied hydrogen occupies only $14.1\,\text{l}\,\text{kg}^{-1}$. Hydrogen has a high energy content, $120\,\text{MJ}\,\text{kg}^{-1}$ ($33.3\,\text{kWh}\,\text{kg}^{-1}$), compared to $44.4\,\text{MJ}\,\text{kg}^{-1}$ ($12.4\,\text{kWh}\,\text{kg}^{-1}$) for gasoline.

Combustion of hydrogen with pure oxygen generates only water as a byproduct:

$$H_2(g) + \frac{1}{2}O_2(g) \rightarrow H_2O(l) + 286 \text{ kJ mol}^{-1} \tag{14.1}$$

Combustion with air, however, produces also nitrogen oxides, NO_x, because nitrogen, which constitutes 78% of air, is involved in the high temperature combustion

Table 14.1 Comparison of the properties of hydrogen with those of other fuels.

	Hydrogen	Gasoline	Methanol	Methane	Propane	Ammonia
Boiling point (K)	20.3	350–400	337	112	231	240
Liquid density ($\text{kg}\,\text{m}^{-3}$)	71	702	797	425	507	771
Gas density ($\text{kg}\,\text{m}^{-3}$) (at 0 °C, 1 atm)	0.08	4.68	–	0.66	1.87	0.69
Lower heating value ($\text{MJ}\,\text{kg}^{-1}$) (mass)	120.0	44.4	20.1	50.0	46.4	18.6
Lower heating value ($\text{MJ}\,\text{m}^{-3}$) (liquid)	8960	31170	16020	21250	23520	14350
Lower flammability limit (vol.% in air)	4	1	7	5	2	15
Upper flammability limit (vol.% in air)	75	6	36	15	10	28

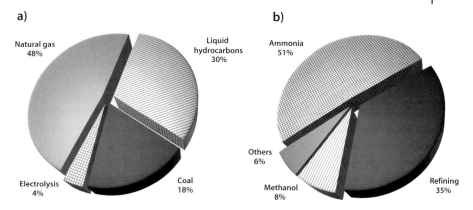

Figure 14.1 (a) Sources of hydrogen production currently used in the world; total production is about 50 million tons. (b) The main hydrogen-consuming sectors in the world.

reaction. In fuel cells, combination of hydrogen and oxygen generates electricity, heat, and water. In the US Space Shuttle missions, hydrogen powered fuel cells operated shuttle electrical systems and produced drinking water for the crew.

Today, about 50 million tons of hydrogen are produced yearly worldwide [111], which correspond to less than 2% of the world's primary energy demand. In most part, hydrogen is used as a feedstock in the chemical and petrochemical industries, to produce principally ammonia, refined ("upgraded") fossil fuels by hydrocracking, and a variety of chemicals. Except as a propellant for rockets, hydrogen is rarely used today as a transportation fuel.

Currently, more than 95% of hydrogen is produced from fossil fuels (Figure 14.1) in a variety of processes that generate, besides hydrogen, carbon dioxide. For example, the extraction of hydrogen from steam reforming of methane, which is today the less expensive process, takes place, via C and CO as intermediate products, according to the following overall reaction:

$$CH_4 + 2H_2O \rightarrow 4H_2 + CO_2 \tag{14.2}$$

A very minor fraction (about 4%) of hydrogen is produced by electrolysis of water, a much costlier process used only when high-purity hydrogen is needed:

$$H_2O + \text{electric energy} \rightarrow H_2 + \frac{1}{2}O_2 \tag{14.3}$$

14.3
Hydrogen as an Energy Carrier: The Scale of the Task

The dream of the hydrogen economy would be that of replacing fossil fuels, starting from those used for transportation [880]. Before discussing how this hydrogen could be produced, let us briefly consider how much of it would be needed. For simplicity, and also because the data are well known, we can make reference to the situation in the US, which uses about 1 million tons of gasoline per day. To replace this fuel alone, the amount of hydrogen produced would need to be 0.34 million tons per day, representing a volume of 3.8 billion m^3, or more than 1200 times the volume of NASA's Kennedy Space Center Vehicle Assembly Building, still the largest building in the world by volume [879]. As discussed below, producing such a large amount of hydrogen, which however would only replace the US gasoline demand for one day, is currently impossible and will be very difficult also in the future, especially if we want to obtain it from renewable energy.

14.4
Methods for Producing Hydrogen

14.4.1
"Clean Coal" Technology

While producing hydrogen as a transportation fuel from natural gas or oil is not convenient for both economic and practical reasons, coal could be used to supply significant amounts of hydrogen [111]. The technology under development to achieve this goal is the integrated gasification combined cycle (IGCC), also called "clean coal" technology (Section 6.7), introduced by the US project FutureGen (Figure 6.4), a public–private partnership announced in 2003 by President Bush. The aim of the project was the construction of near zero emission coal-fueled power plants to co-produce hydrogen and electricity. However, the costs were inaccurately estimated and, in retrospect, FutureGen appears to have been just a public relations ploy for the Bush administration, to make it appear to the world that US was doing something to address global warming despite its refusal to ratify the Kyoto Protocol [881]. In 2008 the Department of Energy decided to restructure the project because of the rising costs and promoted a joint plan with industry to open multiple plants by 2015. Other changes were announced in March 2009 by the current US government, together with a generic intention to support the project. In the meantime, it has become clear that the planned sequestration of CO_2 produced in large amounts from coal is very expensive, might be very dangerous, and involves several complex and so far unsolved problems (Section 6.6).

14.4.2
Biomass

Biomass includes a large variety of materials generated by photosynthesis, such as agricultural residues from farming and wood processing, dedicated bioenergy crops such as switchgrass, and even algae in the sea. As discussed in Section 11.3, biomass can be used for the production of liquid fuels such as ethanol, methanol, and biodiesel. Biomass can also be converted into hydrogen by gasification or pyrolysis coupled with steam reforming. All the economic, ecological, and land-use limitations discussed for the production of liquid fuels from biomass (Section 11.3) apply to hydrogen production, with the further disadvantage deriving from the difficulty of handling hydrogen compared to liquid fuels.

14.4.3
Water Electrolysis

14.4.3.1 General Concepts
Cleaving water into its components needs energy and produces gaseous hydrogen and oxygen. When the energy is supplied in the form of electricity, the process is called electrolysis (Equation 14.3). Today, the production of hydrogen by electrolysis is more expensive than from natural gas and therefore it is used only when high purity hydrogen is needed. However, if we wish to go towards a carbon-free hydrogen economy, electrolysis is potentially the best method since electricity can be obtained from primary energy sources different from fossil fuels, namely renewable energy and nuclear energy. Freshwater is a precious resource for mankind and therefore should not be wasted. This, however, is not a major problem. For instance, a hydrogen-based transportation system in the US using freshwater as feedstock would only employ a small fraction (~1%) of the domestic water use [882].

Figure 14.2 shows a scheme that evidences the use of electricity and hydrogen as parallel and interchangeable energy carriers. As we will see later, in the hydrogen economy the most important role of hydrogen would be that of powering fuel cells to produce electric power. Apparently it is irrational to use electrolysis (i.e.

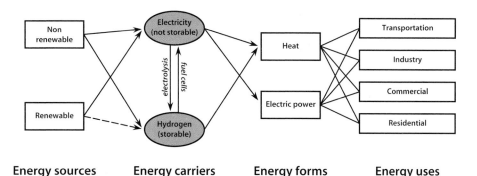

Hydrogen and electricity as energy vectors

Energy sources   Energy carriers   Energy forms   Energy uses

Figure 14.2 Hydrogen and electricity as parallel and interchangeable energy carriers. The dashed arrow represents the water splitting processes under development (e.g. artificial photosynthesis).

electric power) to generate hydrogen, that will then be used in fuel cells to make electric power again. The only reason to proceed in this way is related to the capacity of hydrogen to be stored (Section 14.5), whereas storage of electric power is a difficult problem (Section 13.6).

Electrolysis is a mature technology. In current market conditions, about 50 kWh of electricity are consumed to produce 1 kg of (compressed) hydrogen. A typical commercial electrolyzer has an efficiency of about 70–80%, but a higher efficiency can be obtained with more elevated temperature water or steam electrolysis. Electrolysis of 1 t of water results in 111.5 kg of hydrogen (1237 m^3 at ambient conditions) and 888.5 kg of oxygen (618 m^3). Industrial uses of oxygen are limited. In case of massive hydrogen production, the oxygen co-produced is likely to have limited commercial value and would simply be released into the atmosphere.

Basically, any energy source capable of producing electricity can be used to produce hydrogen as a transportation fuel. The problem, however, is to find a sufficiently large power generating capacity. As we have seen above, just to replace the gasoline component of the US energy economy we would need to produce about 0.34 million tons per day of hydrogen. The production of such an amount of hydrogen will require about 850 GW of power generating capacity and will consume over 20 000 GWh of energy [879]. This is roughly equivalent to the total US generating capacity (including reserves). It can also be noted that 850 GW represents the power that would be generated by 38 Chinese Three Gorges Dam power plants (22.5 GW each). These comparisons emphasize the difficulty in developing a hydrogen economy, but also stresses that the energy consumption of transportation in the US is beyond any reasonable limit.

If hydrogen has to replace gasoline, the US transportation system has to be drastically modified. Perhaps American people should travel less; surely, they should use less fuel consuming vehicles. Furthermore, use of personal cars should be strongly reduced, and public transportation systems based on moderate electricity consumption should be developed. Apparently these concepts inspired the Obama administration which supported the merging of Chrysler with the Italian company Fiat, an automaker specialized in the production of small, low consumption cars, and announced a $8 billion project as a first step to developing networks of high speed electric trains like in Europe, Japan, and China.

14.4.3.2 Hydroelectric Power

Currently, the largest renewable electricity source is hydroelectric power, which in 2008 supplied about 3100 TWh, or 16% of the world's electricity [754]. Most of the hydroelectric power, 750 GW, comes from large plants (Section 12.1). In most countries expansion of hydroelectric energy is currently based on low-cost small hydro plants (less than 10 MW) that use part of the stream flow of a river and have a low environmental impact. In 2009 small hydro installed capacity amounted to about 60 GW worldwide [544]. Increases in the exploitation of hydroelectric power can make only a small contribution towards a hydrogen economy.

14.4.3.3 **Wind Electric Power**

Wind power has an enormous potential, which is estimated to be about 75 TW (Section 12.2). At the end of 2008 the global wind power reached 120 GW. In the US, it has been estimated that to produce by a 70% efficient electrolytic process enough hydrogen (ca 40 million tons) to power 50% of the light-duty fleet based on fuel cell vehicles at twice the 2004 average efficiency would require 555 GW of wind [882]. Comparison with the current US installed levels (35 GW) shows the difficulty of the task. The intermittent nature of wind energy is also a drawback. Nevertheless, wind could certainly make an important contribution towards the development of a clean hydrogen economy.

14.4.3.4 **Solar Photovoltaic and Photoelectrochemical Electricity**

Electricity produced by solar radiation in photovoltaic or photoelectrochemical cells (Chapters 10 and 11) can be used to produce hydrogen. However, it can be estimated [882] that to obtain the 850 GW of power generating capacity required to produce by electrolysis the amount of hydrogen needed for replacing the gasoline component of the US energy economy would require covering 5700 km^2, an area slightly larger than Rhode Island, with 15% efficient solar PV panels. Of course, there would be no need to cover useful land since there are plenty of industrial roofs, highways, roads, and desert areas that could be used.

Unlike wind, however, solar is still an expensive way to generate electricity. Improved efficiencies and/or the use of thin-film technologies, nanostructured films, organic polymers [883], and concentrators [655] (Chapter 10) will certainly reduce the cost of solar electricity [884, 885] and thus of hydrogen production. Indeed, solar PV can make an important contribution to the development of a hydrogen economy, but costs have to be further reduced [886].

14.4.3.5 **Solar Thermal Electricity**

With current technology, storage of heat is much cheaper and more efficient than storage of electricity. Therefore, in CSP plants heat is sometimes stored before conversion to electricity (Section 10.3). In this way, the CSP plant can produce electricity day and night. If the CSP site has predictable solar radiation, then the CSP plant becomes a reliable power plant. Reliability can be further improved by installing a back-up system that uses fossil energy. Electrolyzers for hydrogen production coupled with solar thermal farms can operate continuously at full capacity. CSP plants have no fuel cost, do not generate pollution, and can be placed in unused desert areas. Indeed, construction of CSP plants in desert areas might contribute to the development of a hydrogen economy. Renewable energy proponents are also planning a massive electricity grid in which CSP plants in North Africa and the Middle East would feed power as far as northern Europe (DESERTEC project) [887]. The system would also hook in solar PV systems and other renewable-energy resources [888].

In principle, the heat generated by a CSP plant could also be used for the direct thermal decomposition of water. This process is impractical since it requires temperatures higher than 2000 °C, but water splitting can be achieved at much lower temperatures by using thermochemical cycles as described in Section 10.3.7.

This technology could make an important contribution to the development of a hydrogen economy, but once again there are substantial economic problems.

14.4.4
Photoelectrochemical and Photochemical Water Splitting

Harnessing sunlight directly as a hydrogen fuel in photoelectrochemical and photochemical devices circumvents the need to couple a solar PV device with a water electrolyzer. Since the discovery of the first photoelectrochemical system based on a semiconductor that can absorb only a very limited fraction of the solar spectrum [742], significant advances in basic science have been attained [748] (Section 11.5), but these systems are still far from practical applications.

14.4.5
Nuclear Energy

In 2009, production of nuclear electricity was 2558 TWh, i.e. 13.5% of the generated total [471]. Unfortunately, comparison with the power that would be needed for promoting a hydrogen economy is again disappointing. It has been estimated that to produce by a 70% efficient electrolytic process enough hydrogen (ca 40 million tons) to power 50% of the US light-duty fleet based on fuel cell vehicles at twice the 2004 average efficiency would require 216 GW of nuclear power [882]. By comparison, the current US installed level is 92 GW. This evaluation shows once again that it will not be possible in a hydrogen economy to maintain the huge energy consumption level characteristic of the affluent countries and, in particular, of US society. A contribution to the development of a hydrogen economy, however, could certainly come from nuclear plants since they work at a constant level, whereas the consumption of electric power changes very much during 24 h. Using off-peak periods for producing hydrogen through electrolysis would enable a greater utilization of the plants. Expansion of nuclear energy, however, will cause all the problems discussed in Section 8.2.

As in the case of CSP plants, the heat generated by a nuclear plant could be used for the thermal decomposition of water by using thermochemical cycles. The Japan Atomic Energy Agency (JAEA) has conducted successful experiments with the sulfur–iodine (S–I) cycle. with the intention of using nuclear high-temperature generation IV reactors to produce hydrogen. With these thermochemical cycles, however, there are severe corrosion problems that have still to be solved and, in the US, start-up of research on very high-temperature reactors for hydrogen production have been postponed to 2018–2020 [58].

In one way or another, nuclear energy could certainly make an important contribution towards a hydrogen economy. The hydrogen produced by nuclear energy, however, could hardly be defined as a clean fuel because it would be accompanied by the generation of radioactive waste (Section 8.2).

14.5
Hydrogen Storage

14.5.1
A Difficult Problem

Hydrogen storage is a key factor for the successful introduction of hydrogen as a transportation fuel, which is the major application for the hydrogen economy. The primary task of any on-board fuel storage system is to provide a safe containment and handling system for delivery of sufficient fuel to satisfy the driving range requirement of the vehicle. These objectives must be achieved at low cost, without adding excessive weight, and without compromising the interior volume that is needed for passengers and baggage. Liquid hydrocarbons are almost perfect fuels in all these respects. They have high energy density (gasoline, $8.07\,kWh\,l^{-1}$; diesel, $10.9\,kWh\,l^{-1}$) and can be contained in simple metal or plastic tanks of any form, that can be rapidly filled.

Compared with hydrocarbons, hydrogen gas has a good energy density by weight ($33.3\,Wh\,kg^{-1}$), but poor energy density by volume ($2.5\,Wh\,l^{-1}$). Under normal conditions, hydrogen requires about 3000 times more space than gasoline for an equivalent amount of energy (Table 14.1). Increasing gas pressure improves the energy density by volume, allowing the tank to be smaller, but not lighter. Alternatively, to increase its volumetric energy density, liquid hydrogen may be used. However, liquid hydrogen boils at around $-253\,°C$. Therefore, its liquefaction imposes a large energy loss and the storage tanks must be well insulated to prevent boil off. Assuming all of that is solvable, the density problem remains. Liquid hydrogen ($2.5\,kWh\,l^{-1}$) has a worse energy density *by volume* than hydrocarbon fuels such as gasoline, by a factor of approximately 3.5. It should be noted that there is about 55% more hydrogen in $1\,l$ of gasoline (about 110 g of hydrogen) than there is in $1\,l$ of pure liquid hydrogen (71 g of hydrogen).

On-board hydrogen storage in the range of approximately $5–13\,kg$ is required to enable a driving range greater than $480\,km$ for vehicles using fuel cell power plants. The DOE targets are a $1.8\,kWh\,kg^{-1}$ system (5.5 wt% hydrogen) and $1.3\,kWh\,l^{-1}$ (0.040 kg of hydrogen per liter) in 2015 and the ultimate targets are a $2.5\,kWh\,kg^{-1}$ system (7.5 wt% hydrogen) and $2.3\,kWh\,l^{-1}$ (0.070 kg of hydrogen/per liter). Low-cost, energy-efficient off-board storage of hydrogen will also be needed for stationary and portable applications and throughout the hydrogen delivery infrastructure. For example, storage is required at hydrogen production sites, hydrogen refueling stations, and stationary power generation sites. Temporary storage may also be required at terminals and/or intermediate storage locations.

In an attempt to find a convenient way to store hydrogen in vehicles, both physical storage (liquefied or compressed hydrogen) and chemical storage (absorption by metals or other materials) are being intensively investigated.

14.5.2
Liquid Hydrogen

After helium, hydrogen is the most difficult gas to liquefy and it can be produced only by a complex and expensive multi-stage cooling procedure. About 30–40% of the energy content of the hydrogen is required for its liquefaction. Insulation for liquid hydrogen tanks is usually expensive and delicate. Any liquid hydrogen storage system loses hydrogen gas over time by evaporation. The rate of loss is larger for small quantities of liquid hydrogen. A big problem is that hydrogen leaks can result in major safety hazards.

14.5.3
Compressed Hydrogen

Hydrogen compression is currently the preferred solution used in most hydrogen fuel cell-powered prototype cars, but it is still far from being satisfactory. Hydrogen can now be held under 350 or 700 atm in tanks made from new composite light-weight materials such as carbon fiber with metal (aluminum or steel) or polymer (thermoplastic) liners. Such high-tech materials, however, are very expensive. Even at 700 atm, hydrogen has energy content per unit volume 4.6 times lower than that of gasoline, which means that the hydrogen tank must be much larger. Furthermore, about 10–15% of the energy content of hydrogen is required for compression. A complicated problem concerns the on-board location of the tank. In contrast to the tanks for liquid hydrocarbons, the hydrogen tank cannot adopt the shape that best suits the vehicle, since compressed hydrogen tanks have to be cylindrical to ensure their integrity under high pressure. Therefore they are difficult to integrate in the vehicle.

As previously mentioned, hydrogen is able to diffuse through many materials, including metals. Since the overall fuel system contains several metallic components, material failure could cause leaks with a major safety hazard because hydrogen is a highly flammable and explosive gas.

14.5.4
Metal Hydrides

Many metals and metallic alloys have the ability to absorb hydrogen, like a sponge, to form hydrides [889]. The amount of hydrogen that can be inserted depends on the chemical properties of the system but, in any case, is a small fraction by weight since only one or a few hydrogen atoms can be bound by each metal atom. As a result, in order to store 5 kg of hydrogen, the storing system needs to weigh several hundreds kilograms. The advantage of metal hydrides is their compactness, requiring less space than compressed hydrogen. Since they work at moderate pressure, hydride tanks can also be shaped more freely and are thus easier to integrate in the vehicle.

Ideally, as much hydrogen as possible should be rapidly absorbed and all of the absorbed hydrogen should be released on command, on a suitable time scale. As could be expected, however, when hydrogen is absorbed rapidly, it is released slowly, and vice versa. Both uptake and release are controlled by pressure and temperature and can be accelerated by catalysts. Today, research is mainly focused on light compounds based on alkali and alkaline earth elements (NaH, MgH_2) and complex hydrides such as $NaAlH_4$, $LiBH_4$, $NaBH_4$, and $LiNH_2$ [890–892]. The maximum material (not system) gravimetric capacity of 5.5 wt% hydrogen for $NaAlH_4$ is much below the DOE target for 2015 (5.5 wt%), which refers to the entire system (including tank, valves, regulators, and other hardware). It should also be noted that the maximum theoretical capacity for $NaAlH_4$ is 7.4 wt%. Reversibility of these and other materials for over 1000 cycles has not yet been demonstrated.

14.5.5
Other Systems

Organic materials offer important advantages over inorganic systems [879, 893]. They are light, cheap, and "green," but they also have some major drawbacks. Physisorption takes place on most porous organic solids like activated carbon, graphene, carbon nanotubes, fullerenes, and polymers with intrinsic microporosity or suitably functionalized [894]. Since physisorption is governed by van der Waals interactions, hydrogen is absorbed only as a monolayer at temperatures higher than its boiling point (20.4 K). Therefore, it is quite important to use materials with specific surface areas as large as possible. The maximum capacity of any carbon-based sorbent is 7.7 wt% for a 1:1 ratio between carbon and hydrogen atoms. Microporous metal organic framework (MOF) compounds [895] have recently emerged as some of the most promising candidate materials for hydrogen storage because their pore structure can be chemically tuned.

In order to preserve the important advantages of liquid fuels like gasoline and diesel, namely fast refilling times and high energy density, it would be of great interest to achieve hydrogen storage in liquid materials. An ideal molecular storage liquid system for hydrogen should be capable of storing as much hydrogen as possible and of releasing hydrogen between 80 and 150 °C, which is the working temperature of polymer membrane fuel cells. The dehydrogenation process, of course, should be reversible. Several pairs of hydrogenated and dehydrogenated molecules have been proposed, for example, CH_3OH–CO_2, $HCOOH$–CO_2, C_2H_6–C_2H_4, and NH_3–N_2 [893]. Ammonia is a particularly interesting system: its hydrogen storage capacity is 18 wt% and its synthesis is a well established industrial process, but it is too stable to release hydrogen easily. Much interest is also focused on compounds like ammonia borane (NH_3BH_3) [896] and lithium amidoborane ($LiNH_2BH_3$), which releases nearly 11 wt% of hydrogen at 90 °C [897].

14.6
Hydrogen Transportation and Distribution

If hydrogen is to become the energy source of the future, it would need to be easily available anywhere at an affordable price and its distribution should be safe and user-friendly, as happens today for fuels based on hydrocarbons. There is, however, a fundamental difference. Since hydrogen is not a primary energy source but an energy carrier, a key feature of a hydrogen economy is that in mobile applications (primarily vehicular transport), energy generation and use are decoupled. The primary energy source needs no longer traveling with the vehicle, as it currently does with hydrocarbon fuels. Therefore, two options are possible concerning hydrogen supply to the end users: centralized or decentralized distribution.

14.6.1
Centralized Distribution

In the centralized distribution approach, hydrogen should be produced in big plants and stored in large amounts in relatively few locations. Then it should be transported and stored in lower quantities in several places, and finally it should be distributed to customers. Hydrogen transportation, however, is not an easy task. Large amounts of hydrogen gas could be transported by pipelines, but natural gas pipelines are not suited for hydrogen transport because high-pressure hydrogen leaks easily through the smallest of holes and also embrittles the type of mild steel used for gas pipeline construction. Pipelines for hydrogen would require different materials, different welding procedures, and different designs (for valves, compressors, sensors, and safety devices) than those used for natural gas. It has been estimated [879] that about 1.2% of the hydrogen energy will be used for every 150 km to power the compressors, compared with 0.3% for natural gas. An additional source of energy loss is leakage, which is about 1–3% for natural gas and will likely be much higher for hydrogen.

Similar problems will be encountered for truck or rail transportation of compressed hydrogen gas, except that higher costs will be incurred as material handling losses and round-trip fuel costs increase substantially. A 40 000 kg tube truck pressurized to 200 atm carries only 320 kg of hydrogen, less than 1% of its dead weight. In comparison, a similar truck could deliver about 26 t of gasoline, containing over 20 times more energy than the compressed hydrogen tube.

Today, limited amounts of hydrogen are transported as a cryogenic liquid. Cryogenic trailers, however, are very expensive because they must be double hulled and vacuum insulated. It should also be recalled that up to 40% of the energy content of the shipped hydrogen is consumed by its liquefaction, making the process very expensive from both energetic and economic viewpoints.

In conclusion, hydrogen's properties raise major issues regarding the distribution of hydrogen starting from centralized production plants.

14.6.2
Decentralized Distribution

Since electricity can be easily transported and hydrogen can be produced by electrolysis, it could be convenient to distribute electricity and to prepare hydrogen on-site, where it is needed. While attractive on paper, this option presents many hurdles in practice. Consider, for example, a very busy bus station (2000 vehicles per day) [879]. It would be equipped to receive and handle substantial amounts of electric power, transform it, and convert it from AC to DC. To produce the hydrogen needed (up to 45 t per day) such an on-site facility should electrolyze about 400 m^3 of water, consuming more than 100 MW of electric power, which corresponds to the power consumed by 20 000 small homes. The total energy needed to generate and compress hydrogen at the station would exceed the heating value of the hydrogen generated by at least 50%, partly because small scale compression technology for hydrogen is currently inefficient. Safety problems related to local manufacture of such an explosive gas should also be considered.

In a decentralized hydrogen delivery system based on electricity, the ecological consequences of energy use are decoupled, as happens in the centralized distribution scheme: no problem where energy is used because electrolysis is a clean process and the hydrogen produced is a clean energy carrier, but problems remain where electricity is produced. Whether the resulting energy system is sustainable depends on the method used for producing electricity.

14.7
End Uses of Hydrogen Fuel

14.7.1
Fuel Cells: General Concepts

In a hydrogen economy it is expected that most of the hydrogen will be used to produce electricity by fuel cells [898]. A fuel cell is a device that produces electricity from the reaction of a reductant (fuel) with an oxidant (usually oxygen) in the presence of an electrolyte [899]. The reactants flow into the cell, and the reaction products flow out of it, while the electrolyte remains within it and electricity is produced. Fuel cells are thermodynamically open systems that convert the chemical energy of a fuel directly into electrical energy. They can operate virtually continuously as long as the necessary flows are maintained. Fuel cells, in principle, can be built based on any spontaneous chemical reaction. A hydrogen fuel cell uses hydrogen as a reductant and oxygen (usually from air) as oxidant.

In contrast to internal combustion engines, fuel cells do not involve the generation of heat by burning the fuel and the subsequent conversion of heat into mechanical energy. Therefore, the efficiency of a fuel cell is not limited by the Carnot cycle. The theoretical efficiency of a fuel cell (η) is determined by the ratio

Table 14.2 Theoretical reversible cell potentials ($E°$, V) and theoretical maximum efficiencies (η) of fuel cell reactions under standard conditions (20°C and 1 atm); n, $\Delta G°$ (kJ mol^{-1}), and $\Delta H°$ (kJ mol^{-1}) are the number of exchanged electrons, the free energy change, and the enthalpy change, respectively.

Fuel	Reaction	n	$-\Delta H°$	$-\Delta G°$	$E°$	$\eta(\%)$
Hydrogen	$H_2 + 0.5O_2 \rightarrow H_2O(l)$	2	286	237	1.23	83
Methane	$CH_4 + 2O_2 \rightarrow CO_2 + 2H_2O(l)$	8	891	818	1.06	92
Methanol	$CH_3OH + 1.5O_2 \rightarrow CO_2 + 2H_2O(l)$	6	727	703	1.21	97
Formic acid	$HCOOH + 0.5O_2 \rightarrow CO_2 + H_2O(l)$	2	270	286	1.48	106
Ammonia	$NH_3 + 0.75O_2 \rightarrow 0.5N_2 + 1.5H_2O(l)$	3	383	338	1.17	88

between the free energy change ($\Delta G°$) and the enthalpy change ($\Delta H°$) of the chemical reaction, and the theoretical reversible potential ($E°$) is determined by $\Delta G°$ (Table 14.2). For hydrogen fuel cells, which are based on the reaction in Equation 14.4, $\Delta G° = -237$ kJ mol^{-1} and $\Delta H° = -286$ kJ mol^{-1}, which yields theoretical efficiency $\eta = 83\%$, and cell voltage $E° = 1.23$ V, obtained from Equation 14.5, where n is the number of exchanged electrons ($n = 2$) and F is the Faraday constant, 96 485 C mol^{-1} [5]:

$$H_2(g) + \frac{1}{2}O_2(g) \rightarrow H_2O(l) \tag{14.4}$$

$$-\Delta G° = nFE° \tag{14.5}$$

The working efficiency of a fuel cell is dependent on the amount of power drawn from it. The more power (current) is drawn, the lower is the efficiency.

14.7.2
Proton Exchange Membrane (PEM) Hydrogen Fuel Cells

In hydrogen-based fuel cells the electrolyte can be made by a variety of ion conductor materials, from polymers to ceramics.

In proton exchange membrane (PEM) fuel cells [900] (Figure 14.3), hydrogen entering the fuel cell is split with the help of a platinum catalyst on the anode side into electrons and protons. The protons migrate to the cathode through a polymer electrolyte membrane and the electrons are forced to travel through an external circuit generating electric power. At the cathode, protons and electrons are recombined with oxygen of the air in a platinum catalyzed process to form water as a waste product. The core of the fuel cell is a thin polymer film (usually a perfluorosulfonic acid), which is permeable to protons when saturated with water, but does not conduct electrons. PEM fuel cells have high power density, operate at fairly low temperature (70–85°C), and thus can start up almost immediately. They are safe, quiet, and easy to operate. For all these reasons, PEM fuel cells are considered the most suitable candidate for automotive applications. Currently, however, hydrogen fuel cells are expensive and somewhat fragile.

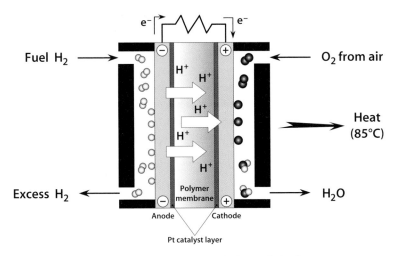

Fuel H$_2$

O$_2$ from air

Heat
(85°C)

Excess H$_2$

Polymer membrane

H$_2$O

Anode   Cathode

Pt catalyst layer

Figure 14.3 Proton exchange membrane (PEM) hydrogen fuel cell.

A typical fuel cell produces a voltage from 0.6 to 0.7 V, so that many cells must be stacked together to generate sufficient power to propel an automobile. A cell running at 0.7 V has an efficiency of about 50%, with the remaining 50% of the energy content of the hydrogen being converted into heat. The current density at 0.7 V is ca 0.5 A cm^{-2}, and the power output ($P = IE$) is about 0.35 W cm^{-2}. In terms of weight, a fuel cell system (that includes fuel and oxidant controls, cooling and produced water removal) will provide 4.46 kW m^{-3} and about 88 W kg^{-1} [879]. Thus, to provide the 150 kW generated by an average American four or six cylinder gasoline motor (weight 180 kg) one would need fuel cells with an overall area of polymer electrolyte membrane about 43 m^2 and a weight of 1700 kg. Although there is no need to drive a 150 kW power car, it is clear that there is still some way to go concerning weight reduction. Research on hydrogen fuel cells is very active particularly in the directions of finding more efficient and cheaper catalysts [901].

14.7.3
Other Types of Hydrogen Fuel Cells

There are many other types of hydrogen based fuel cells [904]. The nature of the electrolyte determines many of the fuel cell's properties, including the temperature of operation, which is a very important parameter for possible applications.

The alkaline fuel cell (AFC) has been used by NASA in the Apollo missions to produce electricity and potable water. In AFCs the two electrodes are separated by a porous matrix saturated with an aqueous alkaline solution, such as potassium hydroxide (KOH). Since aqueous alkaline solutions react with carbon dioxide (CO_2) which is present in air, the fuel cell can become "poisoned" through the conversion of KOH to potassium carbonate (K_2CO_3). Therefore, alkaline fuel cells typically operate on pure oxygen.

Regenerative PEM Cells

A PEM cell can, if designed for the purpose, be run in reverse. In other words, it can use electric power to electrolyze water and produce hydrogen and oxygen. This dual function system is known as a regenerative fuel cell or unitized regenerative fuel cell (URFC) [902]. Lighter than a separate electrolyzer and generator, a URFC is an excellent energy source in situations where weight is a concern. A URFC uses bifunctional electrodes (oxidation and reduction electrodes that reverse roles when switching from charge to discharge, as with a rechargeable battery) and cathode-feed electrolysis (water is fed from the hydrogen side of the cell). This technology has been employed by NASA to power unmanned, solar-powered aircraft (Pathfinder and Helios) used for high-altitude surveillance, communications, and atmospheric sensing as part of the Strategic Defense Initiative.

Waiting for application in real-word transportation, there are clean energy education kits on the market (e.g. Hydrocar [903]), consisting of a toy car which contains a dual function fuel cell. First, the fuel cell is operated in reverse to produce hydrogen by water electrolysis powered by a photovoltaic panel. Then the hydrogen and oxygen produced, which are stored in small plastic containers, are used by the fuel cell to power an electric motor which moves the car.

Dual function cells are under development also for static applications such as daytime generation of hydrogen from solar-generated electric power so that the hydrogen can be stored for use at night.

Phosphoric acid fuel cells (PAFCs) use liquid phosphoric acid as an electrolyte at an operating range of 150–200 °C. They are employed for stationary applications, mainly for consumers in need of a very stable, reliable, and clean on-site electricity source such as banks, airports, hospitals, and military bases.

Molten-carbonate fuel cells (MCFCs) operate at temperatures of 600 °C and above. Since they operate at very high temperatures non-precious metals can be used as catalysts, reducing costs. The primary disadvantage of current MCFC technology is durability because of the high operational temperature and the corrosive action of the electrolyte.

Solid oxide fuel cells (SOFCs) are promising for stationary power generation, but novel materials are needed to reduce the temperature of operation [905]. Several other types of fuel cells are under investigation for stationary applications.

14.7.4
Reformed Methanol Fuel Cells

In order to overcome the problems deriving from hydrogen storage and distribution, liquids rich in hydrogen like methanol can be used to produce hydrogen via on-board reforming [reformed methanol fuel cells (RMFCs)] [111]. The waste products with these types of fuel cells are carbon dioxide and water. The liquid fuel is easy to store and to handle and contains, on a volume basis, more hydrogen than even liquid hydrogen (98.8 g of hydrogen in 1 l of methanol at room temperature; 70.8 g of hydrogen in liquid hydrogen at −253 °C). On-board reformers, however, add to the complexity, cost, and maintenance demands of fuel cell systems.

RMFCs have several applications besides producing hydrogen for transportation purposes. For example, they have been chosen by the US Army to provide a lightweight (less than 1 kg), energy dense (500 Wh) solution, with low acoustic and thermal signatures, for the power needed by a soldier during a digitized 72 h mission [906].

14.7.5
Direct Methanol Fuel Cells

In direct methanol fuel cells (DMFCs) methanol fuel is not reformed, but fed directly to the fuel cell operating at a temperature of ca 90–120 °C (methanol boiling point: 64.7 °C). The waste products are carbon dioxide and water (Figure 14.4). The anode is exposed to a methanol–water mixture supplied by an external container. Methanol and water are adsorbed on a catalyst usually made of platinum and ruthenium particles, and lose protons until carbon dioxide is formed. Hydrogen ions (H^+) are transported across the proton exchange membrane, often made from Nafion, to the cathode where they react with oxygen to produce water. Electrons are transported through an external circuit from anode to cathode, providing power to connected devices. The reactions involved are as follows:

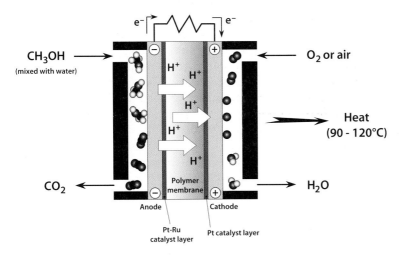

Figure 14.4 Direct methanol fuel cell (DMFC). For technical reasons, pure methanol cannot be used and a methanol–water mixture is supplied.

Anode reaction:

$$CH_3OH + H_2O \rightarrow CO_2 + 6H^+ + 6e^- \tag{14.6}$$

Cathode reaction:

$$1.5O_2 + 6H^+ + 6e^- \rightarrow 3H_2O \tag{14.7}$$

Overall reaction:

$$CH_3OH + 1.5O_2 \rightarrow CO_2 + 2H_2O \tag{14.8}$$

Pure methanol cannot be used because of the high permeation of methanol through the membrane materials (methanol crossover), which decreases the efficiency of the cell.

Current DMFCs are limited in the power they can produce, but can store a high energy content in a small space. This makes them unsuitable at present for powering vehicles, but ideal for equipment that requires low power over a long period of time, like mobile phones, digital cameras, and laptops.

14.8
Hydrogen Powered Vehicles

There is at least one example of any kind of vehicle, from bicycle to Space Shuttle, powered by hydrogen. Liquid hydrogen probably remains the most convenient way to power hydrogen vehicles, but there are at least three very difficult problems unsolved. Hydrogen supplies three times the energy per kilogram of gasoline, but it has only one-tenth the density. Therefore, the tank should be three times the size of a gasoline tank and it must also be insulated. Hydrogen is a dangerous and explosive gas and leaks are difficult to avoid and to detect. Some people believe that employing hydrogen in road vehicles is seriously irresponsible [879]. Since the insulation cannot be perfect, the hydrogen will gradually evaporate. Daily loses of between 1 and 5% of the tank content can be expected.

Hydrogen as a transportation fuel can be used to power internal combustion engines (ICEs) or the fuel cells of electric vehicles (FCEVs) [5]. The hydrogen ICE is a slightly modified version of the traditional gasoline ICE. Demonstrations of hydrogen ICE vehicles using compressed hydrogen gas or liquid hydrogen have been experimented by several companies, but they are not efficient [907]. Certainly, the right way to fight global warming is not that of using a hydrogen-fueled Hummer H2H (Figure 14.5), a vehicle that in the normal version needs 1l of gasoline for less than 5 km.

Most of the attention is focused on FCEVs. In the past few years, following the Hydrogen Fuel Initiative promoted by President Bush (Section 14.1), several companies have committed to develop FCEVs, but in the following years doubts have arisen about the viability of hydrogen fuel cells for mass-market production in the near term. Many FCEVs are currently available in demonstration models, but they are not ready for general public use. According to a July 2008 study by the US

Figure 14.5 California Governor and Humvee fan Arnold Schwarzenegger fuels up a hydrogen-powered Hummer.

National Research Council [908], bringing the technology of hydrogen FCEVs to market viability will require substantial time and additional investment.

Waiting for the development of the infrastructure for hydrogen delivery and distribution, only large vehicles like buses, trucks, and fishing boats might enter the commercial market. Such vehicles, used for local travel over fixed daily work periods, can refuel overnight at relatively large refueling stations and can also take advantage of their ability to hold a large volume of hydrogen fuel for daily consumption, and sufficient room for large fuel-cell stacks. Iceland, where hydrogen is produced at low cost because of the abundance of electric energy from geothermal and hydroelectric facilities, has a plan for powering its fishing fleet. Several dozen advanced engineering versions of fuel-cell buses are under way in demonstration projects in cities all over the world, but we are far from the day on which a significant number of hydrogen FCEVs will be introduced into the market.

The effect of the recent crisis on hydrogen FCEV expansion is not encouraging. While Honda Motors has begun to lease about 200 FCX Clarity FCEVs, primarily in southern California, at a 3-year cost of $600 per month, most of the automobile companies do not seem very interested in near-term plans on hydrogen vehicles. Hydrogen skeptics point out not only the large capital costs associated with the production, transportation, and storage of hydrogen, but also the availability of far more viable alternatives [907]. The above mentioned US National Research Council report on fuel cell development was substantially disclaimed by Secretary of Energy Stephen Chu. In May 2009 he cut funds for the development of hydrogen vehicles since they will not be practical over the next 10–20 years [909]. Bush's 2003 prediction that "the first car driven by a child born today could be powered by hydrogen, and pollution-free" [910] will hardly come true.

14.9
Towards a Hydrogen Economy?

Hydrogen is not an alternative fuel, but a secondary form of energy, like electricity. Although the proper use of hydrogen is not expected to cause big environmental problems, one cannot say that hydrogen is a "clean" form of energy. In fact, hydrogen is "clean" or "dirty" depending of the primary energy form used to produce it. Hydrogen can be easily obtained by on-board reforming of gasoline or natural gas, but this will not help to solve the problems of diminishing oil and gas resources and avoiding production of greenhouse gases. Centralized hydrogen production from coal by integrated gasification combined cycle with capture and sequestration of CO_2 would be acceptable, but such a coal gasification technique is very expensive, might be dangerous, and involves many complex and so far unsolved problems. Furthermore, even coal reserves are depleting, so that this method of producing hydrogen would only be a temporary solution. Artificial photosynthesis, photobiological methods to produce hydrogen from water by using algae, and water splitting at high temperatures obtained by nuclear or concentrated solar power plants are still far from practical applications. Therefore, the development of the hydrogen economy should rely mostly, if not only, on water electrolysis. This task, however, presents noticeable difficulties from several viewpoints and, as we have seen in Section 14.3, the electric power needed would be enormous. Producing electricity by burning fossil fuels, of course, cannot be a solution, whereas hydroelectric power can make only a modest contribution. Therefore, it will be necessary to produce large amounts of electric power by nuclear energy of by renewable energies like wind, PV, or concentrated solar power [5]. In some countries (e.g. France and the US), a large fraction of the electricity is already generated by atomic power plants that, even if "almost" emission free, are not ecologically friendly (Chapter 8). A hydrogen economy based on electricity produced by atomic power plants would imply the construction of thousands of reactors, thereby magnifying all the problems related to the use of nuclear energy. Indeed, production of hydrogen by water electrolysis is an ecologically friendly process only if the electricity used comes from renewable energy sources. In principle, wind, PV, and CSP plants have the potential to produce the enormous amounts of electric power needed, but, except for wind, such technologies are by far too underdeveloped and/or expensive to tackle such a big task in a short time period. Talking about a hydrogen economy before having developed exploitation on a large scale of renewable energy sources is, at least in part, a futile exercise.

Leaving aside the problem of hydrogen production, there are several other scientific and technological hurdles to overcome before arriving at a hydrogen-based economy. Hydrogen transportation and storage are energy consuming and technically difficult processes, and handling hydrogen leads to severe safety issues. As long as there is no adequate hydrogen distribution infrastructure, there will only be a limited demand for hydrogen-powered cars and other applications. On the other hand, there is no incentive for investing large amounts of money in a hydrogen infrastructure unless there is a sustained demand for it. Therefore, we are far from the day on which hydrogen can be filled into the tanks of our cars as easily and safely as gasoline or natural gas are today.

The current evolution of the automobile market could accelerate the growth of hydrogen-powered vehicles. In the meantime, however, interest in grid-powered electrical vehicles is rapidly expanding (Section 13.7). Both hydrogen and electricity can be profitably used as energy carriers (Figure 14.2), but it is irrational to use electrolysis (i.e. electric power) to generate hydrogen that will then be used in fuel cells to make electric power again. The only reason to proceed in this way is related to the capacity of hydrogen to store electricity, even that generated from intermittent sources, much better than batteries. It could happen, however, that storing electric energy in hydrogen becomes much less important if a smart grid for electricity distribution (Section 13.8) develops faster than fuel cell technology. Indeed, most automakers are engaged in building plug-in hybrid and full electric-powered vehicles rather than hydrogen-powered FCEVs.

Nobody knows whether, in the long term, grid-powered EVs or FCEVs will win in the market, but in any case affluent countries will need an exceedingly large amount of additional low-cost electric energy to detach the present transportation system from fossil fuel dependency. In principle, they (more likely, only a few of them) could succeed in doing that by constructing solar–nuclear energy parks located in remote areas. We should not forget, however, that this is only part of the problem because, for several reasons, it will not be possible to supply with personal FCEVs or grid-powered EVs more than a small percentage of the world's population. In the long term, in the affluent countries mobility has to be thoroughly reorganized and substantially reduced. Governments and citizens should begin to think seriously about how to solve this looming problem.

Part Six
Scenarios for a Sustainable Future

Energy for a Sustainable World: From the Oil Age to a Sun-Powered Future. Nicola Armaroli and Vincenzo Balzani
© 2011 WILEY-VCH Verlag GmbH & Co. KGaA, Weinheim
ISBN: 978-3-527-32540-5

15
The Challenge Ahead

"The energy challenge should be considered
the moral equivalent of a war."

Jimmy Carter

15.1
Reflection on the State of Our Planet: Now We Know

The view from space (Figures 1.1 and 1.2) has allowed us to observe the entire Earth as a planet. In the Earth-at-day images from space, national boundaries are invisible and this may strengthen the consciousness of the collective human responsibility for the future of our planet. In contrast, the Earth-at-night images (Figure 15.1) show boundaries: those between highly populated and scarcely populated regions, those between affluent and poor areas. Earth-at-night images also report that most of the light we produce is wasted into the outer space, tropical forests are often burning and gas extracted from oil fields is intensively flared. Inspecting in detail, several other signatures of the heavy human presence materialize (cities, industries, highways, mines).

In the last few decades, we have gathered several important pieces of information on the state of our planet and we have realized that the Earth is an extremely fragile entity sustained by an intricate network of physical, chemical, and biological processes [911]. We have observed the melting of glaciers, the sea level rise, how ocean circulation works, and we have estimated how much carbon moves through ocean, land, atmosphere, and lithosphere plants every year through the carbon cycle [536]. Definitely, spending money on investigating the Earth is much more useful and motivating than going again to the Moon.

As the knowledge of the state of our planet increases, we may have two contrasting reactions. On the one hand, we feel excited to learn the details of the miracle of a living Earth. On the other hand, we feel disconcerted to realize that mankind causes damage that endangers the vitality of the planet. For a long time, some of this damage was unexpected or overlooked and little has been done to stop wounding the Earth. Now we know, but we are not yet prepared to change our way of life and to deal with the complex problems that we have created [57].

Energy for a Sustainable World: From the Oil Age to a Sun-Powered Future. Nicola Armaroli and Vincenzo Balzani
© 2011 WILEY-VCH Verlag GmbH & Co. KGaA, Weinheim
ISBN: 978-3-527-32540-5

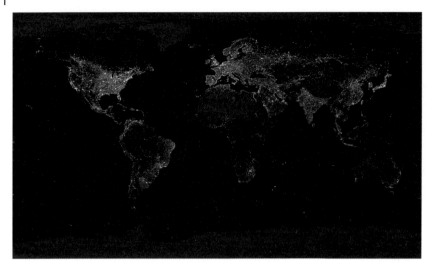

Figure 15.1 Composite image of the Earth at night.

Besides being worried by the data received from space satellites on the precarious health of our planet, we are almost daily distressed by media news that we would prefer to ignore. We find out that the vast disparity between poor and rich is growing in human society. Too many people do not have education, medical care, a house, food, and water, while enormous amounts of money and natural resources are mobilized or employed in military activities worldwide. However, we wish to have quiet travel in the infinity of the Universe, we must take care not only of the state of our spaceship, but also of the well-being of all passengers [912].

Now we know all of this, besides the vast amount of technical information summarized in the previous chapters on the various energy resources and technologies. Accordingly, we can try to answer to the unavoidable questions listed at the end of Chapter 1.

15.2
Energy Demand and Supply

It is simply not possible to address the energy problem by relentlessly increasing supply and this is immediately clear if we make some simple calculations. The rate of consumption of the 2.3 billion people populating the Earth in 1950 was 2.85 TW, corresponding to about 1.1 TW per billion people [52]. In 2010, the world population reached 6.8 billion and the rate of consumption was about 2.2 TW per billion, yielding an overall consumption of about 15 TW. Continuing along this trend, in 2050 the rate of consumption will be around 5 TW per billion people [52]. With a population of over 9 billion, the overall consumption should exceed 40 TW [52].

To cope with such a huge increase of about 24 TW (ignoring replacements), between now and 2050 we should build the equivalent of 48 000 carbon-burning power plants (500 MW each), or 24 000 nuclear power plants (1 GW), or 150 000 km^2 of PV modules (half the area of Italy), that is, more than three carbon-burning power plants, or almost two nuclear plants, or 10 km^2 of PV modules per day. Such an enterprise is simply unachievable because (i) we must drastically reduce CO_2 emissions, (ii) we do not yet have any viable solution for the problem of nuclear waste, and (iii) there are obvious limitations on territory and resource availability.

Clearly, the only possible answer to the expansion of energy demand in affluent countries can no longer be an increase of energy supply, but rather a decrease of energy consumption.

15.3
Energy and the Quality of Life

15.3.1
A Focusing Illusion

Since more energy yields more goods and more services, one is tempted to believe that the quality of life is linearly related to energy consumption. This, however, is a "focusing illusion," like in the case of increasing income [913]. In fact, higher experienced quality of life compensates for lower wages [914]. Quality of life and energy consumption are strictly correlated during basic economic development of a country, but they are not correlated in affluent nations. Annual per capita energy consumption around 1.2–1.6 toe is necessary for satisfaction of essential physical needs, respect for basic individual rights, and widespread opportunities for intellectual advancement [52]. However, when the annual primary energy consumption reaches a value of about 2.6 toe (~100 GJ) per capita, all key indicators (e.g. food availability, infant mortalities, life expectancy, United Nations Human Development Index) show that there is no additional improvement in the quality of life with further increase in energy consumption [52]. Currently, the average EU consumption is about 3.3 toe/capita [7].

Several studies suggest that people living in countries with very high energy consumption are often less happy than people living in countries with lower energy use [3]. Once the basic physical needs are satisfied, the quality of life is a multidimensional concept that embraces intellectual aspects and human relationships not related to energy availability. Well-being and happiness, indeed, cannot be measured by the number of energy slaves that we can painfully learn to master [915]. The energy crisis is actually related to the scarcity of fodder for these voracious slaves. But do we really need all of them?

We have to change the paradigm that governs our energy policy. It is necessary to identify the threshold of real energy needs and to decide when enough is enough. It is not only a problem of material sustainability, but also of mental wisdom. The way in which the energy crisis can be defeated in wealthy countries is to throw out the illusion that well-being depends on ever increasing energy consumption.

15.3.2
Energy, Obesity, Iniquity

The amount of food required by a person depends on several physical and cultural factors. For each person, over a given threshold, food consumption causes obesity that reduces the quality of personal life, endangers social relations, and boost medical expenses. Calories are both biologically and socially healthy only as long as they stay within the narrow range that separates enough from too much.

It has been pointed out that there is a strict parallelism between overconsumption of food and energy [915]. Indeed, the nation with the highest number of overweight or obese people (the US, with more than 60%) [239] is also the one with the highest energy consumption per capita (7.1 toe). Curiously, Barack Obama is trying to decrease the per capita energy consumption of US adults to keep oil imports under control, whereas Michelle Obama has launched her "Let's Move" campaign to get childhood obesity under control within the next generation [916].

As in the case of food, excessive energy uptake is detrimental. It has been estimated that in 2007 the 90 largest metropolitan areas of the US experienced 3.6 billion vehicle-hours of delay, resulting in 9.5 billion liters of wasted fuel and $75.8 billion of congestion costs [917]. In the US there were over 6 million car accidents in 2008, with OVER 37 000 people killed, more than 2.5 million injured, and financial costs higher than $200 billion [918]. In Europe, road accidents caused 34 500 victims in 2009 [919].

A generally overlooked aspect of the energy problem is that equity and energy can grow concurrently only to a point. Before a threshold of per capita wattage, machines improve the conditions for social progress; above that threshold, technology is no longer under our control. Eventually, we become conditioned by technology and technical processes dictate social relations: energy grows at the expense of equity and the growing disparity destabilizes society. Energy overconsumption acts on society like a drug that, besides being physically dangerous, is also psychically enslaving [915].

15.4
Saving the Climate

As discussed in Chapter 7, the accumulation of anthropogenic CO_2 and other greenhouse gases in the atmosphere has modified the Earth's heat balance. From the vast outpouring of data from satellites and other monitoring technologies, scientists have elaborated computer-based climate models [920]. Indeed, long-range forecasts may fail for various reasons, but for sure the dramatic signatures of the greenhouse effect are becoming increasingly evident.

Recent re-examination of the five "reasons for concern" presented by the 2001 IPCC Working Group has led to the conclusion that we are drifting more and more rapidly towards dangerous interference with the climate system [381]. Recent studies have also shown that the time scale for corrective action is not set by the increase in the global mean temperature, but by powerful feedbacks controlling the flow of heat in the climate system [921]. We should stabilize greenhouse emissions as soon as possible and in the long term we should decrease them, hoping that we will be able to regain control of the feedbacks that we have allowed to progress. One more reason for taking action is that climate change will cause the greatest damage in poor countries [922].

Definitely, there is the need for international cooperation [406] and strong leadership to move on to the right direction quickly. Further delay will increase the cost of meeting any given climate target.

15.5
Phasing Out Fossil Fuels

In the last century, particularly in the past 50 years, exploitation of fossil fuels has offered mankind plenty of energy at low price. Fossil fuels have been the engine of human progress in the affluent regions of our planet and today's high technology societies mostly rely on them. We know, however, that oil, natural gas, and carbon are finite resources poised to be exhausted within decades (Chapters 4–6). Our energy bonanza has several costs, difficult to quantify, that are not paid directly by individuals or by energy companies, but by society as a whole (Chapter 7). Last but not least, the massive exploitation of fossil fuels have indirectly contributed to establish disparities and iniquities in human society, at both the international and domestic levels.

For all of these reasons, and particularly for the climate change issue, we must phase out fossil fuels. A sudden stop in their use, in fact, is unrealistic because their infrastructures permeate the entire society and, in the absence of comparable amounts of alternative energy resources, this would cause collapse of the world's economy.

15.6
Avoiding Nuclear Energy

As discussed in Chapter 8, many factors militate against the expansion of nuclear power, including difficulties of finding the huge capital needed, long and uncertain construction times, uncertainties on availability and cost of uranium, limited industrial capacity, skilled worker shortage, threat of nuclear terrorism and proliferation, anti-nuclear sentiment in many nations, future cost competition with the growing market of renewable energies, and two big unsolved problems that had been completely overlooked in the past: disposal of nuclear waste and decommissioning. The world financial crisis of 2009 has exacerbated these problems, particularly in "newcomer" countries. In the US there is an attempt to revive the nuclear industry but there are several technical, economic, and political hurdles to overtake [923].

Nuclear energy can play only a minor role in decreasing greenhouse gas emission in the near future because the deployment of new facilities, if any, will be very slow. Furthermore, construction of nuclear plants has consequences that extend too far into the future: 7–10 years for construction, 40–60 years of operation, 100 years before dismantling, tens of hundreds years of waste to manage. Nuclear energy is indeed a risky mortgage that would pass on to the shoulders of several unacquainted future generations.

This technology is characterized by so many uncertainties, unsolved problems, and hazardous features as to endanger crucial values such as health, environmental sustainability, social stability, and international relationships. Fossil fuels have already shown that in a restricted and highly populated place like our planet, opportunities exploited by a generation can cause severe damage to the subsequent ones. Trying to tackle the energy and climate crises with the nuclear option would be not only an impossible mission, but also a repetition of an unsuccessful strategy.

15.7
Ecological Sustainability

15.7.1
Natural Capital

Life and the human economy depend on the planet's natural capital. The most important element of this capital is sunlight. Another part of the natural capital, the renewable resources (e.g. oxygen, water, fertile soil, fish, trees), are regenerated for free by the direct and indirect action of sunlight on the biosphere. Humanity's footprint (Chapter 1) corresponded to 70% of the capacity of the global biosphere in 1961, and grew to 100% in the 1980s, and to 140% in 2005 [924]; therefore, we are now living above our possibilities and are depleting our capital stock. Recently, nine interlinked planetary boundaries have been identified and three of them – climate change, rate of biodiversity loss, and interference with the nitrogen cycle – have already been transgressed [925].

Fossil fuels are not the only resources that we are consuming at an unsustainable pace. We are exploiting increasing stocks of many elements, such as metals [861], mostly by mining them from finite reserves in the Earth's crust. The rate of extraction of many geochemically scarce metals from the lithosphere has increased in excess of 3% per year through the last half century [926]. Some economists, however, claim that global resource depletion will never occur because a commodity can be replaced by another one [92]. For example, copper wires can sometimes be replaced by fiber-optics; nonetheless copper consumption in China and India is growing at the rate of 10% and 6% per year, respectively. It has been estimated that to extend worldwide the current level of service for copper (as well as for zinc and, perhaps, other metals) of developed countries would require conversion of essentially *all* the ore in the lithosphere to stock-in-use materials [926].

There are several other examples indicating our irrational voracity for natural resources and we are not far from the time at which some chemical elements will be considered "endangered species" [927]. Platinum and rhodium are key materials for catalytic converters: without these devices cars cannot be sold nowadays. Unfortunately, there would not be enough platinum and rhodium to equip all the 800 million cars currently used in the world with such catalytic converters [928], let alone the 5.7 billion cars that should be constructed to bring the car density of the entire world at the level of the US. Further complications arise because substitution of one metal with another one is frequently imposed by environmental reasons, like in the case of mercury.

15.7.2
Learning to Say Enough

Up until now we have taken from Nature any kind of resources to increase our well-being. Only a relatively small part of mankind, however, has benefited from them, and it appears that there are insufficient natural resources to bring all people to the level of consumption of affluent countries. For example, almost half of the world's total primary energy supply is consumed by G8 nations (US, Japan, Germany, China, France, Italy, UK, Canada), despite having only 12% of the world's population. The poorest quarter of humanity consumes less than 3%.

The claim for new goods and services is deeply entrenched in Western culture, which sees growth and development as absolutes. Indeed, in the Western world, the pressure made by ceaseless advertisements quickly converts goods and services, originally considered luxuries, into necessities for everyone. We are encouraged to "upgrade" our mobile phone, computer, television, and so on as soon as a newer, better, bigger version appears on the market. We are persuaded to consume at a faster and faster rate, without any understanding of the consequences of that consumption [929]. Notably, this attitude is not accounted by CO_2 inventories; for instance, it has been estimated that more than 30% of CO_2 emissions of wealthy European countries are actually "imported" through goods produced by emerging economies, particularly China [930].

Sooner or later we will be forced to change our lifestyle based on consumerism, that means *produce–sell–buy–use–throw away* regardless of the resource consumed, the real utility of the object made or service supplied, and the kind of waste generated. Economists and politicians should become aware that sustainability requires that we abandon the notion of endless economic expansion, as measured by the gross domestic product (GDP) [931], as the necessary condition for a successful society [932]. We need to enter a logic of sufficiency to attain ecological stability [26].

After all, we are interested in taking advantage of end-use services, regardless of the way in which they are generated. For example, in the context of business and scientific collaboration, our end-use needs are communication, exchanges of ideas, and personal interactions. They can be satisfied by expensive meetings, but also by video conferences, with much less energy consumption and pollution generation, no waste of time and no travel risk [933]. Communication technology will soon yield a video conference almost as pleasant and productive as a real meeting [934].

Even with present technologies, a lot of energy could be saved in the sectors of building [935] and transportation [936], which, however, are characterized by a high inertia due to their life cycle times. Appliances generally last 5–15 years, cars 10–15 years, heating and cooling systems in commercial buildings 20 years, and building structures 50–100 years or more. Therefore, market forces alone cannot drive the adoption of energy-efficiency technologies within the timeframe imposed by the challenges of global warming. Carefully crafted policies are needed, based on incentives, taxes, mandates, and energy labeling of products, including food and houses, but it will take time also for positive effects of legislation to have a wide impact.

As already mentioned, a most important concept has to be emphasized: there is no way to maintain the present level of consumption in the affluent countries and raise at the same level the consumption of the entire mankind without devastating the planet. If civilization is to progress, in affluent countries a culture of uncontrolled growing consumption should change into a culture of a planned descent phase. Perhaps some people do not like such an idea, but the alternative is that events and contingencies will force us to change and this will be much more painful.

15.8
Why We Need to Develop Renewable Energies

In the long term, mankind will have to rely more and more on recycling. Extensive recycling, however, will require large amounts of energy, which reinforces the notion that energy is the most crucial resource for mankind. Accordingly, we need to entrust our future to energy sources that are not only safe and abundant, but also inexhaustible and available in all the countries of the planet, not only in a few regions.

Energy resources based on Sun, wind, water and Earth are inexhaustible and well distributed all over the planet. With the exception of large hydroelectric plants, they are also safe, because of their low power density. This feature, of course, is a drawback for several applications, but it can be remedied by storage, a strategy that can also resolve the other drawback of solar energies, namely local intermittency. But are energy resources based on Sun, wind, water and Earth abundant enough to supply the huge power that would be needed in 2050 by over 9 billion people, even on the undesirable assumption that they will consume over 7 toe/y

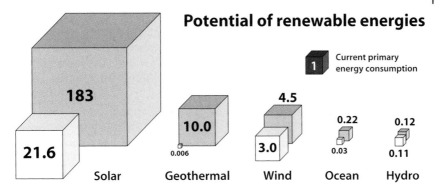

Figure 15.2 Schematic representation of the relative amounts of potentially available (light gray) and technically exploitable (white) renewable energies, compared to the current primary energy consumption (dark gray) [516]. This is a conservative estimation compared to others [937].

per capita, as today in the US? Although it is difficult to estimate the technically exploitable amount of renewable energy resources, there is a general consensus that the answer is "yes" [516, 664, 665, 756, 937] (see, e.g., Figure 15.2). There are, indeed, accurate studies showing that affluent countries in a few decades could at least partially replace fossil fuels with clean and renewable energies without draconian austerity plans, but even continuing to grow their energy consumption. For example, a grand plan that could provide 69% of the US's electricity and 35% of its total energy with solar power alone by 2050, with an annual 1% growth demand, has been proposed [665]. The huge subsidy needed ($420 billion) for the development of the PV and CSP plants, and the related electric grid, would be less than the tax subsidies that have been levied to build the country's high speed telecommunication infrastructure in 35 years. Another study claims that it would be possible to provide 100% of the world's energy need (assumed to be 16.9 TW) by as early as 2030 with already, or closely available, wind, water, and solar technologies [756]. This plan calls for 3.8 million large wind turbines, 90 000 solar plants and numerous geothermal, tidal, and rooftop PV installations worldwide. These are big numbers indeed, but it should be recalled that during WWII, in a few months the US retooled automobile factories to produce 300 000 aircraft, while other countries produced 486 000 more. An obstacle to such a huge development of renewable energies could perhaps be shortages of a few special materials [756], but the greatest barrier is certainly the lack of political will. Unfortunately, most world's leaders have not yet understood that the energy challenge is indeed the moral equivalent of a war.

The above-mentioned studies on the availability of renewable energies are perhaps too optimistic and, more important, do not take into due consideration the problem of current disparities in energy consumption. We believe that the a crucial role in the fossil-to-renewable energy transition should be played by energy saving and efficiency [938]. In affluent countries, any plan for the development of renewable energies should be accompanied or, better, preceded by a plan for a reduction of energy consumption. In fact, if this is substantially reduced, the possibility that renewables may successfully replace fossil fuels is much higher. This is indeed what happens in Europe, according to the 20/20/20 EU Directive [939].

An even more important role in solving the energy crisis should be played by greater awareness of the fact that the quality of life is not necessarily related to energy consumption and that disparities in energy availability are one of the most destabilizing factors in a society (Section 15.3). Therefore, the transitions from fossil fuels to renewable energies will be accelerated not only by governmental decisions and international cooperation [940], but also by educational efforts [941].

Of course, some renewable energies will play only a minor role (Figure 15.2), but we have to exploit as much as possible *all kinds* of renewable energies that are present *in any specific context and territory*. However, we must not forget that also renewable energies can cause problems if they are exploited in a short-sighted fashion, regardless of the benefit to people and the safeguard of the biosphere. A classic example in this regard would be the extensive use of crops to feed cars instead of people (Section 11.3).

Since the construction of the renewable energy infrastructure will require a lot of energy, it would be wise to invest the last portion of fossil fuel reserves to build up such an infrastructure, so that it can be energetically self-sustaining in the next phase, decades ahead.

15.9
Conclusion

Our generation has the crucial duty to become a knowledge-based society capable of achieving sustainable economic growth with a greater social cohesion, as essentially stated by the 2000 declaration of the Council of Europe in Lisbon [942]. Only a new set of ethics and policies, accompanied by decisive changes in attitudes and practices, can prevent the collapse of our civilization [943]. We should take the energy and climate crisis as an opportunity to move away from fossil fuels, reduce disparities, increase international cooperation, and lead humanity to an innovative concept of prosperity.

There is a way out from the energy and climate crisis, that can be based on a few fundamental concepts. What we need is:

- *to acknowledge* that we live in a crowded spaceship that cannot land and cannot dock anywhere to be refueled or repaired, and that there are no places to which our species can migrate;

- *to recognize* that the fossil fuels are causing serious, perhaps irreversible damage to the biosphere, so that we must progressively, but rapidly, phase them out;

- *to convince ourselves* that, over a definite threshold, well-being and happiness are not related to resource consumption, but spring from intellectual and spiritual features, human relationships, cultural development, and values like freedom and peace;

- *to realize* that the resources available on the surface or stored in the holds of our spaceship are limited, and, as a consequence, we must use them with much care and recycle them as much as possible;

- *to understand* that the most abundant and inexhaustible resources that we can trust are renewable energies directly or indirectly related to sunlight, and therefore more research should be devoted to improve energy conversion efficiencies and to develop means that can counter the two intrinsic defects of sunlight, low density and intermittency.

At present, the receipt for our energy future is to follow life's example, which has learnt to set up a circular sustainable cycle (photosynthesis) in which, through the external input of sunlight, is capable of recycling its own exhausts [536]. Likewise, we have to find more efficient ways of extracting free energy from sunlight using our ingenuity. This does not mean, of course, denying the possibility of other extraordinary human achievements capable of changing the frame as well as the content of the above discussion, because we know that the progress of science will never end [45].

Hence, in the future, a most important achievement will likely be artificial photosynthesis: the production of a powerful and clean fuel, hydrogen, starting from water and sunlight (Section 11.5). Until now, mankind has taken from Nature prodigious amounts of resources but, in the long term, scientific research could allow mankind to reverse this trend. Starting from seawater and the fundamental components of our atmosphere (nitrogen, oxygen, and carbon dioxide), by means of sunshine, we will hopefully "fabricate" not only fuels, electricity, and pure water, but also polymers, food, and almost everything else we need [668]. Hopefully, with the extraterrestrial input of solar energy, we will pay back Earth with a capital generated from human intelligence [912, 944].

We should be conscious, however, that science and technology alone will not be able to provide all what we need and ecological sustainability will not be sufficient to allow quiet travel of our spaceship. Social sustainability is even more important: each passenger has the same right to make use of the limited resources of spaceship Earth.

Our spaceship Earth is fragile and living on it denying the concepts of "enough" and "equity" is going to be quite pernicious [77]. Learning to say "enough" is a necessary condition for a sustainable world, whereas establishing equity is a basic need to enforce social sustainability [945]. Earth is not just a space in which we live, but a complex system that we belong to and whose destiny we share [946]. Scientists, regardless of their specialization, should clearly explain these concepts to citizens and especially to policy makers and younger generations.

To live in the third millennium, we need innovative social and economic paradigms, new thinking, and new ways of perceiving the world's problems. Science, but also consciousness, responsibility, compassion, and care, must be the roots of a new knowledge-based society, powered by renewable energies, that we are called to build up in the next few decades.

The alternative, perhaps, will be only barbarism.

Appendix

Did You Know That ... ?

Consumption

- Every second mankind consumes almost 1000 barrels of oil, 93 000 cubic meters of natural gas and 221 tons of coal.

- If we wish to maintain the trend of increasing energy consumption of the past 60 years up to 2050, we need to build *every day* about three carbon-burning power plants, or two nuclear plants, or $10\,km^2$ of PV modules.

- A Second Life Avatar, a digital person that one can create and customize in a virtual world, consumes more electricity than a real person in developing countries.

Disparities

- The G8 nations (12% of the world's population) consume about 50% of the world's total primary energy supply, whereas the poorest 25% of humanity consumes less than 3%.

- As an average, an American consumes about 7.11 toe of energy per year, a quantity approximately equal to that consumed by two Europeans, four Chinese, 17 Indians, and 240 Ethiopians.

- In China, the number of cars per 1000 people is 36, compared to 840 for the US. China is where the US was in 1916 and India (fewer than 20 vehicles per 1000 people) is further behind.

Nuclear Energy

- The effort of the US to build the first permanent repository for high-level radioactive waste at Yucca Mountain (Nevada) has failed after more than 25 years of work and several billion dollars of expenses.

Energy for a Sustainable World: From the Oil Age to a Sun-Powered Future. Nicola Armaroli and Vincenzo Balzani
© 2011 WILEY-VCH Verlag GmbH & Co. KGaA, Weinheim
ISBN: 978-3-527-32540-5

- The initial cost of the Olkiluoto-3 Finnish reactor was €3 billion and the turnkey construction time 4 years starting from September 2005. In November 2009 there was a cost over-run of $2.3 billion and a delay of 3.5 years.

- Electricity from nuclear energy is not "emission free"; it indirectly produces between 40 and 110g CO_2 kWh^{-1}, to be compared with 900g CO_2 kWh^{-1} for coal, 360g CO_2 kWh^{-1} for natural gas, 20–60g CO_2 kWh^{-1} for photovoltaics, and 10g CO_2 kWh^{-1} for wind and CSP.

- In 1948–2000, 59% of US energy R&D expenditure was on nuclear, 23% on fossil fuels, 11% on renewables, and 7% on energy efficiency. In the US, nuclear energy continues to receive the largest proportion of tax breaks (76%), followed by coal at 12%, with renewables trailing at 3%.

Fossil Fuels

Russia is the world giant of fossil energy. It holds the world's largest natural gas reserves, the second largest coal reserves, and the eighth largest oil reserves. Russia is the world's largest exporter of natural gas, the second largest oil exporter, and the third consumer of primary energy. In 2009, after many years of leadership, it was surpassed by the US as the first natural gas producer, but overtook Saudi Arabia as the first world oil producer.

- The US imports about 48% of its oil consumption, Japan 100%, China 47%, and Europe 82%.

- In 2006, the average production of oil wells worldwide was about 100 bbl/well per day. In the US it was only 10 bbl/well per day, with 30% of the wells producing less than one barrel per day. In Saudi Arabia the average daily production was 5700 bbl per well.

- About one-third of the world's oil supply, including three-quarters of all Japan's oil, passes through the Strait of Hormuz, making it one of most important strategic sites of the world.

- China consumes 2.4 barrels per capita annually, about 9 times less than the consumption in US. At the level of US consumption, China would consume over 29 Gbbl, a quantity comparable to the current global annual world oil consumption (30.7 Gbbl).

- The US government provided $72 billion in subsidies to the fossil fuel industry between 2002 and 2008; worldwide the fossil fuel subsidies currently amount to about $330 billion per year.

Renewable Energies

- In 1979 President Carter installed solar panels on the roof of the White House. In 1986 President Reagan took them down.

- China's installed wind energy was 0.34 GW in 2000 and 25.1 GW in 2009. For 2020 China sets national target for wind at more than double that of nuclear.

- The amount of global wind capacity installed 2009 (37 GW) is much higher than the new nuclear power installed in the decade 1999–2009.

- In 2009, wind power accounted for 39% of all new electrical capacity installed in the US and 37% of that installed in Europe.

- An estimated 7 GW of grid-tied photovoltaic capacity was added in 2009, increasing the existing world total by 53 percent to about 21 GW. In the same year, the world nuclear capacity decreased by 1.5 GW.

- On average, covering ~0.6% of the European territory by PV modules would theoretically satisfy its entire electricity demand.

- China has installed a solar collector area of 150 km^2, to be doubled by 2020. The latter target would avoid 60 Mt of CO_2 emissions per year, about the entire CO_2 production of nations such as Portugal or Hungary.

- 1 GW of CSP is under construction, another 14 GW are announced globally through 2014 and a potential global deployment of 25 GW is expected for 2020.

- Photosynthesis uses at least eight photons, some of which have more energy than needed, per O_2 molecule generated and CO_2 molecule "fixed." Therefore, formation of a molecule of glucose, $C_6H_{12}O_6$, requires at least 48 photons.

- A future global fossil fuel consumption of 20 TW (current consumption 12 TW), would correspond to almost three times all cultivable land currently used for agriculture globally, even using the best plants for energy production.

Transportation

- The 50 000 horses required to keep Victorian London's public transport running deposited 1000 tons of manure on the roads every day.

- During 2009 US vehicles traveled about 5 trillion km, that is, 6.3 million round trips from Earth to the Moon.

- In 2007 the 90 largest metropolitan areas of the US experienced 3.6 billion vehicle-hours of delay, resulting in 9.5 billion liters in wasted fuel.

- In the US transportation accounts for about 6% of the *global* primary energy supply.

- A medium-sized car (power 80 kW) running on a highway consumes an amount of energy equivalent to that produced by the work of 1000 human slaves.

Waste and Pollution

- The amount of high-level radioactive nuclear waste worldwide amounts to about 250 000 tons and is increasing by about 12 000 t per year.

- The coal combustion residues generated annually by the 460 US coal-fired power plants would fill a train stretching from New York City to Los Angeles three-and-one-half times over.

- In the US, fossil fuel-based energy production causes $120 billion worth of health and other non-climate-related damages each year, that are not figured into the price of energy; the impact of climate change is estimated to be between $7 billion and $700 billion.

- More than 30% of CO_2 emissions of wealthy European countries are "imported" through goods produced by emerging economies, particularly China.

War

- A US soldier consumed 1 gallon of gasoline per day during World War II, 9 in Vietnam, 15 nowadays.

- In the first air raid in 1914, a single German plane dropped one bomb in Dover, UK. At the beginning of August 1945, 801 planes were used in a single bombing raid on Japan.

- A 1000 MW power plant operating 7000 h per year produces about one-tenth of the energy delivered by the largest H-bomb.

Websites

Main Databases on Energy

US Energy Information Administration:
http://www.eia.doe.gov/
http://tonto.eia.doe.gov/country/index.cfm/

International Energy Agency:
http://www.iea.org/stats/index.asp/
http://www.worldenergyoutlook.org/

World Energy Council:
http://www.worldenergy.org/publications/

European Union:
http://www.energy.eu/

European Environment Agency:
http://www.eea.europa.eu/themes/energy/

BP – British Petroleum:
http://www.bp.com/

General Database and Publications on Energy, Resources, and sustainability

United Nations Development and Environmental Programmes:
http://www.undp.org/
http://geodata.grid.unep.ch/

World Resources Institute:
http://earthtrends.wri.org/

Index Mundi:
http://indexmundi.com/

Global Footprint Network:
http://www.footprintnetwork.org/

Earth Policy Institute:
http://www.earth-policy.org/

Worldwatch Institute:
http://www.worldwatch.org/

Conversion of Energy Units

http://iea.org/stats/unit.asp/

Fossil Fuels

OPEC:
http://www.opec.org/

Naturalgas.org:
http://www.naturalgas.org/

World Coal Institute:
http://www.worldcoal.org/

ASPO, Association for the Study of Peak Oil and Gas:
http://www.peakoil.net/

Nuclear Energy

International Atomic Energy Agency:
http://www.iaea.org/

World Nuclear Association:
http://www.world-nuclear.org/

Nuclear Information and Resource Service:
http://www.nirs.org/

The Generation IV International Forum:
http://www.gen-4.org/

International project on nuclear fusion, ITER:
http://www.iter.org/

Renewable Energies

REN 21 – Renewable Energy Policy Network for the twenty-first century:
http://www.ren21.net/

US National Renewable Energy Laboratories:
http://www.nrel.gov/

European Photovoltaic Industry Association:
http://www.epia.org/

Solarbuzz:
http://www.solarbuzz.com/

European Solar Thermal Industry Federation:
http://www.estif.org/

Global Wind Energy Council:
http://www.gwec.net/

European Wind Energy Association:
http://www.ewea.org/

American Wind Energy Association:
http://www.awea.org/

International Geothermal Association:
http://www.geothermal-energy.org/

International Small Hydro Atlas:
http://www.small-hydro.com/

Climate Change and Related Issues

Intergovernmental Panel on Climate Change:
http://www.ipcc.ch/

Climate science from climate scientists:
http://www.realclimate.org/

NOAA-Mauna Loa Observatory, trends in CO_2 concentration:
http://www.esrl.noaa.gov/gmd/ccgg/trends/

References

Last access to web addresses August 25, 2010.

1. Space Quotations, http://www. spacequotations.com/earth.html.
2. Population Reference Bureau (2009) 2009 World Population Datasheet, www.prb.org/.
3. Smil, V. (2008) *Energy in Nature and Society: General Energetics of Complex Systems*, MIT Press, Cambridge, MA.
4. International Energy Agency (2009) Key World Energy Statistics, www.iea.org/.
5. Kruger, P. (2006) *Alternative Energy Resources: the Quest for Sustainable Energy*, John Wiley & Sons, Inc., Hoboken, NJ.
6. Speth, G.J. (2004) *Red Sky at Morning*, Yale University Press, New Haven, CT.
7. British Petroleum (2010) BP Statistical Review of World Energy, www.bp.com.
8. Baum, R.M. (2009) The energy commons. Chem. Eng. News, 13 November, 3.
9. Hardin, G. (1968) The tragedy of the commons. The population problem has no technical solution; it requires a fundamental extension in morality. *Science*, **162**, 1243.
10. US Census Bureau, International Data Base, www.census.gov/
11. Lutz, W., Sanderson, W., and Scherbov, S. (2008) The coming acceleration of global population ageing. *Nature*, **451**, 716.
12. Wackernagel, M., Schulz, N.B., Deumling, D., *et al.* (2002) Tracking the ecological overshoot of the human economy. *Proc. Natl. Acad. Sci. USA*, **99**, 9266.
13. Global Footprint Network, www.footprintnetwork.org/.
14. Wilson, E.O. (2006) *The Creation. An Appeal to Save Life on Earth*, W. W. Norton, New York.
15. UN Intergovernmental Panel on Climate Change (2007) IPCC Fourth Assessment Report: Climate Change 2007, www.ipcc.ch.
16. Kerr, R.A. (2009) Splitting the difference between oil pessimists and optimists. *Science*, **326**, 1048.
17. Kerr, R.A., and Stone, R. (2010) Two years later, new rumblings over origins of Sichuan quake. *Science*, **327**, 1184.
18. Kirilenko, A.P., and Sedjo, R.A. (2007) Climate change impacts on forestry. *Proc. Natl. Acad. Sci. USA*, **104**, 19697.
19. Johnson, J. (2008) The forever waste. Chem. Eng. News, 5 May, 15.
20. Johnson, J. (2008) A tsunami of electronic waste. Chem. Eng. News, 26 May, 32.
21. Grossman, E. (2006) *High Tech Trash: Digital Devices, Hidden Toxics, and Human Health*, Shearwater/Island Press, Washington, DC.
22. Brown, L.R. (2009) Could food shortages bring down civilization? *Sci. Am.*, **300** (5), 50.
23. February 12 (2010) Special issue on food security. *Science*.
24. Crutzen, P.J. (2002) Geology of mankind. *Nature*, **415**, 23.
25. Zalasiewicz, J., Williams, M., Smith, A., *et al.* (2008) Are we now living

Energy for a Sustainable World: From the Oil Age to a Sun-Powered Future. Nicola Armaroli and Vincenzo Balzani
© 2011 WILEY-VCH Verlag GmbH & Co. KGaA, Weinheim
ISBN: 978-3-527-32540-5

in the Anthropocene? *GSA Today*, **18** (2), 4.

26. Princen, T. (2005) *The Logic of Sufficiency*, MIT Press, Cambridge, MA.

27. United Nations Development Programmne, www.undp.org.

28. EIA, www.eia.doe.gov/.

29. Stone, R. (2009) As China's rare earth R&D becomes ever more rarefied, others tremble. *Science*, **325**, 1336.

30. Rauch, J.N. (2009) Global mapping of Al, Cu, Fe, and Zn in-use stocks and in-ground resources. *Proc. Natl. Acad. Sci. USA*, **106**, 18920.

31. The World Bank, Data Catalog, http://data.worldbank.org/data-catalog.

32. Shah, A. (2010) Poverty Facts and Stats, Global Issues, 28 March 2010, http://www.globalissues.org/article/26/poverty-facts-and-stats.

33. Global Footprint Network, Data Sources. http://www.footprintnetwork.org/en/index.php/GFN/page/data_sources/.

34. Central Intelligence Agency, https://www.cia.gov/library/publications/the-world-factbook/fields/2047.html.

35. Zweig, M. (ed.) (2004) *What's Class Got to Do With It?: American Society in the Twenty-First Century*, Cornell University Press, Ithaca, NY.

36. Economic Policy Institute (2009) The State of Working in America 2008–2009, www.stateofworkingamerica.org/.

37. Barr, D.A. (2008) *Health Disparities in the United States: Social Class, Race, Ethnicity, and Health*, The Johns Hopkins University Press, Baltimore, MD.

38. Godfray, H.C.J., Beddington, J.R., Crute, I.R., *et al.* (2010) Food security: the challenge of feeding 9 billion people. *Science*, **327**, 812.

39. Barnett, T.P., Adam, J.C., and Lettenmaier, D.P. (2005) Potential impacts of a warming climate on water availability in snow-dominated regions. *Nature*, **438**, 303.

40. Brander, K.M. (2007) Global fish production and climate change. *Proc. Natl. Acad. Sci. USA*, **104**, 19709.

41. Millero, F.J. (2007) The marine inorganic carbon cycle. *Chem. Rev.*, **107**, 308.

42. Jorgenson, M.T., Shur, Y.L., and Pullman, E.R. (2006) Abrupt increase in permafrost degradation in Arctic Alaska. *Geophys. Res. Lett.*, **33**, 4.

43. King, D. (2009) The climate in Copenhagen. *Science*, **326**, 1319.

44. Charles, D. (2009) From protest to power: an advocacy group turns 40. *Science*, **323**, 1279.

45. Siegfried, T. (2005) In praise of hard questions. *Science*, **309**, 76.

46. Feynman, R.P. (2005) *Six Easy Pieces: Essentials of Physics by Its Most Brilliant Teacher*, Perseus Books, New York.

47. Atkins, P. (2007) *Four Laws That Drive the Universe*, Oxford University Press, Oxford.

48. Mulder, K., and Hagens, N.J. (2008) Energy return on investment: toward a consistent framework. *Ambio*, **37**, 74.

49. Cleveland, C.J. (2005) Net energy from the extraction of oil and gas in the United States. *Energy*, **30**, 769.

50. Smil, V. (2006) *Energy. A Beginner's Guide*, Oneworld, Oxford.

51. Smil, V. (1999) *Energies. An Illustrated Guide to the Biosphere and Civilization*, MIT Press, Cambridge, MA.

52. Smil, V. (2003) *Energy at the Crossroads: Global Perspectives and Uncertainties*, MIT Press, Cambridge, MA.

53. (2010) What it takes to make that meal. *Science*, **327**, 809.

54. Hoekstra, A.Y., and Chapagain, A.K. (2007) Water footprints of nations: water use by people as a function of their consumption pattern. *Water Resour. Manage.*, **21**, 35.

55. Karbuz, S. (2004) Conversion factors and oil statistics. *Energy Policy*, **32**, 41.

56. International Energy Agency and Organisation for Economic Co-operation and Development (2004) Energy Statistics Manual, www.iea.org/.

57. Armaroli, N., and Balzani, V. (2007) The future of energy supply: challenges and opportunities. *Angew. Chem. Int. Ed.*, **46**, 52.

58. Committee on America's Energy Future (2009) *America's Energy Future: Technology and Transformation*, The National Academies Press, Washington, DC.

59. US Energy Information Administration (2010) Country Energy Profiles, http://tonto.eia.doe.gov/country/index.cfm.

60. Ponting, C. (1991) *A Green History of the World: the Environment and the Collapse of Great Civilizations*, Penguin Books, New York.

61. Montgomery, D.R. (2007) *Dirt: the Erosion of Civilizations*, University of California Press, Berkeley, CA.

62. Odum, H.T. (1971) *Environment, Power and Society*, University of North Carolina Press, Chapel Hill, NC.

63. Diamond, J.M. (2005) *Collapse – How Societies Choose to Fail or Succeed*, Viking Press, New York.

64. Tainter, J.A. (1990) *The Collapse of Complex Societies*, Cambridge University Press, Cambridge.

65. Wikipedia, Paddle Steamer, http://en.wikipedia.org/wiki/Paddle_steamer.

66. Davies, S. (2004) The Great Horse-Manure Crisis of 1894, The Freeman. Ideas On Liberty, September 2004, **54** (7).

67. Dodson, H. (2005) Slavery in the Twenty-First Century, UN Chronicle Online Edition, Issue 3, November 2005, http://www.un.org/wcm/content/site/chronicle/.

68. Encyclopedia of Human Thermodynamics (Wiki) Energy Slave, http://www.eoht.info/page/energy+slave.

69. Yorke, S. (2006) *Windmills and Waterwheels Explained*, Countryside Books, Newbury.

70. Puddle and Pond, Water Wheels. History of the Water Wheel, http://www.puddleandpond.com/water_wheels/water-wheel-history.htm.

71. Smil, V. (1994) *Energy in World History*, Westview Press, Boulder, CO.

72. Dahl, E. (2000) Naval innovation: from coal to oil – cover story. *Joint Force Q.*, Winter.

73. The History of the Automobile. The First Mass Producers of Cars – the Assembly Line, http://inventors.about.com/library/weekly/aacarsassemblya.htm.

74. (1962) *The American People's Encyclopedia*, Grolier, New York.

75. Wikipedia, 1918 in Aviation, http://en.wikipedia.org/wiki/1918_in_aviation.

76. International Technology Roadmap for Semiconductors, 2009 Edition, www.itrs.net/.

77. Balzani, V., Credi, A., and Venturi, M. (2008) *Molecular Devices and Machines, Concepts and Perspectives for the Nanoworld*, Wiley-VCH Verlag GmbH, Weinheim.

78. Fraunhofer IZM (2009) Electricity Consumption by Information Technology is Steadily Rising – "GreenIT" Can Stem the Tide, http://www.izm.fraunhofer.de/.

79. Fettweis, G., and Zimmermann, E. (2008) ICT energy consumption – trends and challenges, 11th International Symposium on Wireless Personal Multimedia Communications (WPMC'08), September, 2008.

80. World Energy Council, London (2007) 2007 Survey of Energy Resources, www.worldenergy.org/.

81. Ball, P. (2004) Heavy computing. *Nat. Mater.*, **3**, 287.

82. United Nations Environment Programme United Nations Environment Programme, Climate Neutral Network: Information and Communication Technologies, http://www.unep.org/climateneutral/.

83. Smil, V. (2008) *Oil. A Beginner's Guide*, Oneworld, Oxford.

84. Hunt, J.M. (1996) *Petroleum Geochemistry and Geology*, Freeman, San Francisco.

85. Kenney, J.F., Kutcherov, V.A., Bendeliani, N.A., *et al.* (2002) The evolution of multicomponent systems at high pressures: VI. The thermodynamic stability of the hydrogen–carbon system: the genesis of hydrocarbons and the origin of petroleum. *Proc. Natl. Acad. Sci. USA*, **99**, 10976.

86. Scott, H.P., Hemley, R.J., Mao, H.K., *et al.* (2004) Generation of methane in the Earth's mantle: *in situ* high pressure–temperature measurements of carbonate reduction. *Proc. Natl. Acad. Sci. USA*, **101**, 14023.

87. Glasby, G.P. (2006) Abiogenic origin of hydrocarbons: an historical overview. *Resour. Geol.*, **56**, 83.

88. Dukes, J.S. (2003) Burning buried sunshine: human consumption of ancient solar energy. *Clim. Change*, **61**, 31.

89. API gravity definition, http://www.engineeringtoolbox.com/

90. Black, B. (2000) *Petrolia, the Landscape of America's First Oil Boom*, Johns Hopkins University Press, Baltimore, MD.

91. Irifune, T., Kurio, A., Sakamoto, S., et al. (2003) Ultrahard polycrystalline diamond from graphite. *Nature*, **421**, 599.

92. Gorelick, S.M. (2009) *Oil Panic and the Global Crisis*, Wiley-Blackwell, Chichester.

93. Hess, G. (2009) Offshore drilling revisited. Chem. Eng. News, 23 March, 30.

94. Baum, R.M. (2010) Tapping a beast. Chem. Eng. News, 14 June, 3.

95. Energy Information Administration (2009) Annual Energy Outlook 2009 – with Projections to 2030, http://www.eia.doe.gov.

96. Deffeyes, K.S. (2005) *Beyond Oil. The View from the Hubbert's Peak*, Hill and Wang, New York.

97. Map Search (2010) Oil & Gas – Electric Power Maps and Atlases, www.mapsearch.com/.

98. T2 Tanker www.t2tanker.org/.

99. Hayler, W.B., and Keever, J.M. (2003) *American Merchant Seaman's Manual*, Cornell Maritime Press, Atglen, PA.

100. Wise, H.L. (2007) *Inside the Danger Zone: the US Military in the Persian Gulf 1987–88*, Naval Institute Press, Annapolis, CA.

101. Parkash, S. (2003) *Refining Processes Handbook*, Elsevier, Amsterdam.

102. Energy Information Administration, http://www.eia.doe.gov/.

103. Strategic Petroleum Reserve, Department of Energy, http://www.spr.doe.gov/dir/dir.html.

104. The China Sourcing Blog (2009) China's Energy Security: Strategic Petroleum Reserve, 10 October, www.chinasourcingblog.org/.

105. Nikiforuk, A. (2008) *Tar Sands: Dirty Oil and the Future of a Continent*, Greystone Books, Vancouver.

106. Alberta Chamber of Resources (2004) Oil Sands Technology Roadmap. Unlocking the Potential, www.acr-alberta.com/.

107. Kean, S. (2009) Eco-alchemy in Alberta. *Science*, **326**, 1052.

108. Kelly, E.N., Short, J.W., Schindler, D.W., et al. (2009) Oil sands development contributes polycyclic aromatic compounds to the Athabasca River and its tributaries. *Proc. Natl. Acad. Sci. USA*, **106**, 22346.

109. US Department of Energy (2004) Strategic Significance of America's Oil Shale Resources. Volume I. Assessment of Strategic Issues, www.fossil.energy.gov/.

110. Matar, S., and Hatch, L.F. (2001) *Chemistry of Petrochemical Processes*, 2nd edn, Elsevier, Amsterdam.

111. Olah, G.A., Goeppert, A., and Prakash, G.K.S. (2006) *Beyond Oil and Gas: the Methanol Economy*, Wiley-VCH Verlag GmbH, Weinheim.

112. OICA, http://oica.net/.

113. China auto sales, output go over 12 million units, 7 December, 2009, http://www.allvoices.com/.

114. (2009) Motoring ahead. more cars are now sold in China than in America. The Economist, 23 October.

115. US Department of Energy (2007) Changes in Vehicles Per Capita around the World, www1.eere.energy.gov/.

116. Jia, H. (2009) Into Africa. Chem. World-UK, January, 48.

117. Greenspan, A. (2007) *The Age of Turbulence: Adventures in a New World*, Penguin, New York.

118. OPEC, www.opec.org/.

119. Hakes, J. (2008) *A Declaration of Energy Independence*, John Wiley & Sons, Inc., Hoboken, NJ.

120. NHTSA, www.nhtsa.dot.gov/.

121. Carter, J. (1977) TV Speech, 18 April, www.pbs.org/.

122. US Energy Information Administration, US Energy Independence and Security Act of 2007: Summary of Provisions, www.eia.doe.gov/.

123. Jackson, J.K. (2009) US Trade Deficit and the Impact of Changing Oil Prices, Congressional Research Service, 15 April, www.cnie.org/.

124. Ernst, R.R. (2003) The responsibility of scientists, a European view. *Angew. Chem. Int. Ed.*, **42**, 4434.

125. National Priorities Project, Cost of War, http://www.nationalpriorities.org/costofwar_home.

126. Reuters (2007) Reuters, US CBO Estimates $2.4 Trillion Long-Term War Costs, 24 October, www.reuters.com/.

127. Stiglitz, J., and Bilmes, L. (2008) The three trillion dollar war. the cost of the Iraq and Afghanistan conflicts have grown to staggering proportions. The Times, 23 February.

128. Antiwar, http://www.antiwar.com/casualties/.

129. Karbuz, S. (2007) US Military Energy Consumption – Facts and Figures, http://karbuz.blogspot.com/.

130. InflationData, http://inflationdata.com/.

131. Shafiq, T. (2009) Iraq's oil prospects face political impediments. *Oil Gas J.*, **107** (3), 46.

132. Delucchi, M.A., and Murphy, J.J. (2008) US military expenditures to protect the use of Persian Gulf oil for motor vehicles. *Energy Policy*, **36**, 2253.

133. Blanchard, R.D. (2005) *The Future of Global Oil Production: Facts, Figures, Trends and Projections, by Region*, McFarland, Jefferson, NC.

134. Campbell, C. (2005) Assessment and importance of oil depletion, in *The Final Energy Crisis* (eds A. McKillop and S. Newman), Pluto Press, London, p. 29.

135. Campbell, C. (2005) Oil and troubled waters, in *The Final Energy Crisis* (eds A. McKillop and S. Newman), Pluto Press, London, p. 133.

136. Society of Petroleum Engineers (2007) Petroleum Resources Management System, www.spe.org/.

137. Hubbert, M.K. (1956) Nuclear energy and the fossil fuels, Spring Meeting of the Southern District. Division of Production, American Petroleum Institute, Shell Development Company, www.hubbertpeak.com/.

138. Peak Oil, www.peakoil.net/.

139. US Geological Survey (2000) USGS World Petroleum Assessment 2000, pubs.usgs.gov/.

140. Lynch, M.C. (2003) The New Pessimism About Petroleum Resources: Debunking the Hubbert Model (and Hubbert Modelers), Minerals & Energy – Raw Materials Report, **18** (1), 21.

141. Daly, M.C. (2007) Peak oil: a metaphor for anxiety. Speech at Geological Society Bicentenery Conference, London, September 10, www.bp.com/.

142. Smil, V. (2006) Peak oil: a catastrophist cult and complex realities. World Watch Mag., January–February, 22.

143. CERA, www.cera.com.

144. Hook, M., Hirsch, R., and Aleklett, K. (2009) Giant oil field decline rates and their influence on world oil production. *Energy Policy*, **37**, 2262.

145. Tsoskounogiou, M., Ayerides, G., and Tritopoulou, E. (2008) The end of cheap oil: current status and prospects. *Energy Policy*, **36**, 3797.

146. Mohr, S.H., and Evans, G.M. (2010) Long term prediction of unconventional oil production. *Energy Policy*, **38**, 265.

147. Campbell, C.J. (2005) *Oil Crisis*, Multiscience Publishing, Brentwood.

148. Heinberg, R. (2003) *The Party's Over: Oil, War, and the Fate of Industrial Societies*, New Society Publishers, Gabriola Island, Canada.

149. Simmons, M.R. (2005) *Twilight in the Desert*, John Wiley & Sons, Inc., Hoboken, NJ.

150. Duncan, R.C. (2001) World energy production, population growth, and the road to the olduvai gorge. *Popul. Environ.*, **22**, 503.

151. UK Energy Research Centre (2009) Global Oil Depletion. An Assessment of the Evidence for a Near-Term Peak in Global Oil Production, www.ukerc.ac.uk/.

152. Smith, P., Martino, D., Cai, Z., et al. (2008) Greenhouse gas mitigation in agriculture. *Philos. Trans. R. Soc. Lond. B Biol. Sci.*, **363**, 789.

153. Proskurowski, G., Lilley, M.D., Seewald, J.S., et al. (2008) Abiogenic

hydrocarbon production at lost city hydrothermal field. *Science*, **319**, 604.

154. The World Bank, Global Gas Flaring Reduction Partnership, http://web.worldbank.org/.

155. Darley, J. (2004) *High Noon for Natural Gas. The New Energy Crisis*, Chelsea Green Publishing, White River Junction, VT.

156. Hoyos, C., and Crooks, E. (2010) A foot on the gas. Financial Times, 12 March, 6.

157. ENI (2008) World Oil and Gas Review, www.eni.it.

158. World Energy Council, London (2007) 2007 Survey of Energy Resources, http://www.worldenergy.org/.

159. Victor, D.G., Jaffe, A.M., and Hayes, M.H. (eds) (2006) *Natural Gas and Geopolitics. From 1970 to 2040*, Cambridge University Press, Cambridge.

160. Farrell, A.E., Zerriffi, H., and Dowlatabadi, H. (2004) Energy infrastructure and security. *Annu. Rev. Environ. Res.*, **29**, 421.

161. Remme, U., Blesl, M., and Fahl, U. (2008) Future european gas supply in the resource triangle of the Former Soviet Union, the Middle East and Northern Africa. *Energy Policy*, **36**, 1622.

162. Bilgin, M. (2009) Geopolitics of European natural gas demand: supplies from Russia, Caspian and the Middle East. *Energy Policy*, **37**, 4482.

163. Sagen, E.L., and Tsygankova, M. (2008) Russian natural gas exports – will Russian gas price reforms improve the European security of supply? *Energy Policy*, **36**, 867.

164. Söderbergh, B., Jakobsson, K., and Aleklett, K. (2009) European energy security: the future of Norwegian natural gas production. *Energy Policy*, **37**, 5037.

165. Goldthau, A. (2008) Rhetoric versus reality: Russian threats to European energy supply. *Energy Policy*, **36**, 686.

166. California Energy Commission, Liquified Natural Gas, www.energy.ca.gov/.

167. Energy Information Administration (2006) Natural Gas Processing: the Crucial Link between Natural Gas Production and Its Transportation to Market, www.eia.doe.gov.

168. Lockhart, T., and Crescenzi, F. (2007) Sour oil and gas management, in *Encyclopaedia of Hydrocarbons – ENI*, vol. **3**, Istituto della Enciclopedia Italiana Fondata da Giovanni Treccani, Rome, p. 237.

169. Wikipedia, For an exhaustive list see http://en.wikipedia.org/wiki/List_of_natural_gas_pipelines.

170. Reshetnikov, A.I., Paramonova, N.N., and Shashkov, A.A. (2000) An evaluation of historical methane emissions from the Soviet gas industry. *J. Geophys. Res.*, **105**, 3517.

171. Leliveld, J., Lechtenbolumer, S., Assonov, S.S., *et al.* (2005) Greenhouse gases: low methane leakage from gas pipelines. *Nature*, **434**, 841.

172. International Association for Natural Gas Vehicles, Statistics, www.iangv.org/.

173. Pasquon, I., and Forzatti, P. (2007) The petrochemical industry, in *Encyclopaedia of Hydrocarbons – ENI*, vol. **2**, Istituto della Enciclopedia Italiana Fondata da Giovanni Treccani, Rome, p. 407.

174. Bellusi, G., and Zennaro, R. (2007) Hydrocarbons from natural gas, in *Encyclopaedia of Hydrocarbons – ENI*, vol. **3**, Istituto della Enciclopedia Italiana Fondata da Giovanni Treccani, Rome, p. 161.

175. Smil, V. (1999) Detonator of the population explosion. *Nature*, **400**, 415.

176. Schlesinger, W.H. (2009) On the fate of anthropogenic nitrogen. *Proc. Natl. Acad. Sci. USA*, **106**, 203.

177. Galloway, J.N., Townsend, A.R., Erisman, J.W., *et al.* (2008) Transformation of the nitrogen cycle: recent trends, questions, and potential solutions. *Science*, **320**, 889.

178. Hogberg, P. (2007) Nitrogen impacts on forest carbon. *Nature*, **447**, 781.

179. Magnani, F., Mencuccini, M., Borghetti, M., *et al.* (2007) The human footprint in the carbon cycle of temperate and boreal forests. *Nature*, **447**, 848.

180. Lawrence Berkeley National Laboratory (2005) Easing the Natural Gas Crisis, www.lbl.gov/.

181. Boyer, C.M., Frantz, J.H., and Jenkins, C.D. (2007) Non-conventional gas, in *Encyclopaedia of Hydrocarbons – ENI*, vol. **3**, Istituto della Enciclopedia Italiana Fondata da Giovanni Treccani, Rome, p. 57.

182. Wylie, G., Eberhard, M., and Mullen, M. (2007) Unconventional gas technology – 1: advances in fracs and fluids improve tight-gas production. *Oil Gas J.*, **105** (47), 39.

183. Tullo, A.A. (2009) Stepping on the gas. Chem. Eng. News, 21 September, 26.

184. Hoyos, C. (2010) Europe the new frontier in shale gas rush. Financial Times, 8 March, 19.

185. Kerr, R. (2010) Natural Gas from Shale Burst into the Scene. *Science*, **328**, 1624.

186. Smith, P., and Hoyos, C. (2010) Australia set for lead export role. Financial Times, 8 March, 19.

187. Kvenvolden, K.A. (1993) Gas hydrates – geological perspective and global change. *Rev. Geophys.*, **31**, 173.

188. Archer, D., Buffett, B., and Brovkin, V. (2009) Ocean methane hydrates as a slow tipping point in the global carbon cycle. *Proc. Natl. Acad. Sci. USA*, **106**, 20596.

189. Borghi, G.P. (2007) Gas hydrates, in *Encyclopaedia of Hydrocarbons – ENI*, vol. 3, Istituto della Enciclopedia Italiana Fondata da Giovanni Treccani, Rome, p. 85.

190. Boswell, R. (2009) Is gas hydrate energy within reach? *Science*, **325**, 957.

191. Kerr, R.A. (2004) Gas hydrate resource: smaller but sooner. *Science*, **303**, 946.

192. Park, Y., Kim, D.Y., Lee, J.W., *et al.* (2006) Sequestering carbon dioxide into complex structures of naturally occurring gas hydrates. *Proc. Natl. Acad. Sci. USA*, **103**, 12690.

193. Kvenvolden, K.A., and Rogers, B.W. (2005) Gaia's breath – global methane exhalations. *Mar. Pet. Geol.*, **22**, 579.

194. Dickens, G.R. (2004) Hydrocarbon-driven warming. *Nature*, **429**, 513.

195. Svensen, H., Planke, S., Malthe-Sorenssen, A., *et al.* (2004) Release of methane from a volcanic basin as a mechanism for initial eocene global warming. *Nature*, **429**, 542.

196. Sowers, T. (2006) Late quaternary atmospheric CH_4 isotope record suggests marine clathrates are stable. *Science*, **311**, 838.

197. US Energy Information Administration, Coals Data, Reports, Analysis and Surveys, www.eia.doe.gov/.

198. Aldhous, P. (2004) Land remediation: Borneo is burning. *Nature*, **432**, 144.

199. Page, S.E., Siegert, F., Rieley, J.O., *et al.* (2002) The amount of carbon released from peat and forest fires in Indonesia during 1997. *Nature*, **420**, 61.

200. Palmer, M.A., Bernhardt, E.S., Schlesinger, W.H., *et al.* (2010) Mountaintop mining consequences. *Science*, **327**, 148.

201. Mitchell, J.G. (2006) When mountains move. Natl. Geogr., March, 105.

202. Fox, J.F., and Campbell, J.E. (2010) Terrestrial carbon disturbance from mountaintop mining increases lifecycle emissions for clean coal. *Environ. Sci. Technol.*, **44**, 2144.

203. Homer, A.W. (2009) Coal mine safety regulation in China and the USA. *J. Contemp. Asia*, **39**, 424.

204. Zheng, I.-P., Feng, C.-G., Ging, G.-X. (2009) A statistical analysis of coal mine accidents caused by coal dust explosions in China. *J. Loss Prevent. Proc.*, **22**, 528.

205. World Coal Institute, Coal and Cement, www.worldcoal.org/.

206. Kerr, R.A. (2008) World oil crunch looming? *Science*, **322**, 1178.

207. Kintisch, E. (2008) The greening of synfuels. *Science*, **320**, 306.

208. America's Energy Future Panel on Alternative Liquid Transportation Fuels (2009) *Liquid Transportation Fuels from Coal and Biomass: Technological Status, Costs, and Environmental Impacts*, The National Academies Press, Washington, DC.

209. Yang, L.H. (2009) Modeling of contaminant transport in underground coal gasification. *Energy Fuels*, **23**, 193.

210. US Energy Information Administration, Country Energy Profiles, www.eia.doe.gov.

211. World Coal Institute, Coal Statistics, www.worldcoal.org/.

212. Kerr, R.A. (2009) How much coal remains? *Science*, **323**, 1420.

213. Orr, F.M. (2009) CO_2 capture and storage: are we ready? *Energy Environ. Sci.*, **2**, 449.

214. Metz, B., Davidson, O., de Coninck, H., *et al.* (eds) (2005) *IPCC Special Report on Carbon Dioxide Capture and Storage*, Cambridge University Press, New York. Available online at www.ipcc.ch.

215. Hileman, B., and Johnson, J. (2007) Driving CO_2 underground. Chem. Eng. News, 24 September, 74.

216. Carbon Monitoring for Action – CARMA, Power Plant Data, www.carma.org/.

217. Kintisch, E. (2007) Making dirty coal plants cleaner. *Science*, **317**, 184.

218. US Energy Information Administration, Energy-Related Emissions Data and Environmental Analyses, www.eia.doe.gov/.

219. Schiermeier, Q. (2006) Putting the carbon back: the hundred billion tonne challenge. *Nature*, **442**, 620.

220. Massachusetts Institute of Technology (2007) The Future of Coal. An Interdisciplinary MIT Study, http://web.mit.edu/coal/.

221. Van Noorden, R. (2010) Buried trouble. *Nature*, **463**, 871.

222. (2009) Saved by sequestration? *Nat. Geosci.*, **2**, 809.

223. Haszledine, R.S. (2009) Carbon capture and storage: how green can black be? *Science*, **325**, 1647.

224. Service, R. (2009) Carbon sequestration. *Science*, **325**, 1644.

225. House, K.Z., Harvey, C.F., Aziz, M.J., *et al.* (2009) The energy penalty of post-combustion CO_2 capture and storage and its implications for retrofitting the US installed base. *Energy Environ. Sci.*, **2**, 193.

226. Thayer, A.M. (2009) Chemicals to help coal come clean. Chem. Eng. News, 13 July, 18.

227. Rochelle, G.T. (2009) Amine scrubbing for CO_2 capture. *Science*, **325**, 1652.

228. Nayar, A. (2009) A lakeful of trouble. *Nature*, **460**, 321.

229. Stone, E.J., Lowe, J.A., and Shine, K.P. (2009) The impact of carbon capture and storage on climate. *Energy Environ. Sci.*, **2**, 81.

230. Charles, D. (2009) Stimulus gives DOE billions for carbon-capture projects. *Science*, **323**, 1158.

231. Orr, F.M., Jr (2009) Onshore geologic storage of CO_2. *Science*, **325**, 1656.

232. Schrag, D.P. (2009) Storage of carbon dioxide in offshore sediments. *Science*, **325**, 1658.

233. Hawkins, D.G., Lashof, D.A., and Williams, R.H. (2006) What to do about coal. *Sci. Am.*, **295** (3), 68.

234. Futuregenalliance, www.futuregenalliance.org/.

235. Neville, A. (2009) IGCC update: are we there yet? *Power*, **153**, 52.

236. European Commission (2003) External Costs – Research Results on Socio-Environmental Damages Due to Electricity and Transport, http://ec.europa.eu/research/energy/pdf/externe_en.pdf.

237. European Commission (2005) Externe – Externalities of Energy – Methodology 2005 Update, www.externe.info/.

238. Committee on Health, Environmental, and Other External Costs and Benefits of Energy Production and Consumption (2009) *Hidden Costs of Energy: Unpriced Consequences of Energy Production and Use*, National Academies Press, Washington, DC.

239. Popkin, B.M. (2007) The world is fat. *Sci. Am.*, **297** (3), 88.

240. Doll, C., and Wietschel, M. (2008) Externalities of the transport sector and the role of hydrogen in a sustainable transport vision. *Energy Policy*, **36**, 4069.

241. Fuglestvedt, J., Berntsen, T., Myhre, G., *et al.* (2008) Climate forcing from the transport sectors. *Proc. Natl. Acad. Sci. USA*, **105**, 454.

242. Berntsen, T., and Fuglestvedt, J. (2008) Global temperature responses to current emissions from the transport sectors. *Proc. Natl. Acad. Sci. USA*, **105**, 19154.

243. Volk, T. (2008) *CO_2 Rising. The World's Greatest Environmental Challange*, MIT Press, Cambridge, MA.

244. Schrag, D.P. (2007) Preparing to capture carbon. *Science*, **315**, 812.

245. Le Quere, C., Raupach, M.R., Canadell, J.G., *et al.* (2009) Trends in the sources and sinks of carbon dioxide. *Nat. Geosci.*, **2**, 831.

246. van der Werf, G.R., Morton, D.C., DeFries, R.S., *et al.* (2009) CO_2 emissions from forest loss. *Nat. Geosci.*, **2**, 737.

247. Foley, J.A., DeFries, R., Asner, G.P., *et al.* (2005) Global consequences of land use. *Science*, **309**, 570.

248. Jones, P.D., and Mann, M.E. (2004) Climate over past millennia. *Rev. Geophys.*, **42** (2), RG2002.

249. Luthi, D., Le Floch, M., Bereiter, B., *et al.* (2008) High-resolution carbon dioxide concentration record 650000–800000 years before present. *Nature*, **453**, 379.

250. Keeling, R.F. (2008) Recording Earth's vital signs. *Science*, **319**, 1771.

251. Cocks, F.H. (2009) *Energy Demand and Climate Change. Issues and Resolutions*, Wiley-VCH Verlag GmbH, Weinheim.

252. Pearson, P.N., Foster, G.L., and Wade, B.S. (2009) Atmospheric carbon dioxide through the eocene–oligocene climate transition. *Nature*, **461**, 1110.

253. Revelle, R., and Suess, H.E. (1957) Carbon dioxide exchange between atmosphere and ocean and the question of an increase of atmospheric CO_2 during the past decades. *Tellus*, **9**, 18.

254. Hogue, C. (2010) Climate-change panel under scrutiny. Chem. Eng. News, 15 March, 13.

255. Hulme, M., Zorita, E., Stocker, T.F., *et al.* (2010) IPCC: cherish it, tweak it or scrap it? *Nature*, **463**, 730.

256. Joos, F., and Spahni, R. (2008) Rates of change in natural and anthropogenic radiative forcing over the past 20000 years. *Proc. Natl. Acad. Sci. USA*, **105**, 1425.

257. Stine, A.R., Huybers, P., and Fung, I.Y. (2009) Changes in the phase of the annual cycle of surface temperature. *Nature*, **457**, 435.

258. Swanson, K.L., Sugihara, G., and Tsonis, A.A. (2009) Long-term natural variability and 20th century climate change. *Proc. Natl. Acad. Sci. USA*, **106**, 16120.

259. Parker, D.E. (2006) A demonstration that large-scale warming is not urban. *J. Clim.*, **19**, 2882.

260. Foukal, P., Fröhlich, C., Spruit, H., *et al.* (2006) Variations in solar luminosity and their effect on the Earth's climate. *Nature*, **443**, 161.

261. Lockwood, M., and Fröhlich, C. (2007) Recent oppositely directed trends in solar climate forcings and the global mean surface air temperature. *Proc. R. Soc. A Math. Phys.*, **463**, 2447.

262. (2010) Climate of fear. *Nature*, **464**, 141.

263. Frank, D.C., Esper, J., Raible, C.C., *et al.* (2010) Ensemble reconstruction constraints on the global carbon cycle sensitivity to climate. *Nature*, **463**, 527.

264. Lynas, M. (2008) *Six Degrees. Our Future on a Hotter Planet*, Harper Perennial, London.

265. Schneider von Deimling, T., Ganopolski, A., Held, H., *et al.* (2006) How cold was the last glacial maximum? *Geophys. Res. Lett.*, **33**, 5.

266. Thompson, L.G., Brecher, H.H., Mosley-Thompson, E., *et al.* (2009) Glacier loss on Kilimanjaro continues unabated. *Proc. Natl. Acad. Sci. USA*, **106**, 19770.

267. United Nations Framework Convention on Climate Change (2009) Climate Change Science Compendium 2009, www.unfccc.int/.

268. Serreze, M.C., Holland, M.M., and Stroeve, J. (2007) Perspectives on the Arctic's shrinking sea-ice cover. *Science*, **315**, 1533.

269. Kaufman, D.S., Schneider, D.P., McKay, N.P., *et al.* (2009) Recent warming reverses long-term Arctic cooling. *Science*, **325**, 1236.

270. Wang, M.Y., and Overland, J.E. (2009) A sea ice free summer Arctic within 30 years? *Geophys. Res. Lett.*, **36**, L07502.

271. Pritchard, H.D., Arthern, R.J., Vaughan, D.G., *et al.* (2009) Extensive dynamic thinning on the margins of the Greenland and Antarctic ice sheets. *Nature*, **461**, 971.

272. Velicogna, I. (2009) Increasing rates of ice mass loss from the Greenland and Antarctic ice sheets revealed by GRACE. *Geophys. Res. Lett.*, **36**, L19503.

273. Shepherd, A., and Wingham, D. (2007) Recent sea-level contributions of the Antarctic and Greenland ice sheets. *Science*, **315**, 1529.

274. Bamber, J.L., Riva, R.E.M., Vermeersen, B.L.A., *et al.* (2009) Reassessment of the potential sea-level rise from a collapse of the west Antarctic ice sheet. *Science*, **324**, 901.

275. Chao, B.F., Wu, Y.H., and Li, Y.S. (2008) Impact of artificial reservoir water impoundment on global sea level. *Science*, **320**, 212.

276. Barnett, T.P., Pierce, D.W., AchutaRao, K.M., *et al.* (2005) Penetration of human-induced warming into the world's oceans. *Science*, **309**, 284.

277. Halpern, B.S., Walbridge, S., Selkoe, K.A., *et al.* (2008) A global map of human impact on marine ecosystems. *Science*, **319**, 948.

278. Lenton, T.M., Held, H., Kriegler, E., *et al.* (2008) Tipping elements in the Earth's climate system. *Proc. Natl. Acad. Sci. USA*, **105**, 1786.

279. Scheffer, M., Bascompte, J., Brock, W.A., *et al.* (2009) Early-warning signals for critical transitions. *Nature*, **461**, 53.

280. de Vernal, A., and Hillaire-Marcel, C. (2008) Natural variability of Greenland climate, vegetation, and ice volume during the past million years. *Science*, **320**, 1622.

281. Steffensen, J.P., Andersen, K.K., Bigler, M., *et al.* (2008) High-resolution Greenland ice core data show abrupt climate change happens in few years. *Science*, **321**, 680.

282. Schiermeier, Q. (2006) A sea change. *Nature*, **439**, 256.

283. Zeebe, R.E., Zachos, J.C., Caldeira, K., *et al.* (2008) Carbon emissions and acidification. *Science*, **321**, 51.

284. Sabine, C.L., Feely, R.A., Gruber, N., *et al.* (2004) The oceanic sink for anthropogenic CO_2. *Science*, **305**, 367.

285. Archer, D., Kheshgi, H., and Maier-Reimer, E. (1997) Multiple timescales for neutralization of fossil fuel CO_2. *Geophys. Res. Lett.*, **24**, 405.

286. Feely, R.A., Sabine, C.L., Lee, K., *et al.* (2004) Impact of anthropogenic CO_2 on the $CaCO_3$ system in the oceans. *Science*, **305**, 362.

287. Caldeira, K., and Wickett, M.E. (2003) Anthropogenic carbon and ocean pH. *Nature*, **425**, 365.

288. Orr, J.C., Fabry, V.J., Aumont, O., *et al.* (2005) Anthropogenic ocean acidification over the twenty-first century and its impact on calcifying organisms. *Nature*, **437**, 681.

289. Hofmann, M., and Schellnhuber, H.J. (2009) Oceanic acidification affects marine carbon pump and triggers extended marine oxygen holes. *Proc. Natl. Acad. Sci. USA*, **106**, 3017.

290. Shaffer, G., Olsen, S.M., and Pedersen, J.O.P. (2009) Long-term ocean oxygen depletion in response to carbon dioxide emissions from fossil fuels. *Nat. Geosci.*, **2**, 105.

291. Hester, K.C., Peltzer, E.T., Kirkwood, W.J., *et al.* (2008) Unanticipated consequences of ocean acidification: a noisier ocean at lower pH. *Geophys. Res. Lett.*, **35**, 5.

292. Shi, D., Xu, Y., Hopkinson, B.M., *et al.* (2010) Effect of ocean acidification on iron availability to marine phytoplankton. *Science*, **327**, 676.

293. Qiu, J. (2008) The third pole. *Nature*, **454**, 393.

294. Bohannon, J. (2008) The big thaw reaches Mongolia's pristine north. *Science*, **319**, 567.

295. Nayar, A. (2009) When the ice melts. *Nature*, **461**, 1042.

296. Gruber, S., Hoelzle, M., and Haeberli, W. (2004) Permafrost thaw and destabilization of alpine rock walls in the hot summer of 2003. *Geophys. Res. Lett.*, **31**, 4.

297. Walker, G. (2007) A world melting from the top down. *Nature*, **446**, 718.

298. Davidson, E.A., and Janssens, I.A. (2006) Temperature sensitivity of soil carbon decomposition and feedbacks to climate change. *Nature*, **440**, 165.

299. Walter, K.M., Zimov, S.A., Chanton, J.P., *et al.* (2006) Methane bubbling from Siberian thaw lakes as a positive feedback to climate warming. *Nature*, **443**, 71.

300. Trapp, R.J., Diffenbaugh, N.S., Brooks, H.E., *et al.* (2007) Changes in severe

thunderstorm environment frequency during the 21st century caused by anthropogenically enhanced global radiative forcing. *Proc. Natl. Acad. Sci. USA*, **104**, 19719.

301. Webster, P.J., Holland, G.J., Curry, J.A., *et al.* (2005) Changes in tropical cyclone number, duration, and intensity in a warming environment. *Science*, **309**, 1844.

302. Allan, R.P., and Soden, B.J. (2008) Atmospheric warming and the amplification of precipitation extremes. *Science*, **321**, 1481.

303. Stott, P.A., Stone, D.A., and Allen, M.R. (2004) Human contribution to the European heatwave of 2003. *Nature*, **432**, 610.

304. Schar, C., Vidale, P.L., Luthi, D., *et al.* (2004) The role of increasing temperature variability in European summer heatwaves. *Nature*, **427**, 332.

305. de Wit, M., and Stankiewicz, J. (2006) Changes in surface water supply across Africa with predicted climate change. *Science*, **311**, 1917.

306. Kerr, R.A. (2007) Global warming is changing the world. *Science*, **316**, 188.

307. Butt, T.A., McCarl, B.A., Angerer, J., *et al.* (2005) The economic and food security implications of climate change in Mali. *Clim. Change*, **68**, 355.

308. Burke, M.B., Miguel, E., Satyanath, S., *et al.* (2009) Warming increases the risk of civil war in Africa. *Proc. Natl. Acad. Sci. USA*, **106**, 20670.

309. Battisti, D.S., and Naylor, R.L. (2009) Historical warnings of future food insecurity with unprecedented seasonal heat. *Science*, **323**, 240.

310. Schlenker, W., and Roberts, M.J. (2009) Nonlinear temperature effects indicate severe damages to US crop yields under climate change. *Proc. Natl. Acad. Sci. USA*, **106**, 15594.

311. Schmidhuber, J., and Tubiello, F.N. (2007) Global food security under climate change. *Proc. Natl. Acad. Sci. USA*, **104**, 19703.

312. Barnett, T.P., Adam, J.C., and Lettenmaier, D.P. (2005) Potential impacts of a warming climate on water availability in snow-dominated regions. *Nature*, **438**, 303.

313. Westerling, A.L., Hidalgo, H.G., Cayan, D.R., *et al.* (2006) Warming and earlier spring increase western US forest wildfire activity. *Science*, **313**, 940.

314. Qiu, J. (2009) Tundra's burning. *Nature*, **461**, 34.

315. Parmesan, C., and Yohe, G. (2003) A globally coherent fingerprint of climate change impacts across natural systems. *Nature*, **421**, 37.

316. Stafford, N. (2007) The other greenhouse effect. *Nature*, **448**, 526.

317. Long, S.P., Ainsworth, E.A., Leakey, A.D.B., *et al.* (2006) Food for thought: lower-than-expected crop yield stimulation with rising CO_2 concentrations. *Science*, **312**, 1918.

318. Akimoto, H. (2003) Global air quality and pollution. *Science*, **302**, 1716.

319. Gustafsson, O., Krusa, M., Zencak, Z., *et al.* (2009) Brown clouds over South Asia: biomass or fossil fuel combustion? *Science*, **323**, 495.

320. Ramanathan, V., and Carmichael, G. (2008) Global and regional climate changes due to black carbon. *Nat. Geosci.*, **1**, 221.

321. Law, K.S., and Stohl, A. (2007) Arctic air pollution: origins and impacts. *Science*, **315**, 1537.

322. Andreae, M.O., Jones, C.D., and Cox, P.M. (2005) Strong present-day aerosol cooling implies a hot future. *Nature*, **435**, 1187.

323. Goldstein, A.H., Koven, C.D., Heald, C.L., *et al.* (2009) Biogenic carbon and anthropogenic pollutants combine to form a cooling haze over the southeastern United States. *Proc. Natl. Acad. Sci. USA*, **106**, 8835.

324. Ramanathan, V., Ramana, M.V., Roberts, G., *et al.* (2007) Warming trends in Asia amplified by brown cloud solar absorption. *Nature*, **448**, 575.

325. Arneth, A., Unger, N., Kulmala, M., *et al.* (2009) Clean the air, heat the planet? *Science*, **326**, 672.

326. Myhre, G. (2009) Consistency between satellite-derived and modeled estimates of the direct aerosol effect. *Science*, **325**, 187.

327. Quaas, J. (2009) Smoke and climate change. *Science*, **325**, 153.

328. Stevens, B., and Feingold, G. (2009) Untangling aerosol effects on clouds and precipitation in a buffered system. *Nature*, **461**, 607.

329. Schwartz, S.E. (2008) Uncertainty in climate sensitivity: causes, consequences, challenges. *Energy Environ. Sci.*, **1**, 430.

330. Andreae, M.O., Rosenfeld, D., Artaxo, P., *et al.* (2004) Smoking rain clouds over the Amazon. *Science*, **303**, 1337.

331. Rosenfeld, D., Lohmann, U., Raga, G.B., *et al.* (2008) Flood or drought: how do aerosols affect precipitation? *Science*, **321**, 1309.

332. Kerr, R.A. (2007) Pollutant hazes extend their climate-changing reach. *Science*, **315**, 1217.

333. Shindell, D.T., Faluvegi, G., Koch, D.M., *et al.* (2009) Improved attribution of climate forcing to emissions. *Science*, **326**, 716.

334. McConnell, J.R., Edwards, R., Kok, G.L., *et al.* (2007) 20th-century industrial black carbon emissions altered Arctic climate forcing. *Science*, **317**, 1381.

335. Xu, B., Cao, J., Hansen, J., *et al.* (2009) Black soot and the survival of Tibetan glaciers. *Proc. Natl. Acad. Sci. USA*, **106**, 22114.

336. Kintisch, E. (2009) New push focuses on quick ways to curb global warming. *Science*, **324**, 323.

337. Gauss, M., Myhre, G., Isaksen, I.S.A., *et al.* (2006) Radiative forcing since preindustrial times due to ozone change in the troposphere and the lower stratosphere. *Atmos. Chem. Phys.*, **6**, 575.

338. European Environment Agency (2009) Assessment of Ground-Level Ozone in EEA Member Countries, with a Focus on Long-Term Trends, www.eea.europa.eu.

339. Son, S.W., Tandon, N.F., Polvani, L.M., *et al.* (2009) Ozone hole and southern hemisphere climate change. *Geophys. Res. Lett.*, **36**, L15705.

340. Walsh, J.E. (2009) A comparison of Arctic and Antarctic climate change, present and future. *Antarct. Sci.*, **21**, 179.

341. Turner, J., and Overland, J. (2009) Contrasting climate change in the two polar regions. *Polar Res.*, **28**, 146.

342. Cohen, J., Rau, A., and Bruning, K. (2009) Bridging the Montreal–Kyoto Gap. *Science*, **326**, 940.

343. Ravishankara, A.R., Daniel, J.S., and Portmann, R.W. (2009) Nitrous oxide (N_2O): the dominant ozone-depleting substance emitted in the 21st century. *Science*, **326**, 123.

344. Wuebbles, D.J. (2009) Nitrous oxide: no laughing matter. *Science*, **326**, 56.

345. Schiermeier, Q. (2009) Fixing the sky. *Nature*, **460**, 792.

346. Forster, P.M., Bodeker, G., Schofield, R., *et al.* (2007) Effects of ozone cooling in the tropical lower stratosphere and upper troposphere. *Geophys. Res. Lett.*, **34**, 5.

347. Shindell, D. (2008) Climate change: cool ozone. *Nat. Geosci.*, **1**, 85.

348. Andrady, A., Aucamp, P.J., Bais, A.F., *et al.* (2009) Environmental effects of ozone depletion and its interactions with climate change: progress report, 2008. *Photochem. Photobiol. Sci.*, **8**, 13.

349. Solomon, S., Plattner, G.K., Knutti, R., *et al.* (2009) Irreversible climate change due to carbon dioxide emissions. *Proc. Natl. Acad. Sci. USA*, **106**, 1704.

350. Royal Society (2009) Geoengineering the Climate: Science, Governance and Uncertainty, www.royalsociety.org/.

351. Robock, A. (2000) Volcanic eruptions and climate. *Rev. Geophys.*, **38**, 191.

352. Hegerl, G.C., and Solomon, S. (2009) Risks of climate engineering. *Science*, **325**, 955.

353. Bala, G., Duffy, P.B., and Taylor, K.E. (2008) Impact of geoengineering schemes on the global hydrological cycle. *Proc. Natl. Acad. Sci. USA*, **105**, 7664.

354. Crutzen, P.J. (2006) Albedo enhancement by stratospheric sulfur injections: a contribution to resolve a policy dilemma? *Clim. Change*, **77**, 211.

355. Morton, O. (2009) Great white hope. *Nature*, **458**, 1097.

356. Angel, R. (2006) Feasibility of cooling the Earth with a cloud of small spacecraft near the inner Lagrange point (L1). *Proc. Natl. Acad. Sci. USA*, **103**, 17184.

357. Bengtsson, L. (2006) Geo-engineering to confine climate change: is it at all feasible? *Clim. Change*, **77**, 229.

358. Robock, A. (2008) 20 reasons why geoengineering may be a bad idea. *Bull. At. Sci.*, **64**, 14.

359. Matthews, H.D., and Caldeira, K. (2007) Transient climate-carbon simulations of planetary geoengineering. *Proc. Natl. Acad. Sci. USA*, **104**, 9949.

360. Keith, D.W., Parson, E., and Morgan, M.G. (2010) Research on global Sun block needed now. *Nature*, **463**, 426.

361. Robock, A. (2008) Whither geoengineering? *Science*, **320**, 1166.

362. Kintisch, E. (2007) Should oceanographers pump iron? *Science*, **318**, 1368.

363. Ledford, H. (2009) Ocean fertilization: dead in the water? *Nature*, **457**, 520.

364. Pollard, R.T., Salter, I., Sanders, R.J., *et al.* (2009) Southern ocean deep-water carbon export enhanced by natural iron fertilization. *Nature*, **457**, 577.

365. Strong, A., Chisholm, S., Miller, C., *et al.* (2009) Ocean fertilization: time to move on. *Nature*, **461**, 347.

366. Lehman, J., Gaunt, J., and Rondom, M. (2006) Bio-char sequestration in terrestrial ecosystems – a Review. *Mitig. Adapt. Strateg. Glob. Change*, **11**, 403.

367. Marris, E. (2006) Putting the carbon back: black is the new green. *Nature*, **442**, 624.

368. International Biochar Initiative, www.biochar-international.org/.

369. United Nations Framework Convention on Climate Change, Kyoto Protocol, http://unfccc.int/kyoto_protocol/items/2830.php.

370. Victor, D.G., and Cullenward, D. (2007) Making carbon markets work. *Sci. Am.*, **297** (6), 70.

371. Mukerjee, M. (2009) A mechanism of hot air. *Sci. Am.*, **300** (6), 18.

372. Searchinger, T.D., Hamburg, S.P., Melillo, J., *et al.* (2009) Fixing a critical climate accounting error. *Science*, **326**, 527.

373. Schiermeier, Q. (2009) Prices plummet on carbon market. *Nature*, **457**, 365.

374. Tollefson, J. (2008) Carbon-trading market has uncertain future. *Nature*, **452**, 508.

375. Schiermeier, Q. (2008) Europe agrees emissions deal. *Nature*, **456**, 847.

376. European Climate Exchange, www.ecx.eu/.

377. Benkovic, S.R., and Kruger, J. (2001) US sulfur dioxide emissions trading program: results and further applications. *Water Air Soil Pollut.*, **130**, 241.

378. Turner, W.R., Oppenheimer, M., and Wilcove, D.S. (2009) A force to fight global warming. *Nature*, **462**, 278.

379. Tollefson, J. (2009) Counting carbon in the Amazon. *Nature*, **461**, 1048.

380. Nepstad, D., Soares-Filho, B.S., Merry, F., *et al.* (2009) The end of deforestation in the Brazilian Amazon. *Science*, **326**, 1350.

381. Smith, J.B., Schneider, S.H., Oppenheimer, M., *et al.* (2009) Assessing dangerous climate change through an update of the Intergovernmental Panel on Climate Change (IPCC) "Reasons for Concern." *Proc. Natl. Acad. Sci. USA*, **106**, 4133.

382. Kerr, R.A. (2009) Amid worrisome signs of warming, "Climate Fatigue" sets in. *Science*, **326**, 926.

383. Biello, D. (2010) Energy and environment climate numerology. *Sci. Am.*, **302** (1), 14.

384. UN Intergovernmental Panel on Climate Change (2001) IPCC Third Assessment Report: Climate Change 2001, www.ipcc.ch.

385. O'Neill, B.C., and Oppenheimer, M. (2002) Dangerous climate impacts and the Kyoto Protocol. *Science*, **296**, 1971.

386. Hansen, J., Sato, M., Kharecha, P., *et al.* (2008) Target atmospheric CO_2: where should humanity aim? *Open Atmos. Sci. J.*, **2**, 217.

387. Monastersky, R. (2009) A burden beyond bearing. *Nature*, **458**, 1091.

388. See www.350.org/.

389. Pacala, S., and Socolow, R. (2004) Stabilization wedges: solving the climate problem for the next 50 years with current technologies. *Science*, **305**, 968.

390. Broecker, W.S. (2007) CO_2 arithmetic. *Science*, **315**, 1371.

391. England, M.H., Sen Gupta, A., and Pitman, A.J. (2009) Constraining future

greenhouse gas emissions by a cumulative target. *Proc. Natl. Acad. Sci. USA*, **106**, 16539.

392. Zickfeld, K., Eby, M., Matthews, H.D., *et al.* (2009) Setting cumulative emissions targets to reduce the risk of dangerous climate change. *Proc. Natl. Acad. Sci. USA*, **106**, 16129.

393. Meinshausen, M., Meinshausen, N., Hare, W., *et al.* (2009) Greenhouse-gas emission targets for limiting global warming to 2 °C. *Nature*, **458**, 1158.

394. Matthews, H.D., Gillett, N.P., Stott, P.A., *et al.* (2009) The proportionality of global warming to cumulative carbon emissions. *Nature*, **459**, 829.

395. Allen, M.R., Frame, D.J., Huntingford, C., *et al.* (2009) Warming caused by cumulative carbon emissions towards the trillionth tonne. *Nature*, **458**, 1163.

396. Schmidt, G., and Archer, D. (2009) Too much of a bad thing. *Nature*, **458**, 1117.

397. Matthews, H.D., and Caldeira, K. (2008) Stabilizing climate requires near-zero emissions. *Geophys. Res. Lett.*, **35**, L04705.

398. Parry, M., Lowe, J., and Hanson, C. (2009) Overshoot, adapt and recover. *Nature*, **458**, 1102.

399. Stern, N. (2006) *The Economics of Climate Change*, Cambridge University Press, Cambridge.

400. Nayar, A. (2009) Cost of climate change underestimated. *Nature*, **461**, 24.

401. Brand, C., and Boardman, B. (2008) Taming of the few – the unequal distribution of greenhouse gas emissions from personal travel in the UK. *Energy Policy*, **36**, 224.

402. Chakravarty, S., Chikkatur, A., de Coninck, H., *et al.* (2009) Sharing global CO_2 emission reductions among one billion high emitters. *Proc. Natl. Acad. Sci. USA*, **106**, 11884.

403. Allen, M.R., Frame, D.J., and Mason, C.F. (2009) The case for mandatory sequestration. *Nat. Geosci.*, **2**, 813.

404. Johnson, J. (2008) coal plant permit blocked. *Chem. Eng. News*, 24 November, 9.

405. Johnson, J. (2009) EPA finds CO_2 a health threat. Chem. Eng. News, 14 December, 7.

406. Bawa, K.S., Koh, L.P., Lee, T.M., *et al.* (2010) China, India, and the environment. *Science*, **327**, 1457.

407. Brunekreef, B., and Holgate, S.T. (2002) Air pollution and health. *Lancet*, **360**, 1233.

408. van der A, R.J., Eskes, H.J., Boersma, K.F., *et al.* (2008) Trends, seasonal variability and dominant NO_x source derived from a ten year record of NO_2 measured from space. *J. Geophys. Res. Atmos.*, **113**, 12.

409. Parks, J.E. (2010) Less costly catalysts for controlling engine emissions. *Science*, **327**, 1584.

410. European Environment Agency (2008) Energy and Environment Report 2008, www.eea.europa.eu/.

411. World Health Organization (2005) Air Quality Guidelines Global Update 2005. Particulate Matter, Ozone, Nitrogen Dioxide and Sulfur Dioxide, www.euro.who.int.

412. Li, S.P., Matthews, J., and Sinha, A. (2008) Atmospheric hydroxyl radical production from electronically excited NO_2 and H_2O. *Science*, **319**, 1657.

413. Ryerson, T.B., Trainer, M., Holloway, J.S., *et al.* (2001) Observations of ozone formation in power plant plumes and implications for ozone control strategies. *Science*, **292**, 719.

414. World Health Organization (2006) Effects of Air Pollution on Children's Health and Development. A Review of Evidence, www.euro.who.int.

415. World Health Organization (2008) Health Risks of Ozone from Long-Range Transboundary Air Pollution, www.euro.who.int.

416. Jonson, J.E., Simpson, D., Fagerli, H., *et al.* (2006) Can we explain the trends in European ozone levels? *Atmos. Chem. Phys.*, **6**, 51.

417. Poschl, U. (2005) Atmospheric aerosols: composition, transformation, climate and health effects. *Angew. Chem. Int. Ed.*, **44**, 7520.

418. Fine, P.M., Sioutas, C., and Solomon, P.A. (2008) Secondary particulate matter in the United States: insights from the particulate matter supersites program and related studies. *J. Air Waste Manage.*, **58**, 234.

419. Jimenez, J.L., Canagaratna, M.R., Donahue, N.M., *et al.* (2009) Evolution of organic aerosols in the atmosphere. *Science*, **326**, 1525.

420. Nemmar, A., Hoet, P.H.M., Vanquickenborne, B., *et al.* (2002) Passage of inhaled particles into the blood circulation in humans. *Circulation*, **105**, 411.

421. Chow, J.C., Watson, J.G., Mauderly, J.L., *et al.* (2006) Health effects of fine particulate air pollution: lines that connect. *J. Air Waste Manage.*, **56**, 1368.

422. Dominici, F., Peng, R.D., Bell, M.L., *et al.* (2006) Fine particulate air pollution and hospital admission for cardiovascular and respiratory diseases. *JAMA*, **295**, 1127.

423. Ezzati, M., Bailis, R., Kammen, D.M., *et al.* (2004) Energy management and global health. *Annu. Rev. Environ. Resour.*, **29**, 383.

424. Nel, A. (2005) Air pollution-related illness: effects of particles. *Science*, **308**, 804.

425. Fischer, P.H., Brunekreef, B., and Lebret, E. (2004) Air pollution related deaths during the 2003 heat wave in The Netherlands. *Atmos. Environ.*, **38**, 1083.

426. Stedman, J.R. (2004) The predicted number of air pollution related deaths in the UK during the August 2003 heatwave. *Atmos. Environ.*, **38**, 1087.

427. Marufu, L.T., Taubman, B.F., Bloomer, B., *et al.* (2004) The 2003 North American electrical blackout: an accidental experiment in atmospheric chemistry. *Geophys. Res. Lett.*, **31**, L13106.

428. Ezzati, M., and Kammen, D.M. (2002) Household energy, indoor air pollution and health in devleoping countries: knowledge base for effective interventions. *Annu. Rev. Energy Environ.*, **27**, 233.

429. World Health Organization (2007) Indoor Air Pollution: National Burden of Disease Estimates, www.who.int.

430. Likens, G.E., Driscoll, C.T., and Buso, D.C. (1996) Long-term effects of acid rain: response and recovery of a forest ecosystem. *Science*, **272**, 244.

431. Stoddard, J.L., Jeffries, D.S., Lukewille, A., *et al.* (1999) Regional trends in aquatic recovery from acidification in North America and Europe. *Nature*, **401**, 575.

432. Palmer, S.M., and Driscoll, C.T. (2002) Decline in mobilization of toxic aluminium. *Nature*, **417**, 242.

433. Likens, G.E., Weathers, K.C., Butler, T.J., *et al.* (1998) Solving the acid rain problem. *Science*, **282**, 1991.

434. Kerr, R.A. (1998) Acid rain control: success on the cheap. *Science*, **282**, 1024.

435. Larssen, T., Lydersen, E., Tang, D.G., *et al.* (2006) Acid rain in China. *Environ. Sci. Technol.*, **40**, 418.

436. Cooke, C.A., Balcom, P.H., Biester, H., *et al.* (2009) Over three millennia of mercury pollution in the Peruvian Andes. *Proc. Natl. Acad. Sci. USA*, **106**, 8830.

437. McConnell, J.R., and Edwards, R. (2008) Coal burning leaves toxic heavy metal legacy in the Arctic. *Proc. Natl. Acad. Sci. USA*, **105**, 12140.

438. Faïn, X., Ferrari, C.P., Dommergue, A., *et al.* (2009) Polar firn air reveals large-scale impact of anthropogenic mercury emissions during the 1970s. *Proc. Natl. Acad. Sci. USA*, **106**, 16114.

439. Hong, S.M., Lee, K., Hou, S.G., *et al.* (2009) An 800-year record of atmospheric As, Mo, Sn, and Sb in central Asia in high-altitude ice cores from Mt. Qomolangma (Everest), Himalayas. *Environ. Sci. Technol.*, **43**, 8060.

440. Mercury, http://www.epa.gov/mercury/.

441. Reisch, M.S. (2008) Getting rid of mercury. *Chem. Eng. News*, 24 November, 22.

442. Eckelman, M.J., Anastas, P.T., and Zimmerman, J.B. (2008) Spatial assessment of net mercury emissions from the use of fluorescent bulbs. *Environ. Sci. Technol.*, **42**, 8564.

443. Rauch, S., Hemond, H.F., Barbante, C., *et al.* (2005) Importance of automobile exhaust catalyst emissions for the depostion of platinum, palladium, and rhodium in the Northern Hemisphere. *Environ. Sci. Technol.*, **39**, 8156.

444. Zereini, F., Wiseman, C., and Puttmann, W. (2007) Changes in palladium, platinum, and rhodium concentrations, and their spatial distribution in soils along a major highway in Germany from 1994 to 2004. *Environ. Sci. Technol.*, **41**, 451.

445. Chen, C., Sedwick, P.N., and Sharma, M. (2009) Anthropogenic osmium in rain and snow reveals global-scale atmospheric contamination. *Proc. Natl. Acad. Sci. USA*, **106**, 7724.

446. Epstein, P.R., and Selber, J. (eds) (2002) *Oil: a Life Cycle Analysis of Its Health and Environmental Impact*, Center for Health and the Global Environment, Harvard Medical School, Boston, MA.

447. O'Rourke, D., and Connolly, S. (2003) Just oil? The distribution of environmental and social impacts of oil production and consumption. *Annu. Rev. Environ. Resour.*, **28**, 587.

448. Marshall, A. (2008) Drowning in mud. Natl. Geogr., January, 58.

449. American Association of Petroleum Geologists, www.aapg.org/.

450. Syvitski, J.P.M., Kettner, A.J., Overeem, I., *et al.* (2009) Sinking deltas due to human activities. *Nat. Geosci.*, **2**, 681.

451. Bohannon, J. (2010) The Nile Delta's sinking future. *Science*, **327**, 1444.

452. Bragg, J.R., Prince, R.C., Harner, E.J., *et al.* (1994) Effectiveness of bioremediation for the Exxon-Valdez oil-spill. *Nature*, **368**, 413.

453. Peterson, C.H., Rice, S.D., Short, J.W., *et al.* (2003) Long-term ecosystem response to the Exxon Valdez oil spill. *Science*, **302**, 2082.

454. Guterman, L. (2009) Exxon Valdez turns 20. *Science*, **323**, 1558.

455. Short, J.W., Lindeberg, M.R., Harris, P.M., *et al.* (2004) Estimate of oil persisting on the beaches of Prince William Sound 12 years after the Exxon Valdez oil spill. *Environ. Sci. Technol.*, **38**, 19.

456. Short, J.W., Irvine, G.V., Mann, D.H., *et al.* (2007) Slightly weathered Exxon Valdez oil persists in Gulf of Alaska beach sediments after 16 years. *Environ. Sci. Technol.*, **41**, 1245.

457. Torrice, M. (2009) Science lags on saving the Arctic from oil spills. *Science*, **325**, 1335.

458. Lucas, Z., and MacGregor, C. (2006) Characterization and source of oil contamination on the beaches and seabird corpses, Sable Island, Nova Scotia, 1996–2005. *Mar. Pollut. Bull.*, **52**, 778.

459. Johnson, J. (2009) The foul side of "clean coal." Chem. Eng. News, 23 February, 44.

460. European Environment Agency (2006) EN10 – Residues from Combustion of Coal for Energy Production, www.ims.eionet.europa.eu/.

461. Gruber, N., and Galloway, J.N. (2008) An Earth-system perspective of the global nitrogen cycle. *Nature*, **451**, 293.

462. Smil, V. (2008) *Global Catastrophes and Trends. The Next 50 Years*, MIT Press, Cambridge, MA.

463. Townsend, A.R., and Howarth, R.W. (2010) Fixing the global nitrogen problem. *Sci. Am.*, **302** (2), 64.

464. Hore-Lacy, I. (2007) *Nuclear Energy in the 21st Century*, Elsevier, Amsterdam.

465. Garwin, R.L., and Charpak, G. (2001) *Megawatts and Megatons*, Knopf, New York.

466. Hahn, O., and Strassmann, F. (1939) Über den Nachweis und das Verhalten bei der Bestrahlung des Urans Mittels Neutronen entstehenden Erdalkalimetalle (On the detection and characteristics of the alkaline earth metals formed by irradiation of uranium with neutrons). *Naturwissenschaften*, **27** (1), 11.

467. International Atomic Energy Agency (2004) From Obninsk Beyond: Nuclear Power Conference Looks to the Future, www.iaea.org.

468. Greens–European Free Alliance (2007) The World Nuclear Industry Status Report 2007, www.greens-efa.org/.

469. International Atomic Energy Agency www.iaea.org.

470. German Federal Ministry of Environment, Nature Conservation and Reactor Safety (2009) The World Nuclear Industry Status Report 2009, with Particular Emphasis on Economic Issues, www.bmu.de/.

471. World Nuclear Association, www.world-nuclear.org/.

472. Energy Watch Group (2006) Uranium Resources and Nuclear Energy, www.peakoil.net/.

473. ISA, The University of Sydney (2006) Life-Cycle Energy Balance and Greenhouse Gas Emissions of Nuclear Energy in Australia, http:// www.isa.org.usyd.edu.au/.

474. Nuclear Energy Agency, www.nea.fr/.

475. Massachusetts Institute of Technology (2009) Update of the MIT 2003 Future of Nuclear Power, www.mit.edu/.

476. The Oil Drum (2009) The Future of Nuclear Energy: Facts and Fiction, www.europe.theoildrum.com/.

477. Generation IV International Forum, www.gen-4.org/.

478. (2006) Recycling the past. *Nature*, **439**, 509.

479. Johnson, J. (2008) The forever waste. Chem. Eng. News, 5 May, 15.

480. von Hippel, F.N. (2008) Rethinking neuclear fuel recycling. *Sci. Am.*, **298** (5), 88.

481. Jacoby, M. (2009) Reintroducing thorium. Chem. Eng. News, 16 November, 44.

482. Allen, G. (2008) US gives nuclear power a second look. National Public Radio, 28 March, www.npr.org/.

483. Hamilton, T. (2009) $26B cost killed nuclear bid. The Star, 14 July.

484. Kanter, J. (2009) Nuclear power renaissance stalls in Finland. International Herald Tribune, 29 May.

485. Electricity Forum (2009) Shadow Cast over Areva's EPR Nuclear Reactor, November, www.electricityforum.com/.

486. Moody's Corporate Finance (2008) New Nuclear Generating Capacity: Potential Credit Implications for US Investor Owned Utilities.

487. Synapse Energy Economics (2008) Nuclear Power Plant Construction Costs, www.synapse-energy.com/.

488. Climate Progress (2009) Business Risks and Costs of New Nuclear Power, www.climateprogress.org/.

489. Federation of American Scientists, www.fas.org/.

490. Bohannon, J. (2007) Profile: Tariq Rauf – treading the nuclear fuel cycle minefield. *Science*, **315**, 791.

491. (2009) Struggling to hold up a bank. Economist, 6 August.

492. Harrell, E. (2010) Bomb chasers. Time – Europe Edition, 19 April, 14.

493. (2008) Nuclear security undervalued. *Nature*, **451**, 745.

494. Williams, D., and Baverstock, K. (2006) Too soon for a final diagnosis. *Nature*, **440**, 993.

495. Bohannon, J. (2005) Panel puts eventual Chernobyl death toll in thousands. *Science*, **309**, 1663.

496. Abbotts, J. (1979) Radioactive-waste – technical solution. *Bull. At. Sci.*, **35**, 12.

497. Johnson, J. (2009) DOE drops Yucca Mountain. Chem. Eng. News, 23 March, 35.

498. Ewing, R.C., and von Hippel, F.N. (2009) Nuclear waste management in the United States – starting over. *Science*, **325**, 151.

499. (2008) Yucca Mountain cost estimate rises to $96 billion. World Nucl. News, 6 August, www.world-nuclear-news.org/.

500. Morgan, S. (2006) Going underground. *Chem. World UK*, **3**, 46.

501. Vandenbosch, R., and Vandenbosch, S.E. (2007) *Nuclear Waste Stalemate: Political and Scientific Controversies*, University of Utah Press, Salt Lake City, UT.

502. Giles, J. (2006) When the price is right. *Nature*, **440**, 984.

503. Caldicott, H. (2006) *Nuclear Power Is Not the Answer*, The New Press, New York.

504. Wikipedia, http://en.wikipedia.org/ wiki/Brennilis_Nuclear_Power_Plant.

505. Yankee Rowe, www.yankeerowe.com/.

506. Connecticut Yankee, www.connyankee.com/.

507. Nuclear Decommissioning Authority, www.nda.gov.uk/.

508. NRC, www.nrc.gov/.

509. Cyranoski, D. (2008) Japanese nuclear plant in quake risk. *Nature*, **453**, 704.

510. Hightower, M., and Pierce, S.A. (2008) The energy challenge. *Nature*, **452**, 285.

511. Diehl, P. (2007) Uranium Mining Overview, 27 April, www.nonuclear.se/.

512. WISE, www.wise-uranium.org/.

513. Nuclear Energy Institute (1998) Energy Department's FY99 Budget Request Recognizes Nuclear Energy's Value as Carbon-Free Electricity Source. Nuclear Energy Institute News Release, 2 February, nei.org/.

514. Sovacool, B.K. (2008) Valuing the greenhouse gas emissions from nuclear power: a critical survey. *Energy Policy*, **36**, 2950.

515. Beerten, J., Laes, E., Meskens, G., *et al.* (2009) Greenhouse gas emissions in the nuclear life cycle: a balanced appraisal. *Energy Policy*, **37**, 5056.

516. Jacobson, M.Z. (2009) Review of solutions to global warming, air pollution, and energy security. *Energy Environ. Sci.*, **2**, 148.

517. Johnson, J. (2010) Obama sings a nuclear tune. Chem. Eng. News, 22 February, 8.

518. Johnson, J. (2010) New jolt for nuclear power. Chem. Eng. News, 8 March, 31.

519. International Atomic Energy Agency (2008) Energy, Electricity and Nuclear Power Estimates for the Period up to 2030, www-pub.iaea.org/.

520. Energy Information Administration (2008) International Energy Outlook 2008 – Highlights, www.eia.doe.gov.

521. International Energy Agency (2008) World Energy Outlook 2008, www.worldenergyoutlook.org/.

522. ITER, www.iter.org/.

523. Clery, D. (2009) ITER blueprints near completion, but financial hurdles lie ahead. *Science*, **326**, 932.

524. Brumfiel, G. (2009) Fusion dreams delayed. *Nature*, **459**, 488.

525. Clery, D. (2010) Budget red tape in Europe brings new delay to ITER. *Science*, **327**, 1434.

526. Starr, C., Hirsch, R.L., Dieckamp, H., *et al.* (2006) Reexamining fusion power. *Science*, **313**, 170.

527. National Ignition Facility, lasers.llnl.gov/.

528. Rochau, G.E., Morrow, C.W., and Pankuch, P.J. (2003) A concept for containing inertial fusion energy pulses in a Z-pinch-driven power plant. *Fusion Sci. Technol.*, **43**, 447.

529. Fleischmann, M., and Pons, S. (1989) Electrochemically induced nuclear-fusion of deuterium. *J. Electroanal. Chem.*, **261**, 301.

530. US Department of Energy (2004) Report of the Review of Low Energy Nuclear Reactions, www.lenr-canr.org/.

531. Schmitt, H.H. (2003) Private enterprise approach to lunar base activation. *Adv. Space Res.*, **31**, 2441.

532. Atomic Bomb: Decision. Documents on the Decision to Use the Atomic Bombs on the Cities of Hiroshima and Nagasaki, http://www.dannen.com/decision/index.html.

533. Walker, J.S. (2005) Recent literature on Truman's atomic bomb decision: a search for middle ground. *Diplomatic Hist.*, **29**, 311.

534. Sthol, M. (ed.) (1988) *National Interest and State Terrorism. The Politics of Terrorism*, 3rd edn, Marcel Dekker, New York.

535. Buchen, L. (2010) What will the next solar cycle bring? *Nature*, **463**, 414.

536. Morton, O. (2007) *Eating the Sun: How Plants Power the Planet*, Fourth Estate, London.

537. See for an example the United Nations Environmental Programme Website, http://maps.grida.no/go/graphic/natural-resource-solar-power-potential.

538. Solar4Power, http://www.solar4power.com/solar-power-insolation-window.html.

539. Lindley, D. (2010) The energy storage problem. *Nature*, **463**, 18.

540. Sartori, I., and Hestnes, A.G. (2007) Energy use in the life cycle of conventional and low-energy buildings: a review article. *Energy Build.*, **39**, 249.

541. Thirugnanasambandam, M., Iniyan, S., and Goic, R. (2010) A review of solar thermal technologies. *Renew. Sust. Energy Rev.*, **14**, 312.

542. Kalogirou, S.A. (2004) Solar thermal collectors and applications. *Prog. Energy Combust.*, **30**, 231.

543. Ardente, F., Beccali, G., Cellura, M., *et al.* (2005) Life cycle assessment of a solar thermal collector: sensitivity analysis, energy and environmental balances. *Renew. Energy*, **30**, 109.

544. REN 21 – Renewable Energy Policy Network for the 21st Century (2010) Renewables Global Status Report – 2010, www.ren21.net/.

545. Han, J., Mol, A.P.J., and Lu, Y. (2010) Solar water heaters in China: a new day dawning. *Energy Policy*, **38**, 383.

546. Celik, A.N., Muneer, T., and Clarke, P. (2009) A review of installed solar photovoltaic and thermal collector capacities in relation to solar potential for the EU-15. *Renew. Energy*, **34**, 849.

547. European Solar Thermal Industry Federation (2009) Potential of Solar Thermal in Europe, www.estif.org/.

548. Weiss, W., and Mauthner, F. (2010) Solar Heat Worldwide. Markets and Contribution to the Energy Supply 2008 – Edition 2010, Solar Heating and Cooling Programme, International Energy Agency, Graz.

549. Fischetti, M. (2008) Warming and cooling. *Sci. Am.*, **299** (2), 104.

550. Yang, W., Zhou, J., Xu, W., *et al.* (2010) Current status of ground-source heat pumps in China. *Energy Policy*, **38**, 323.

551. Service, R.F. (2004) Temperature rises for devices that turn heat into electricity. *Science*, **306**, 806.

552. Saqr, K.M., and Musa, M.N. (2009) Critical review of thermoelectrics in modern power generation applications. *Thermal Sci.*, **13**, 165.

553. Quarez, E., Hsu, K.F., Pcionek, R., *et al.* (2005) Nanostructuring, compositional fluctuations, and atomic ordering in the thermoelectric materials $AgPb_mSbTe_{2+m}$. The myth of solid solutions. *J. Am. Chem. Soc.*, **127**, 9177.

554. Crabtree, G.W., and Lewis, N.S. (2007) Solar energy conversion. Phys. Today, March, 37.

555. Venkatasubramanian, R., Siivola, E., Colpitts, T., *et al.* (2001) Thin-film thermoelectric devices with high room-temperature figures of merit. *Nature*, **413**, 597.

556. US Department of Energy Office of Science (2005) Basic Research Needs for Solar Energy Utilization, www.er.doe.gov/.

557. Whitesides, G.M., and Crabtree, G.W. (2007) Don't forget long-term fundamental research in energy. *Science*, **315**, 796.

558. Patel, M.R. (2005) *Spacecraft Power Systems*, CRC Press, Boca Raton, FL.

559. Schwarzer, K., and da Silva, M.E.V. (2008) Characterisation and design methods of solar cookers. *Solar Energy*, **82**, 157.

560. Knudson, B. (2004) State of the Art of Solar Cooking: a Global Survey of Practices and Promotion Programs, www.she-inc.org/.

561. Solar Cookers International, solarcookers.org/.

562. Wentzel, M., and Pouris, A. (2007) The development impact of solar cookers: a review of solar cooking impact research in South Africa. *Energy Policy*, **35**, 1909.

563. Sharma, A., Chen, C.R., and Lan, N.V. (2009) Solar-energy drying systems: a review. *Renew. Sust. Energy Rev.*, **13**, 1185.

564. Mercer, D.G. (2008) Solar drying in developing countries: possibilities and pitfalls, in *Using Food Science and Technology to Improve Nutrition and Promote National Development: Selected Case Studies* (eds G.L. Robertson and J.R. Lupien), International Union of Food Science and Technology, Oakville, ON, Canada, Chapter 4.

565. Mills, A.A., and Clift, R. (1992) Reflections of the "Burning Mirrors of Archimedes". With a consideration of the geometry and intensity of sunlight reflected from plane mirrors. *Eur. J. Phys.*, **13**, 268.

566. Kodama, T., and Gokon, N. (2007) Thermochernical cycles for high-temperature solar hydrogen production. *Chem. Rev.*, **107**, 4048.

567. Segal, A., and Epstein, M. (2000) The optics of the solar tower reflector. *Solar Energy*, **69**, 229.

568. Mills, D. (2004) Advances in solar thermal electricity technology. *Solar Energy*, **76**, 19.

569. Tsoutsos, T., Frantzeskaki, N., and Gekas, V. (2005) Environmental impacts from the solar energy technologies. *Energy Policy*, **33**, 289.

570. (2009) Global Concentrated Solar Power Markets and Strategies,

2009–2020, Emerging Energy Research, Cambridge, MA.

571. Patel, S. (2009) Interest in solar tower technology rising. Power News, 1 May, www.powermag.com/.

572. Hang, Q., Jun, Z., Xiao, Y., *et al.* (2008) Prospect of concentrating solar power in China – the sustainable future. *Renew. Sust. Energy Rev.*, **12**, 2505.

573. Hamdan, L.K., Zarei, M., Chianelli, R.R., *et al.* (2008) Sustainable water and energy in Gaza Strip. *Renew. Energy*, **33**, 1137.

574. Trieb, F., and Muller-Steinhagen, H. (2008) Concentrating solar power for seawater desalination in the Middle East and North Africa. *Desalination*, **220**, 165.

575. Pitz-Paal, R. (2008) How the sun gets into the power plant, in *Renewable Energy. Sustainable Energy Concepts for the Future* (eds R. Wengenmayr and T. Bührke), Wiley-VCH Verlag GmbH, Weinheim, p. 26.

576. Fthenakis, V., and Kim, H.C. (2009) Land use and electricity generation: a life-cycle analysis. *Renew. Sust. Energy Rev.*, **13**, 1465.

577. Solar Millennium (2008) The Parabolic Trough Power Plants Andasol 1 to 3, Erlangen, Germany, www.solarmillennium.de/.

578. Mills, D.R., and Morrison, G.L. (2000) Compact linear Fresnel reflector solar thermal powerplants. *Solar Energy*, **68**, 263.

579. Stanford University Global Climate and Energy Project (2006) An Assessment of Solar Energy Conversion Technologies and Research Opportunities, gcep.stanford.edu/.

580. Pacheco, J.E. (2001) Demonstration of solar-generated electricity on demand: the Solar Two project. *J. Solar Energy Eng. Trans. ASME*, **123**, 5.

581. Kolb, G.J., Alpert, D.J., and Lopez, C.W. (1991) Insights from the operation of Solar One and their implications for future central receiver plants. *Solar Energy*, **47**, 39.

582. Mancini, T., Heller, P., Butler, B., *et al.* (2003) Dish-Stirling systems: an overview of development and status. *J. Solar Energy Eng. Trans. ASME*, **125**, 135.

583. (2009) The other kind of solar power. Economist, 4 June

584. EnviroMission, www.enviromission.com.au/.

585. Schlaich, J., Bergermann, R., Schiel, W., *et al.* (2005) Design of commercial solar updraft tower systems – utilization of solar induced convective flows for power generation. *J. Solar Energy Eng Trans. ASME*, **127**, 117.

586. Schlaich, J., Bergermann, R., and Weinrebe, G. (2008) Electric power from hot air, in *Renewable Energy. Sustainable Energy Concepts for the Future* (eds R. Wengenmayr and T. Bührke), Wiley-VCH Verlag GmbH, Weinheim, p. 68.

587. Halper, M. (2010) Sunshine's cloudy days. Time, 25 January, p. 43.

588. Ringel, M. (2006) Fostering the use of renewable energies in the European Union: the race between feed-in tariffs and green certificates. *Renew. Energy*, **31**, 1.

589. Erge, T., Hoffmann, V.U., and Kiefer, K. (2001) The German experience with grid-connected PV-systems. *Solar Energy*, **70**, 479.

590. del Rio, P., and Gual, M.A. (2007) An integrated assessment of the feed-in tariff system in Spain. *Energy Policy*, **35**, 994.

591. Jefferson, M. (2008) Accelerating the transition to sustainable energy systems. *Energy Policy*, **36**, 4116.

592. Eilperin, J. (2009) G20 leaders agree to phase out fossil fuel subsidies. Washington Post, 25 September, www.washingtonpost.com/.

593. Skylakakis, T. (2010) EU's message to the world on fossil fuel subsidies. New Europe, 17 January, www.neurope.eu/.

594. Storchmann, K. (2005) The rise and fall of German hard coal subsidies. *Energy Policy*, **33**, 1469.

595. Dahmen, N., Dinjus, E., and Henrich, E. (2008) Synthetic fuels from the biomass, in *Renewable Energy. Sustainable Energy Concepts for the Future* (eds R. Wengenmayr and T. Bührke), Wiley-VCH Verlag GmbH, Weinheim, p. 61.

596. Goyal, H.B., Seal, D., and Saxena, R.C. (2008) Bio-fuels from thermochemical conversion of renewable resources: a review. *Renew. Sust. Energy Rev.*, **12**, 504.

597. Service, R.F. (2009) Biomass fuel starts to see the light. *Science*, **326**, 1474.

598. Service, R.F. (2009) Sunlight in your tank. *Science*, **326**, 1472.

599. Steinfeld, A., and Palumbo, R. (2001) Solar thermochemical process technology, in *Encyclopedia of Physical Science and Technology*, vol. **15** (ed. R.A. Meyers), Academic Press, New York, p. 237.

600. Becquerel, A.E. (1839) On electric effects under the influence of solar radiation. *C. R. Acad. Sci.*, **9**, 711.

601. Adams, W.G., and Day, R.E. (1876) The action of light on selenium. *Proc. R. Soc. London*, **25**, 113.

602. Fritts, C.E. (1883) On a new form of selenium photocell. *Am. J. Sci.*, **26**, 465.

603. Shevaleevskiy, O. (2008) The future of solar photovoltaics: a new challenge for chemical physics. *Pure Appl. Chem.*, **80**, 2079.

604. Chapin, D.M., Fuller, C.S., and Pearson, G.L. (1954) A new silicon p–n junction photocell for converting solar radiation into electrical power. *J. Appl. Phys.*, **5**, 676.

605. Rappaport, P. (1961) Photoelectricity. *Proc. Natl. Acad. Sci. USA*, **47**, 1303.

606. Flood, D.J. (2001) Space photovoltaics – history, progress and promise. *Mod. Phys. Lett. B*, **15**, 561.

607. Swanson, R.M. (2006) A vision for crystalline silicon photovoltaics. *Prog. Photovoltaics Res. Appl.*, **14**, 443.

608. Shockley, W., and Queisser, H.J. (1961) Detailed balance limit of efficiency of p–n junction solar cells. *J. Appl. Phys.*, **32**, 510.

609. Goetzberger, A., Hebling, C., and Schock, H.-W. (2003) Photovoltaic materials, history, status and outlook. *Mater. Sci. Eng. R*, **40**, 1.

610. Reisch, M. (2007) Solar energy advances. Chem. Eng. News, 30 July, 15.

611. Tao, M. (2008) Inorganic photovoltaic solar cells: silicon and beyond. Electrochem. Soc. Interface, Winter, 30.

612. Swanson, R.M. (2009) Photovoltaics power up. *Science*, **324**, 891.

613. Sherwani, A.F., and Usmani, J.A. (2010) Life cycle assessment of solar PV based electricity generation systems: a review. *Renew. Sust. Energy Rev.*, **14**, 540.

614. Mason, J.E., Fthenakis, V.M., Hansen, T., *et al.* (2006) Energy payback and life-cycle CO_2 emissions of the BOS in an optimized 3.5 MW PV installation. *Prog. Photovoltaics Res. Appl.*, **14**, 179.

615. Fthenakis, V.M., Kim, H.C., and Alsema, E. (2008) Emissions from photovoltaic life cycles. *Environ. Sci. Technol.*, **42**, 2168.

616. Voith, M. (2009) Solar on sale. Chem. Eng. News, 9 November, 28.

617. Short, P.L. (2009) Sun's warmth cushions Wacker. Chem. Eng. News, 6 April, 18.

618. Lewis, N.S. (2007) Toward cost-effective solar energy use. *Science*, **315**, 798.

619. Public Renewables Partnership, http://www.repartners.org/solar/pvcost.htm.

620. For an updated pricing of solar components, including past trends, see www.solarbuzz.com/.

621. Yoon, J., Baca, A.J., Park, S.I., *et al.* (2008) Ultrathin silicon solar microcells for semitransparent, mechanically flexible and microconcentrator module designs. *Nat. Mater.*, **7**, 907.

622. Green, M.A. (2003) Crystalline and thin-film silicon solar cells: state of the art and future potential. *Solar Energy*, **74**, 181.

623. Voith, M. (2009) Sneaky solar. Chem. Eng. News, 15 June, 17.

624. Chopra, K.L., Paulson, P.D., and Dutta, V. (2004) Thin-film solar cells: an overview. *Prog. Photovoltaics Res. Appl.*, **12**, 69.

625. (2010) Business watch. *Nature*, **464**, 13.

626. Miles, R.W., Zoppi, G., and Forbes, I. (2007) Inorganic photovoltaic cells. *Mater. Today*, **10**, 20.

627. First Solar, www.firstsolar.com/.

628. Butler, D. (2008) Thin films: ready for their close-up? *Nature*, **454**, 558.

629. Wu, X.Z. (2004) High-efficiency polycrystalline CdTe thin-film solar cells. *Solar Energy*, **77**, 803.

630. Aberle, A.G. (2009) Thin-film solar cells. *Thin Solid Films*, **517**, 4706.

631. Nanosolar (2009) Ultra-Low-Cost Solar Electricity Cells. An Overview of Nanosolar's Cell Technology Platform, www.nanosolar.com/.

632. Fthenakis, V., Wang, W., and Kim, H.C. (2009) Life cycle inventory analysis of the production of metals used in photovoltaics. *Renew. Sust. Energy Rev.*, **13**, 493.

633. USGS, USGS Mineral Information Database, minerals.usgs. gov/.

634. Fthenakis, V.M., Fuhrmann, M., Heiser, J., *et al.* (2005) Emissions and encapsulation of cadmium in CdTe PV modules during fires. *Prog. Photovoltaics Res. Appl.*, **13**, 713.

635. Wadia, C., Alivisatos, A.P., and Kammen, D.M. (2009) Materials availability expands the opportunity for large-scale photovoltaics deployment. *Environ. Sci. Technol.*, **43**, 2072.

636. Hoppe, H., and Sariciftci, N.S. (2004) Organic solar cells: an overview. *J. Mater. Res.*, **19**, 1924.

637. Spanggaard, H., and Krebs, F.C. (2004) A brief history of the development of organic and polymeric photovoltaics. *Solar Energy Mater. Solar Cells*, **83**, 125.

638. Ameri, T., Dennler, G., Lungenschmied, C., *et al.* (2009) Organic tandem solar cells: a review. *Energy Environ. Sci.*, **2**, 347.

639. November (2009) Special issue on Organic Photovoltaics, *Acc. Chem. Res.*

640. Po, R., Maggini, M., and Camaioni, N. (2009) Polymer solar cells: recent approaches and achievements. *J. Phys. Chem. C*, **114**, 695.

641. Bhosale, R., Misek, J., Sakai, N., *et al.* (2010) Supramolecular n/p-heterojunction photosystems with oriented multicolored antiparallel redox gradients (OMARG-SHJs). *Chem. Soc. Rev.*, **39**, 138.

642. Thompson, B.C., and Frechet, J.M.J. (2008) Organic photovoltaics – polymer–fullerene composite solar cells. *Angew. Chem. Int. Ed.*, **47**, 58.

643. Accorsi, G., and Armaroli, N. (2010) Taking advantage of the electronic excited states of [60]-fullerenes. *J. Phys. Chem. C*, **114**, 1385.

644. Li, G., Shrotriya, V., Huang, J.S., *et al.* (2005) High-efficiency solution processable polymer photovoltaic cells by self-organization of polymer blends. *Nat. Mater.*, **4**, 864.

645. Yongye, L., Zheng, X., Jiangbin, X., *et al.* (2010) For the bright future – bulk heterojunction polymer solar cells with power conversion efficiency of 7.4%. *Adv. Mater.*, **22**, 1.

646. Chen, H.-Y., Hou, J., Zhang, S., *et al.* (2009) Polymer solar cells with enhanced open-circuit voltage and efficiency. *Nat. Photon.*, **3**, 649.

647. Park, S.H., Roy, A., Beaupre, S., *et al.* (2009) Bulk heterojunction solar cells with internal quantum efficiency approaching 100%. *Nat. Photon.*, **3**, 297.

648. Jorgensen, M., Norrman, K., and Krebs, F.C. (2008) Stability/degradation of polymer solar cells. *Solar Energy Mater. Solar Cells*, **92**, 686.

649. Solarmer Energy, www.solarmer.com/.

650. Konarka, www.konarka.com/.

651. Royne, A., Dey, C.J., and Mills, D.R. (2005) Cooling of photovoltaic cells under concentrated illumination: a critical review. *Solar Energy Mater. Solar Cells*, **86**, 451.

652. Cotal, H., Fetzer, C., Boisvert, J., *et al.* (2009) III–V multijunction solar cells for concentrating photovoltaics. *Energy Environ. Sci.*, **2**, 174.

653. Geisz, J.F., Friedman, D.J., Ward, J.S., *et al.* (2008) 40.8% efficient inverted triple-junction solar cell with two independently metamorphic junctions. *Appl. Phys. Lett.*, **93**, 3.

654. Concentrix Solar, www.concentrix-solar.de/ or www.solfocus.com/.

655. Currie, M.J., Mapel, J.K., Heidel, T.D., *et al.* (2008) High-efficiency organic solar concentrators for photovoltaics. *Science*, **321**, 226.

656. van Sark, W., Barnham, K.W.J., Slooff, L.H., *et al.* (2008) Luminescent solar concentrators – a review of recent results. *Opt. Express*, **16**, 21773.

657. Huynh, W.U., Dittmer, J.J., and Alivisatos, A.P. (2002) Hybrid nanorod-polymer solar cells. *Science*, **295**, 2425.

658. Gur, I., Fromer, N.A., Geier, M.L., *et al.* (2005) Air-stable all-inorganic nanocrystal solar cells processed from solution. *Science*, **310**, 462.

659. Kamat, P.V. (2008) Quantum dot solar cells. Semiconductor nanocrystals as light harvesters. *J. Phys. Chem. C*, **112**, 18737.

660. Imahori, H., and Umeyama, T. (2009) Donor–acceptor nanoarchitecture on semiconducting electrodes for solar energy conversion. *J. Phys. Chem. C*, **113**, 9029.

661. Fehrenbacher, K. (2009) First Solar to Build World's Largest PV Solar Farm for China, 8 September, http://earth2tech.com/.

662. Bradford, T. (2006) *Solar Revolution*, MIT Press, Cambridge, MA.

663. Martinot, E., Dienst, C., Weiliang, L., *et al.* (2007) Renewable energy futures: targets, scenarios, and pathways. *Annu. Rev. Environ. Resour.*, **32**, 205.

664. Resch, G., Held, A., Faber, T., *et al.* (2008) Potentials and prospects for renewable energies at global scale. *Energy Policy*, **36**, 4048.

665. Fthenakis, V., Mason, J.E., and Zweibel, K. (2009) The technical, geographical, and economic feasibility for solar energy to supply the energy needs of the US. *Energy Policy*, **37**, 387.

666. Súri, M., Huld, T.A., Dunlop, E.D., *et al.* (2007) Potential of solar electricity generation in the European Union member states and candidate countries. *Solar Energy*, **81**, 1295.

667. Nemet, G.F. (2009) Net radiative forcing from widespread deployment of photovoltaics. *Environ. Sci. Technol.*, **43**, 2173.

668. Gray, H.B. (2009) Powering the planet with solar fuel. *Nat. Chem.*, **1**, 7.

669. Barber, J. (2009) Photosynthetic energy conversion: natural and artificial. *Chem. Soc. Rev.*, **38**, 185.

670. Everts, S. (2009) Divining the spliceosome. Chem. Eng. News, 4 May, 46.

671. Blankenship, R.E. (2002) *Molecular Mechanisms of Photosynthesis*, Blackwell Science, Oxford.

672. McDermott, G., Prince, S.M., Freer, A.A., *et al.* (1995) Crystal-structure of an integral membrane light-harvesting complex from photosynthetic bacteria. *Nature*, **374**, 517.

673. Deisenhofer, J., Epp, O., Miki, K., *et al.* (1985) Structure of the protein subunits in the photosynthetic reaction center of *Rhodopseudomonas viridis* at 3 Å resolution. *Nature*, **318**, 618.

674. Balzani, V., Credi, A., and Venturi, M. (2008) Photochemical conversion of solar energy. *ChemSusChem*, **1**, 26.

675. Marcus, R.A., and Sutin, N. (1985) Electron transfers in chemistry and biology. *Biochim. Biophys. Acta*, **811**, 265.

676. Lubitz, W., Reijerse, E.J., and Messinger, J. (2008) Solar water-splitting into H_2 and O_2: design principles of photosystem II and hydrogenases. *Energy Environ. Sci.*, **1**, 15.

677. Hammes-Schiffer, S. (2009) Theory of proton-coupled electron transfer in energy conversion processes. *Acc. Chem. Res.*, **42**, 1881.

678. Siegbahn, P.E.M. (2009) Structures and energetics for O_2 formation in photosystem II. *Acc. Chem. Res.*, **42**, 1871.

679. Dau, H., and Zaharieva, I. (2009) Principles, efficiency, and blueprint character of solar-energy conversion in photosynthetic water oxidation. *Acc. Chem. Res.*, **42**, 1861.

680. Gallezot, P. (2008) Catalytic conversion of biomass: challenges and issues. *ChemSusChem*, **1**, 734.

681. Goldemberg, J. (2008) The challenge of biofuels. *Energy Environ. Sci.*, **1**, 523.

682. Inderwildi, O.R., and King, D.A. (2009) *Quo vadis* biofuels? *Energy Environ. Sci.*, **2**, 343.

683. Luque, R., Herrero-Davila, L., Campelo, J.M., *et al.* (2008) Biofuels: a technological perspective. *Energy Environ. Sci.*, **1**, 542.

684. Fedoroff, N. (2008) Seeds of a perfect storm. *Science*, **320**, 425.

685. Sheehan, J. (2009) Engineering direct conversion of CO_2 to biofuel. *Nat. Biotechnol.*, **27**, 1128.

686. Ritter, S.K. (2007) Biofuel Bonanza. Chem. Eng. News, 25 June, 15.

687. Hess, G. (2007) Renewable Fuels Face Bumpy Road. Chem. Eng. News, 17 September, 28.

688. European Commission Energy, http://ec.europa.eu/energy/index_en.htm.

689. Environmental Protection Agency (2005) US Energy Policy Act, http://www.epa.gov/oust/fedlaws/publ_109-058.pdf.

690. Hill, J., Nelson, E., Tilman, D., *et al.* (2006) Environmental, economic, and energetic costs and benefits of biodiesel and ethanol biofuels. *Proc. Natl. Acad. Sci. USA*, **103**, 11206.

691. US Department of Energy, US DOE Biomass Program, http://www1.eere.energy.gov/biomass/recovery.html.

692. OECD, http://www.oecd.org/home/.

693. Johnson, J. (2008) Food versus fuel fight heats up. Chem. Eng. News, 12 May, 11.

694. Mitchell, D. (2008) A Note on Rising Food Prices, Development Prospects Group, The World Bank, July, www-wds.worldbank.org/.

695. Rabalais, N.N., Turner, R.E., and Wiseman, W.J. (2002) Gulf of Mexico hypoxia, aka "The Dead Zone". *Annu. Rev. Ecol. Syst.*, **33**, 235.

696. Anderson, L.G. (2009) Ethanol fuel use in Brazil: air quality impacts. *Energy Environ. Sci.*, **2**, 1015.

697. Crutzen, P.J., Mosier, A.R., Smith, K.A., *et al.* (2008) N_2O release from agro-biofuel production negates global warming reduction by replacing fossil fuels. *Atmos. Chem. Phys.*, **8**, 389.

698. Goldemberg, J. (2007) Ethanol for a sustainable energy future. *Science*, **315**, 808.

699. Charles, D. (2009) Biofuels corn-based ethanol flunks key test. *Science*, **324**, 587.

700. Fargione, J., Hill, J., Tilman, D., *et al.* (2008) Land clearing and the biofuel carbon debt. *Science*, **319**, 1235.

701. MacKay, D.J.C. (2009) *Sustainable Energy – Without the Hot Air*, UIT Cambridge, Cambridge.

702. Qiu, J. (2008) A sustainable generation? Chem. World-UK, June, 48.

703. United Nations Developement Programme (2003) World Energy Assesment: Energy and the Challenge of Sustainability, unp.un.org/.

704. Lestari, S., Maki-Arvela, P., Beltramini, J., *et al.* (2009) Transforming triglycerides and fatty acids into biofuels. *ChemSusChem*, **2**, 1109.

705. Rinaldi, R., and Schuth, F. (2009) Acid hydrolysis of cellulose as the entry point into biorefinery schemes. *ChemSusChem*, **2**, 1096.

706. Crossley, S., Faria, J., Shen, M., *et al.* (2010) Solid nanoparticles that catalyze biofuel upgrade reactions at the water/oil interface. *Science*, **327**, 68.

707. Milmo, S. (2007) The green fuel myth. Chem. World-UK, October, 48.

708. Jeffries, E. (2008) See the wood for the fuel. Chem. World-UK, June, 44.

709. Righelato, R., and Spracklen, D.V. (2007) Carbon mitigation by biofuels or by saving and restoring forests? *Science*, **317**, 902.

710. Bonan, G.B. (2008) Forests and climate change: forcings, feedbacks, and the climate benefits of forests. *Science*, **320**, 1444.

711. Sanderson, K. (2009) Wonder weed plans fail to flourish. *Nature*, **461**, 328.

712. Ghirardi, M.L., Dubini, A., Yu, J.P., *et al.* (2009) Photobiological hydrogen-producing systems. *Chem. Soc. Rev.*, **38**, 52.

713. Wilkinson, M. (2008) The promise of algae. Chem. World-UK, December.

714. Service, R.F. (2009) Exxonmobil fuels Venter's efforts to run vehicles on algae-based oil. *Science*, **325**, 379.

715. Huber, G.W., and Corma, A. (2007) Synergies between bio- and oil refineries for the production of fuels from biomass. *Angew. Chem. Int. Ed.*, **46**, 7184.

716. Ritter, S.K. (2008) Genes to gasoline. Chem. Eng. News, 25 June, 10.

717. Tollefson, J. (2008) Energy: not your father's biofuels. *Nature*, **451**, 880.

718. Rubin, E.M. (2008) Genomics of cellulosic biofuels. *Nature*, **454**, 841.

719. Farrell, A.E. (2006) Ethanol can contribute to energy and environmental goals. *Science*, **312**, 1748.

720. Johnson, J. (2009) Supporting biofuels. Chem. Eng. News, 1 May, 8.

721. Kwok, R. (2009) Cellulosic ethanol hits roadblocks. *Nature*, **461**, 582.

722. Sanderson, K. (2009) From plant to power. *Nature*, **461**, 710.

723. US Department of Energy (2009) Affordable, Low-Carbon Diesel Fuel from Domestic Coal and Biomass, www.netl.doe.gov.

724. Ciamician, G. (1912) The photochemistry of the future. *Science*, **36**, 385.

725. Balzani, V., Moggi, L., Manfrin, M.F., *et al.* (1975) Solar-energy conversion by water photodissociation. *Science*, **189**, 852.

726. Hautala, R.R., King, R.B., and C. Kutal (eds) (1979) *Solar Energy: Chemical Conversion and Storage*, Humana Press, Clifton, NJ.

727. Collings, A.F., and Critchley, C. (eds) (2005) *Artificial Photosynthesis*, Wiley-VCH Verlag GmbH, Weinheim.

728. Klan, P., and Wirz, J. (eds) (2009) *Photochemistry of Organic Compounds*, John Wiley & Sons, Inc., Hoboken, NJ, USA.

729. Rand, D.A.J., and Dell, R.M. (2008) *Hydrogen Energy: Challenges and Prospects*, RSC Publishing, Cambridge.

730. Aresta, M., and Dibenedetto, A. (2007) Utilisation of CO_2 as a chemical feedstock: opportunities and challenges. *Dalton Trans.*, 2975.

731. Morris, A.J., Meyer, G.J., and Fujita, E. (2009) Molecular approaches to the photocatalytic reduction of carbon dioxide for solar fuels. *Acc. Chem. Res.*, **42**, 1983.

732. Listorti, A., Durrant, J., and Barber, J. (2009) Artificial photosynthesis solar to fuel. *Nat. Mater.*, **8**, 929.

733. Kanan, M.W., and Nocera, D.G. (2008) *In situ* formation of an oxygen-evolving catalyst in neutral water containing phosphate and Co^{2+}. *Science*, **321**, 1072.

734. Kodis, G., Terazono, Y., Liddell, P.A., *et al.* (2006) Energy and photoinduced electron transfer in a wheel-shaped artificial photosynthetic antenna–reaction center complex. *J. Am. Chem. Soc.*, **128**, 1818.

735. Concepcion, J.J., Jurss, J.W., Brennaman, M.K., *et al.* (2009) Making oxygen with ruthenium complexes. *Acc. Chem. Res.*, **42**, 1954.

736. Dismukes, G.C., Brimblecombe, R., Felton, G.A.N., *et al.* (2009) Development of bioinspired Mn_4O_4–cubane water oxidation catalysts: lessons from photosynthesis. *Acc. Chem. Res.*, **42**, 1935.

737. Magnuson, A., Anderlund, M., Johansson, O., *et al.* (2009) Biomimetic and microbial approaches to solar fuel generation. *Acc. Chem. Res.*, **42**, 1899.

738. Yin, Q., Tan, J.M., Besson, C., *et al.* (2010) A fast soluble carbon-free molecular water oxidation catalyst based on abundant metals. *Science*, **328**, 342.

739. Huynh, M.H.V., and Meyer, T.J. (2007) Proton-coupled electron transfer. *Chem. Rev.*, **107**, 5004.

740. Nam, Y.S., Magyar, A.P., Lee, D., *et al.* (2010) Biologically templated photocatalytic nanostructures for sustained light-driven water oxidation. *Nat. Nanotech.*, **5**, 340.

741. Navarro Yerga, R.M., Alvarez Galvan, M.C., del Valle, F., *et al.* (2009) Water splitting on semiconductor catalysts under visible-light irradiation. *ChemSusChem*, **2**, 471.

742. Fujishima, A., and Honda, K. (1972) Electrochemical photolysis of water at a semiconductor electrode. *Nature*, **238**, 37.

743. Inoue, Y. (2009) Photocatalytic water splitting by RuO_2-loaded metal oxides and nitrides with d^0- and d^{10}-related electronic configurations. *Energy Environ. Sci.*, **2**, 364.

744. Abe, R., Shinmei, K., Hara, K., *et al.* (2009) Robust dye-sensitized overall water splitting system with two-step photoexcitation of coumarin dyes and metal oxide semiconductors. *Chem. Commun.*, 3577.

745. Hardin, B.E., Hoke, E.T., Armstrong, P.B., *et al.* (2009) Increased light harvesting in dye-sensitized solar cells with energy relay dyes. *Nat. Photonics*, **3**, 406.

746. O'Regan, B., and Grätzel, M. (1991) A low-cost, high-efficiency solar-cell based on dye-sensitized colloidal TiO_2 films. *Nature*, **353**, 737.

747. Grätzel, M. (2009) Recent advances in sensitized mesoscopic solar cells. *Acc. Chem. Res.*, **42**, 1788.

748. Youngblood, W.J., Lee, S.H.A., Kobayashi, Y., *et al.* (2009) Photoassisted overall water splitting in a visible light-absorbing dye-sensitized photoelectrochemical cell. *J. Am. Chem. Soc.*, **131**, 926.

749. Youngblood, W.J., Lee, S.-H.A., Maeda, K., *et al.* (2009) Visible light water splitting using dye-sensitized oxide semiconductors. *Acc. Chem. Res.*, **42**, 1966.

750. Regalado, A. (2010) Race for cellulosic fuels spurs Brazilian research program. *Science*, **327**, 928.

751. Lewis, N.S., and Nocera, D.G. (2006) Powering the planet: chemical challenges in solar energy utilization. *Proc. Natl. Acad. Sci. USA*, **103**, 15729.

752. Sternberg, R. (2010) Hydropower's future, the environment, and global electricity systems. *Renew. Sust. Energy Rev.*, **14**, 713.

753. Bartle, A. (2002) Hydropower potential and development activities. *Energy Policy*, **30**, 1231.

754. World Energy Council (2009) Survey of Energy Resources Interim Update 2009, www.worldenergy.org/.

755. Marks, J.C. (2007) Down go the dams. *Sci. Am.*, **296** (3), 66.

756. Jacobson, M.Z., and Delucchi, M.A. (2009) A path to sustainable energy by 2030. *Sci. Am.*, **301** (5), 58.

757. Stone, R. (2010) Severe drought puts spotlight on Chinese dams. *Science*, **327**, 1311.

758. Normile, D. (2010) Restoration or devastation? *Science*, **327**, 1568.

759. Dalton, R. (2005) Floods fail to save Canyon beaches. *Nature*, **438**, 10.

760. Schilt, C.R. (2007) Developing fish passage and protection at hydropower dams. *Appl. Anim. Behav. Sci.*, **104**, 295.

761. Stone, R. (2008) Three Gorges Dam: into the unknown. *Science*, **321**, 628.

762. Kerr, R.A., and Stone, R. (2009) A human trigger for the great quake of Sichuan? *Science*, **323**, 322.

763. Sovacool, B.K. (2008) The costs of failure: a preliminary assessment of major energy accidents, 1907–2007. *Energy Policy*, **36**, 1802.

764. Lehner, B., Czisch, G., and Vassolo, S. (2005) The impact of global change on the hydropower potential of Europe: a model-based analysis. *Energy Policy*, **33**, 839.

765. Guerin, F., Abril, G., Richard, S., *et al.* (2006) Methane and carbon dioxide emissions from tropical reservoirs: significance of downstream rivers. *Geophys. Res. Lett.*, **33**, L21407.

766. European Small Hydropower Association, www.esha.be/.

767. Sørensen, B. (1995) History of, and recent progress in, wind-energy utilization. *Annu. Rev. Energy Environ.*, **20**, 387.

768. European Wind Energy Association (2009) Wind Energy – the Facts, www.wind-energy-the-facts.org/.

769. Lu, X., McElroy, M.B., and Kiviluoma, J. (2009) Global potential for wind-generated electricity. *Proc. Natl. Acad. Sci. USA*, **106**, 10933.

770. Huleihil, M. (2009) Maximum windmill efficiency in finite time. *J. Appl. Phys.*, **105**, 104908.

771. Archer, C.L., and Jacobson, M.Z. (2005) Evaluation of global wind power. *J. Geophys. Res.*, **110**, D12110.

772. Archer, C.L., and Jacobson, M.Z. (2007) Supplying baseload power and reducing transmission requirements by interconnecting wind farms. *J. Appl. Meteorol. Climatol.*, **46**, 1701.

773. European Wind Energy Association (2009) Pure Power – Wind Energy Targets for 2020 and 2030, www.ewea.org/.

774. European Wind Energy Association (2010) Wind in Power – 2009 European Statistics, www.wind-energy-the-facts.org/.

775. European Environment Agency (2009) ENER27 Electricity Production by Fuel, www.eea.europa.eu/.

776. Fairless, D. (2007) Energy-go-round. *Nature*, **447**, 1046.

777. American Wind Energy Association (2010) 4th Quarter 2009 Market Report, www.awea.org/.

778. Cyranoski, D. (2009) Renewable energy: Beijing's windy bet. *Nature*, **457**, 372.

779. Global Wind Energy Council, www.gwec.net/.

780. McElroy, M.B., Lu, X., Nielsen, C.P., *et al.* (2009) Potential for wind-generated electricity in China. *Science*, **325**, 1378.

781. Wagner, H.J., and Epe, A. (2009) Energy from wind – perspectives and research needs. *Eur. Phys. J. Spec. Top.*, **176**, 107.

782. Crawford, R.H. (2009) Life cycle energy and greenhouse emissions analysis of wind turbines and the effect of size on energy yield. *Renew. Sust. Energy Rev.*, **13**, 2653.

783. Curry, A. (2009) Deadly flights. *Science*, **325**, 386.

784. Pryor, S.C., and Barthelmie, R.J. (2010) Climate change impacts on wind energy: a review. *Renew. Sust. Energy Rev.*, **14**, 430.

785. Keith, D.W., DeCarolis, J.F., Denkenberger, D.C., *et al.* (2004) The influence of large-scale wind power on global climate. *Proc. Natl. Acad. Sci. USA*, **101**, 16115.

786. Archer, C.L., and Caldeira, K. (2009) Global assessment of high-altitude wind power. *Energies*, **2**, 307.

787. Vance, E. (2009) High hopes. *Nature*, **460**, 564.

788. Canale, M., Fagiano, L., and Milanese, M. (2009) Kitegen: a revolution in wind energy generation. *Energy*, **34**, 355.

789. Kite Gen Research, www.kitegen.com/.

790. Sky WindPower, www.skywindpower.com/.

791. Roberts, B.W., Shepard, D.H., Caldeira, K., *et al.* (2007) Harnessing high-altitude wind power. *IEEE Trans. Energy Conver.*, **22**, 136.

792. Mazumder, R., and Arima, M. (2005) Tidal rhythmites and their implications. *Earth Sci. Rev.*, **69**, 79.

793. Rourke, F.O., Boyle, F., and Reynolds, A. (2010) Tidal energy update 2009. *Appl. Energy*, **87**, 398.

794. Clery, D. (2008) UK ponders world's biggest tidal power scheme. *Science*, **320**, 1574.

795. Mueller, M., and Wallace, R. (2008) Enabling science and technology for marine renewable energy. *Energy Policy*, **36**, 4376.

796. Marine Current Turbines, www.marineturbines.com/.

797. Schiermeier, Q., Tollefson, J., Scully, T., *et al.* (2008) Electricity without carbon. *Nature*, **454**, 816.

798. Callaway, E. (2007) To catch a wave. *Nature*, **450**, 156.

799. Scruggs, J., and Jacob, P. (2009) Harvesting ocean wave energy. *Science*, **323**, 1176.

800. Pelc, R., and Fujita, R.M. (2002) Renewable energy from the ocean. *Mar. Policy*, **26**, 471.

801. Avery, W.H., and Wu, C. (1994) *Renewable Energy from the Ocean. A Guide to OTEC*, Oxford University Press, New York.

802. Nihous, G.C. (2007) A preliminary assessment of ocean thermal energy conversion resources. *J. Energy Resour. ASME*, **129**, 10.

803. Barbier, E. (2002) Geothermal energy technology and current status: an overview. *Renew. Sust. Energy Rev.*, **6**, 3.

804. Kleiner, K. (2007) Promise boiling over. *Nature*, **450**, 934.

805. Stefansson, V. (1997) Geothermal reinjection experience. *Geothermics*, **26**, 99.

806. Hammons, T.J. (2004) Geothermal power generation worldwide: global perspective, technology, field experience, and research and development. *Electr. Power Compon. Syst.*, **32**, 529.

807. Lund, J.W., Freeston, D.H., and Boyd, T.L. (2005) Direct application of geothermal energy: 2005 worldwide review. *Geothermics*, **34**, 691.

808. REN 21 – Renewable Energy Policy Network for the 21st Century (2008) Renewables 2007 – Global Status Report, www.ren21.net/.

809. Omer, A.M. (2008) Ground-source heat pumps systems and applications. *Renew. Sust. Energy Rev.*, **12**, 344.

810. Bertani, R. (2005) World geothermal power generation in the period 2001–2005. *Geothermics*, **34**, 651.

811. Huang, S.P., and Liu, J.Q. (2010) Geothermal energy stuck between a rock and a hot place. *Nature*, **463**, 293.

812. Giardini, D. (2009) Geothermal quake risks must be faced. *Nature*, **462**, 848.

813. US Department of Energy Idaho National Laboratory (2006) The Future of Geothermal Energy – Impact of Enhanced Geothermal Systems (EGS) on the United States in the 21st Century, geothermal.inel.gov/.

814. Majer, E.L., Baria, R., Stark, M., *et al.* (2007) Induced seismicity associated with enhanced geothermal systems. *Geothermics*, **36**, 185.

815. Gallup, D.L. (2009) Production engineering in geothermal technology: a review. *Geothermics*, **38**, 326.

816. Schewe, P.F. (2007) *The Grid: a Journey Through the Heart of Our Electrified World*, Joseph Henry Press, Washington, DC.

817. International Energy Agency (2006) Light's Labour's Lost – Policies for Energy-Efficient Lighting, www.iea.org/.

818. Johnson, J. (2007) The end of the light bulb. Chem. Eng. News, **85**, 46.

819. Appell, D. (2007) Toxic bulbs. *Sci. Am.*, **297** (4), 30.

820. Crawford, M.H. (2009) LEDs for solid-state lighting: performance challenges and recent advances. *IEEE J. Sel. Top. Quantum Electron.*, **15**, 1028.

821. Humphreys, C.J. (2008) Solid-state lighting. *MRS Bull.*, **33**, 459.

822. Phillips, J.M., Coltrin, M.E., Crawford, M.H., *et al.* (2007) Research challenges to ultra-efficient inorganic solid-state lighting. *Laser Photon. Rev.*, **1**, 307.

823. So, F., Kido, J., and Burrows, P. (2008) Organic light-emitting devices for solid-state lighting. *MRS Bull.*, **33**, 663.

824. Müllen, K., and Scherf, U. (eds) (2006) *Organic Light Emitting Devices: Synthesis, Properties, and Applications*, Wiley-VCH Verlag GmbH, Weinheim.

825. US Energy Information Administration (2009) International Energy Outlook 2009, www.eia.doe.gov/.

826. Terna, www.terna.it.

827. Kammen, D.M., and Pacca, S. (2004) Assessing the costs of electricity. *Annu. Rev. Environ. Resour.*, **29**, 301.

828. Rojey, A. (2009) *Energy and Climate: How to Achieve a Successful Energy Transition*, John Wiley & Sons, Ltd, Chichester.

829. Dell, R.M., and Rand, D.A.J. (2004) *Clean Energy*, Royal Society of Chemistry, Cambridge.

830. See www.world-nuclear.org/info/nshare.html.

831. International Atomic Energy Agency (2010) Power Reactor Information System. Nuclear Renaissance, http://iaea.org/programmes/a2.

832. European Energy Forum(2007) World Energy Technology Outlook to 2050, www.europeanenergyforum.eu/.

833. ENTSO-E, www.entsoe.eu/.

834. Johnson, S. (2000) California officials investigate electric "megawatt laundering". *San Jose Mercury News*, http://www.accessmylibrary.com/article-1G1-121918279/california-officials-investigate-electric.html.

835. (2002) Testimony of S. David Freeman before the Subcommittee on Consumer Affairs, Foreign Commerce and Tourism of the Senate Committee on Commerce, Science and Transportation, 15 May, http://en.wikipedia.org/wiki/California_electricity_crisis.

836. Johnston, D.C. (2007) Competitively priced electricity costs more. New York Times, 6 November.

837. (2009) Windpower Monthly Mag., June www.windpowermonthly.com/.

838. Energy Watch Group/Ludwig-Boelkow-Foundation (2008) Renewable Energy Outlook 2030. Energy Watch Group Global Renewable Energy Scenarios, www.energywatchgroup.org/.

839. Oswald, J., Raine, M., and Ashraf-Ball, H. (2008) Will British weather provide reliable electricity? *Energy Policy*, **36**, 3212.

840. Cliburn, J.K. (2009) Going for solar gigawatts at utilities. Solar Today, June, 5.

841. Solar Electric Power Association, www.solarelectricpower.org/.

842. Electricity Storage Association, www.electricitystorage.org/.

843. Flybrid Energy Systems, The Flybrid Kinetic Energy Recovery System (KERS), http://www.flybridsystems.com/F1System.html.

844. Wolsky, A.M. (2002) The status and prospects for flywheels and SMES that incorporate HTS. *Physica C*, **372**, 1495.

845. Battery Council International (2009) Battery Recycling, www.batterycouncil.org/.

846. PowerGenix, PowerGenix Nickel–Zinc High Discharge Battery Technology Overview, http://www.powergenix.com/service.php.

847. PR-inside (2009) ReVolt, BASF Team for Rechargeable Zinc–Air Battery, 9 July, www.pr-inside.com/.

848. Nikkei Electronics Asia (2008) Can Batteries Save Embattled Wind Power?, techon.nikkeibp.co.jp/.

849. Zebra Batteries http://atea.it/pdf/Zebra-Battery-Z57.pdf.

850. Teki, R., Datta, M.K., Krishnan, R., *et al.* (2009) Nanostructured silicon anodes for lithium ion rechargeable batteries. *Small*, **5**, 2236.

851. Cao, F.F., Guo, Y.G., Zheng, S.F., *et al.* (2010) Symbiotic coaxial nanocables: facile synthesis and an efficient and elegant morphological solution to the lithium storage problem. *Chem. Mater.*, **22**, 1908.

852. Armand, M., and Tarascon, J.M. (2008) Building better batteries. *Nature*, **451**, 652.

853. A123 Systems Enabling: a New Era of Sustainable Transportation, www.a123systems.com/.

854. Voith, M. (2009) Powerful stuff. Chem. Eng. News, 7 September, 32.

855. Mahindra Reva, Reva Electric Car, www.revaindia.com/.

856. Tesla Motors, www.teslamotors.com/.

857. HybridCars 2009 Hybrid Cars – Year in Review, http://www.hybridcars.com/2009-hybrid-cars.

858. Rocky Mountain Institute www.rmi.org/.

859. Stafford, N. (2009) Germany plugs electric cars. Chem. World-UK, October, 1.

860. National Renewable Energy Laboratory, Plug-in Hybrid Electric Vehicles, http://www.nrel.gov/vehiclesandfuels/vsa/plugin_hybrid.html.

861. Ritter, S.K. (2009) Future of metals. Chem. Eng. News, 8 June, 53.

862. US Energy Information Administration, http://www.eia.doe.gov/emeu/international/electricitycapacity.html.

863. Marris, E. (2008) Upgrading the grid. *Nature*, **454**, 570.

864. Charles, D. (2009) Energy renewables test IQ of the grid. *Science*, **324**, 172.

865. US Department of Energy, Smart Grid, http://www.oe.energy.gov/smartgrid.htm.

866. Wikipedia, Electranet, http://en.wikipedia.org/wiki/Electranet.

867. IBM – a Smarter Planet, http://www.ibm.com/ibm/ideasfromibm/us/smartplanet/20081106/index2.shtml?&re=sp2.

868. (2008) The greener grid. *Nature*, **454**, 551.

869. America's Energy Future Panel on Electricity from Renewable Resources (2010) *Electricity from Renewable Resources: Status, Prospects, and Impediments*, National Academies Press, Washington, DC.

870. Rifkin, J. (2003) *The Hydrogen Economy: the Creation of the Worldwide Energy Web and the Redistribution of Power on Earth*, Penguin Putnam, New York.

871. Momirlan, M., and Veziroglu, T.N. (2005) The properties of hydrogen as fuel tomorrow in sustainable energy system for a cleaner planet. *Int. J. Hydrogen Energy*, **30**, 795.

872. Züttel, A., Borgschulte, A., and Schlapbach, L. (eds) (2008) *Hydrogen as a Future Energy Carrier*, Wiley-VCH Verlag GmbH, Weinhem.

873. Congressional Research Service (CRS) (2007) Hydrogen and Fuel Cell Vehicle R&D: Freedomcar and the President's Hydrogen Fuel Initiative, http://ncseonline.org/NLE/CRSreports/07May/RS21442.pdf.

874. Kennedy, D. (2004) The hydrogen solution. *Science*, **305**, 917.

875. European Commission, The Fuel Cell and Hydrogen Joint Technology Initiative, http://ec.europa.eu/research/fch/index_en.cfm.

876. Crabtree, G.W., and Dresselhaus, M.S. (2008) The hydrogen fuel alternative. *MRS Bull.*, **33**, 421.

877. Sartbaeva, A., Kuznetsov, V.L., Wells, S.A., *et al.* (2008) Hydrogen nexus in a sustainable energy future. *Energy Environ. Sci.*, **1**, 79.

878. Coontz, R., and Hanson, B. (2004) Not so simple. *Science*, **305**, 957.

879. Wilson, J.R., and Burgh, G. (2008) *Energizing Our Future*, John Wiley & Sons, Inc., Hoboken, NJ.

880. US Department of Energy (2006) Hydrogen Posture Plan. An Integrated Research, Development and Demonstration Plan, http://hydrogen.energy.gov/.

881. Trager, R. (2009) (Mar 17: 2009) Maths mistake sidelined Futuregen project. Chem. World-UK, 17 March, 9.

882. Turner, J.A., Williams, M.C., and Rajeshwar, K. (2004) Hydrogen economy based on renewable energy sources. Electrochem. Soc. Interface, Fall, 24.

883. Pagliaro, M., Ciriminna, R., and Palmisano, G. (2008) Flexible solar cells. *ChemSusChem*, **1**, 880.

884. Johnson, J. (2008) US solar energy heats up. Chem. Eng. News, 20 October, 40.

885. Service, R.F. (2008) Can the upstarts top silicon? *Science*, **319**, 718.

886. Levene, J.I., Mann, M.K., Margolis, R.M., *et al.* (2007) An analysis of hydrogen production from renewable electricity sources. *Solar Energy*, **81**, 773.

887. Desertec Foundation, www.desertec.org.

888. Feresin, E. (2007) Europe looks to draw power from Africa. *Nature*, **450**, 595.

889. Graetz, J. (2009) New approaches to hydrogen storage. *Chem. Soc. Rev.*, **38**, 73.

890. Jacoby, M. (2008) Bottling hydrogen in solids. Chem. Eng. News, 28 January, 67.

891. Liu, J.J., Han, Y., and Ge, Q.F. (2009) Effect of doped transition metal on reversible hydrogen release/uptake from $NaAlH_4$. *Chem. Eur. J.*, **15**, 1685.

892. Yao, X.D., and Lu, G.Q. (2008) Magnesium-based materials for hydrogen storage: recent advances and future perspectives. *Chin. Sci. Bull.*, **53**, 2421.

893. Makowski, P., Thomas, A., Kuhn, P., *et al.* (2009) Organic materials for hydrogen storage applications: from physisorption on organic solids to chemisorption in organic molecules. *Energy Environ. Sci.*, **2**, 480.

894. Svec, F., Germain, J., and Frechet, J.M.J. (2009) Nanoporous polymers for hydrogen storage. *Small*, **5**, 1098.

895. Murray, L.J., Dinca, M., and Long, J.R. (2009) Hydrogen storage in metal–organic frameworks. *Chem. Soc. Rev.*, **38**, 1294.

896. Kang, X.D., Ma, L.P., Fang, Z.Z., *et al.* (2009) Promoted hydrogen release from ammonia borane by mechanically milling with magnesium hydride: a new destabilizing approach. *Phys. Chem. Chem. Phys.*, **11**, 2507.

897. Xiong, Z.T., Yong, C.K., Wu, G.T., *et al.* (2008) High-capacity hydrogen storage in lithium and sodium amidoboranes. *Nat. Mater.*, **7**, 138.

898. Vielstich, W. (ed.) (2009) *Handbook of Fuel Cells*, vols **1–6**, Wiley-VCH Verlag GmbH, Weinheim.

899. Fuel Cells 2000, www.fuelcells.org.

900. Norskov, J.K., Bligaard, T., Rossmeisl, J., *et al.* (2009) Towards the computational design of solid catalysts. *Nat. Chem.*, **1**, 37.

901. Léfevre, M., Proietti, E., Jaouen, F., *et al.* (2009) Iron-based catalysts with improved oxygen reduction activity in polymer electrolyte fuel cells. *Science*, **324**, 71.

902. Lawrence Livermore National Laboratory, The Unitized Regenerative Fuel Cell, http://www.llnl.gov/str/Mitlit.html.

903. Horizon Fuel Cell Technologies, Think Big, Start Small, www.horizonfuelcell.com.

904. Vielstich, W. (ed.) (2009) *Handbook of Fuel Cells: Fuel Cell Technology and Applications*, vol. **4**, Wiley-VCH Verlag GmbH, Weinheim.

905. Selman, J.R. (2009) Poison-tolerant fuel cells. *Science*, **326**, 52.

906. Bostic, E. (2006) Soldier power fuel cell development XX25 contract awarded. RDECOM Mag., June–July, http://www.rdecom.army.mil/.

907. Romm, J. (2008) The last car you would ever buy – literally. *Technol. Rev.*, June 18.

908. Committee on Assessment of Resource Needs for Fuel Cell and Hydrogen

Technologies (2008) *Transitions to Alternative Transportation Technologies – A Focus on Hydrogen*, National Academies Press, Washington, DC.

909. Service, R.F. (2009) Hydrogen cars: fad or the future? *Science*, **324**, 1257.

910. Bush, G.W. (2003) State of the Union Address, 28 January, http://millercenter.org/scripps/archive/speeches/detail/4541.

911. McCarthy, J.J. (2009) Reflections on: our planet and its life, origins, and futures. *Science*, **326**, 1646.

912. Brown, L.R. (2009) *Plan B 4.0. Mobilizing to Save Civilization*, W.W. Norton, New York.

913. Kahneman, D., Krueger, A.B., Schkade, D., *et al.* (2006) Would you be happier if you were richer? A focusing illusion. *Science*, **312**, 1908.

914. Layard, R. (2010) Measuring subjective well-being. *Science*, **327**, 534.

915. Illich, I. (1974) *Energy and Equity*, Trinity Press, Worcester.

916. Let's Move, www.letsmove.gov/.

917. Texas Transportation Institute, Components of the Congestion Problem, 2005 Urban Area Totals, http://en.wikipedia.org/wiki/Traffic_congestion.

918. National Traffic Highway Administration (2010), www.nhtsa.gov/

919. European Commission, Road Safety – Trends, http://ec.europa.eu/transport/road_safety/specialist/statistics/trends/index_en.htm.

920. Heffernan, O. (2010) Earth science: the climate machine. *Nature*, **463**, 1014.

921. Anderson, J.G. (2009) Strategic choices for global energy: constraints from feedbacks in the climate system. *ChemSusChem*, **2**, 369.

922. Inman, M. (2009) Hot, flat, crowded and preparing for the worst. *Science*, **326**, 662.

923. Johnson, J. (2010) New jolt for nuclear power. *Chem. Eng. News*, 8 March, 8.

924. Global Footprint Network, www.globalfootprint.org.

925. Rockstrom, J., Steffen, W., Noone, K., *et al.* (2009) A safe operating space for humanity. *Nature*, **461**, 472.

926. Gordon, R.B., Bertram, M., and Graedel, T.E. (2006) Metal stocks and sustainability. *Proc. Natl. Acad. Sci. USA*, **103**, 1209.

927. Service, R.F. (2010) Nations move to head off shortages of rare earths. *Science*, **327**, 1596.

928. Tollefson, J. (2007) Worth its weight in platinum. *Nature*, **450**, 334.

929. Ogunseitan, O.A., Schoenung, J.M., Saphores, J.D.M., *et al.* (2009) The electronics revolution: from E-wonderland to E-wasteland. *Science*, **326**, 670.

930. Davis, S.J., and Caldeira, K. (2010) Consumption-based accounting of CO_2 emissions. *Proc. Natl. Acad. Sci. USA*, **107**, 5687.

931. (2010) Progressive thinking. *Nature*, **463**, 849.

932. European Commission, DG Environment, Beyond GDP International Initiative, http://www.beyond-gdp.eu/.

933. Stohl, A. (2008) The travel-related carbon dioxide emissions of atmospheric researchers. *Atmos. Chem. Phys.*, **8**, 6499.

934. Welch, C.J., Ray, S., Melendez, J., *et al.* (2010) Virtual conferences becoming a reality. *Nat. Chem.*, **2**, 148.

935. Gershenfeld, N., Samouhos, S., and Nordman, B. (2010) Intelligent infrastructure for energy efficiency. *Science*, **327**, 1086.

936. American Physical Society (2008) Energy = Future: Think Efficiency, http://www.aps.org/energyefficiencyreport/.

937. Kohl, H. (2008) Renewable energy sources on the rise, in *Renewable Energy. Sustainable Energy Concepts for the Future* (eds R. Wengenmayr and T. Bührke), Wiley-VCH Verlag GmbH, Weinheim, p. 4.

938. Lindley, D. (2009) The energy should always work twice. *Nature*, **458**, 138.

939. European Commission, DG Environment, EU Climate and Energy Package, http://ec.europa.eu/environment/climat/climate_action.htm.

940. IRENA – International Renewable Energy Agency, www.irena.org/.

941. Allcott, H., and Mullainathan, S. (2010) Behavior and energy policy. *Science*, **327**, 1204.

942. European Parliament (2000) Lisbon European Council 23 and 24 March 2000. Presidency Conclusions, http://www.europarl.europa.eu/summits/lis1_en.htm.

943. Odum, H.T., and Odum, E.C. (2006) The prosperous way down. *Energy*, **31**, 21.

944. Roberts, L., Stone, R., and Sugden, A. (2009) The rise of restoration ecology introduction. *Science*, **325**, 555.

945. Des Jardins, J.R. (2006) *Environmental Ethics. An Introduction to Environmental Philosophy*, Thomson-Wadsworth, Toronto.

946. Kates, R.W. (2001) Queries on the human use of the earth. *Annu. Rev. Energy Environ.*, **26**, 1.

Index

Energy for a Sustainable World: From the Oil Age to a Sun-Powered Future. Nicola Armaroli and Vincenzo Balzani
© 2011 WILEY-VCH Verlag GmbH & Co. KGaA, Weinheim
ISBN: 978-3-527-32540-5